CW01288180

BATTLE OF THE BIG BANG

BATTLE OF THE BIG BANG

The New Tales of Our Cosmic Origins

NIAYESH AFSHORDI
AND PHIL HALPER

THE UNIVERSITY OF CHICAGO PRESS
Chicago and London

The University of Chicago Press, Chicago 60637
The University of Chicago Press, Ltd., London
© 2025 by Niayesh Afshordi and Phil Halper
All rights reserved. No part of this book may be used or reproduced in any manner whatsoever without written permission, except in the case of brief quotations in critical articles and reviews. For more information, contact the University of Chicago Press, 1427 E. 60th St., Chicago, IL 60637.
Published 2025
Printed in Canada

34 33 32 31 30 29 28 27 26 25   1 2 3 4 5

ISBN-13: 978-0-226-83047-6 (cloth)
ISBN-13: 978-0-226-83048-3 (e-book)
DOI: https://doi.org/10.7208/chicago/9780226830483.001.0001

Library of Congress Cataloging-in-Publication Data

Names: Afshordi, Niayesh, 1978– author. | Halper, Phil, author.
Title: Battle of the Big Bang : the new tales of our cosmic origins / Niayesh Afshordi and Phil Halper.
Description: Chicago : The University of Chicago Press, 2025. | Includes bibliographical references and index.
Identifiers: LCCN 2024042399 | ISBN 9780226830476 (cloth) | ISBN 9780226830483 (ebook)
Subjects: LCSH: Cosmogony. | Cosmogony—History. | Big bang theory. | Cosmology—History.
Classification: LCC QB981 .A385 2025 | DDC 523.1/8—dc23/eng/20241119
LC record available at https://lccn.loc.gov/2024042399

♾ This paper meets the requirements of ANSI/NISO Z39.48-1992 (Permanence of Paper).

CONTENTS

Preface: Before Our First Memory vii

1. The Beginning: Cosmology from Thales to Hawking 1
2. The Multiverse and the Ultimate Free Lunch 31
3. Hawking's U-Turn and a Universe from Nothing 71
4. String Theories and Colliding Branes 99
5. The Big Bounce 127
6. Déjà Vu and the Cyclic Cosmos 153
7. Born from a Black Hole 183
8. Speeds of Light 209
9. Holograms and Missing Dimensions 235
10. Can the Universe Create Itself? 261
11. Science of Religion, Religion of Science 281
12. The End of the Beginning 307

Acknowledgments 329
Appendix: Big Bang Models Cheat Sheet 333
Notes 337 Index 357

Color illustrations follow page 194

PREFACE

BEFORE OUR FIRST MEMORY

The Big Bang theory never was a theory of the bang. It said nothing about what banged, why it banged, or what happened before it banged. ALAN GUTH

"My dear fellow," said Sherlock Holmes . . . "life is infinitely stranger than anything which the mind of man could invent. We would not dare to conceive the things which are really mere commonplaces of existence." ARTHUR CONAN DOYLE

What is the oldest thing you can remember? Mine is a vague outline of the inside of an apartment in a city I can't recognize. Phil recalls pretending he was a cat when he was four. You might remember a time from even earlier in your life. What came before that moment? We have our collective memories, the stories our parents tell and the stories they heard from theirs. We piece together our prehistory from photo albums and other family heirlooms—hidden in dusty attics, revealed, rewritten, and forgotten again.

The Big Bang is science's earliest memory. About 13.8 billion years ago, so the story goes, the universe sprang forth from an infinitely dense inferno. It began to expand and, as we understand now, has been expanding ever since. This current picture of the universe's evolution is relatively new. Parts of the album will seem familiar, including names and concepts you might now remember from other popular accounts: Georges Lemaître and the primeval atom; Henrietta Swan Leavitt and Edwin Hubble and the redshift of the expanding universe; Emmy Noether and Albert Einstein and the mathematics of general relativity; Roger Penrose and Stephen Hawking and the singularity of infinite density, pressure, and temperature.

But at the heart of this book, we hope to give you a glimpse of what remains uncovered in that album in the dusty attic. Cosmologists are now searching for a lost memory, a forgotten time of the universe from before the Big Bang. And just as our relatives might disagree about long-past times, we'll see that scientists also bicker as they try to reconstruct the deep past.

That's where we come in. This book is about the origin of our origins and the need for new physics. We want to guide you through a new landscape of ideas, show the strengths and weaknesses of competing models, and tell the stories of how science is really done. This book is the account of the very human struggle of those who dare to reach for the most ancient forgotten era—through collaboration and consensus, but also through bitter rivalry and conflict.

The story of the Big Bang, like much of modern physics, can be traced back to Albert Einstein. During the darkest days of World War I, he unveiled his greatest masterpiece: the theory of general relativity. Einstein's idea described space and time as one unified whole, a living, breathing dynamic entity. Any mass or energy causes space-time to deform, creating depressions in its fabric that guide every object's path. This warping is what we call gravity. Using this new framework, Einstein realized that the universe itself is unstable. Because gravity is attractive, the combined mass of every planet, star, and galaxy should deform the entire cosmos, causing it to implode. Yet our universe persists, seemingly for eternity, in the face of gravity's fatal attraction. This conundrum led Einstein to commit his "biggest blunder," as he called it: inventing a "cosmological constant" to give the vacuum an energy that would finely balance cosmic gravitation, enabling the universe to be unchanging across time.

Einstein's cosmology started to unravel in the 1920s, when scientists like Alexander Friedmann and Georges Lemaître dared to imagine a universe that was not eternal and static but rather one that expanded in all directions like a cake rising in an oven. Later observations by Edwin Hubble and others confirmed this picture. If we trace the history of this expanding universe backward in time, we are bound to infer an inescapable scenario, one that is supported by dozens of experiments and observations carried out over the past hundred years by thousands of scientists: It all started with a Big Bang.

If drawn, the evolution of the universe resembles an ice cream cone, a pretty and elegant picture. But what lies at the tip of the cone is a paradox: infinitesimal in space and time, and yet it happened everywhere. And as for what happened before the Big Bang, beyond the tip of our cosmic ice cream cone, for decades the answer was: nothing. Mathematical analyses (known as singularity theorems) carried out in the 1960s by giants of cosmology, the Nobel laureate Sir Roger Penrose and the late, great Stephen Hawking, proved that all matter should have emerged out of this singular point, believed to have infinite density, pressure, and temperature. These theorems shut the door to any prehistory of the universe, and thus all space-time histories terminate at the Big Bang. Case closed. Asking what came before the Big Bang was treated as simply a meaningless question, like asking what is north of the North Pole. If time is part of the universe and time began at the Big Bang, then there can be no before, no cause, no explanation for what happened. The Big Bang marks the end of our search for answers. Or so we were told.

We want to debunk this myth—to reveal a gap between what the public understands as the Big Bang and what my colleagues and I think. Today, there is a broad consensus among "early universe" cosmologists that in some sense the theory of the Big Bang cannot be trusted. But I emphasize "in some sense" because the phrase *Big Bang* has more than one meaning and whether we treat the Big Bang as fact or fiction, truth or myth, depends on which definition we adopt. For there is no question that the universe is different today than it was in the past, that it has evolved over billions of years from a fiery hot dense state—the hot Big Bang. We can be confident that this Big Bang happened. But if what we mean by *Big Bang* is a state of infinite density where time stands still and the answers to all our origin questions meet their demise, then no, most cosmologists today do not think that such a Big Bang happened. If you survey my colleagues (as my co-author Phil does for his YouTube series) about their accounts of how our universe looked 13.8 billion years ago, when it was hotter than a few billion degrees, or denser than water, you will get very different answers.

We should then distinguish between the observationally well-established hot Big Bang as a fiery cauldron that forged all sound, sight, and light elements in the cosmos, and the Big Bang singularity, now believed to be a mere mirage of something "infinitely stranger." Whenever you hear the phrase *Big Bang*, keep in mind these two definitions; often you can infer which one is being used by the context, but sometimes cosmologists lazily switch between the two. But always remember it's the hot Big Bang that is well established, not the Big Bang singularity. Hence, even the oft-quoted 13.8-billion-year-age of the universe is nothing but a lower limit on the age of a possibly far more ancient cosmos. Many of the scientists we shall meet in this book think the universe might have always existed. This idea represents an enormous sea change in thinking and, ironically, a return to earlier ideas about the universe. A few decades ago, the idea that the Big Bang was an absolute beginning was dogma, and few would have entertained something as radical as a pre–Big Bang universe. Now, cosmologists are competing to theorize it, using their favorite models.

We are going to dig deeper into our universal history, examine the evidence of what came before the Big Bang, and evaluate the different models for what triggered it that have emerged in the last few years. I have worked on several of these proposals, and this book is partly an account of the wonder, joy, frustration, and struggles of doing so. But of course, I am not alone; there is a community of astronomers and physicists trying to build a picture of how the Big Bang came to be.

Science is a human endeavor. And, as in any human enterprise with rival views and high stakes, competition is fierce. This saga has seen old friends become sworn enemies, and scientists are no strangers to accusations of lies and deception. Sometimes our demons get the best of us. The battle of the Big Bang is a contest to correctly describe the true origins of our expanding cosmos and push beyond the boundaries of what is currently known. The standard formula for popular books that tackle this subject is to have one of its partisans convince you they have the right model. I don't intend to argue in favor of any of the theories in this book, not even the ones I proposed. Instead, the plan is to give a fair description and analysis of the many new Big Bang models—to let you imagine all the different possibilities for the childhood of the cosmos. To get this more balanced view, I have collaborated with someone who has no stake in the game.

That leads me to a brief note on writing styles. You may notice from the cover that this book has two authors. To keep a consistent tone, we have decided that a first-person description should be from my (Niayesh's) point of view and be present throughout the book. Phil's writing brings in his outsider view, informed by years of background research and dozens of exclusive interviews with the leaders in the field, which I blend with my insider stories, insights, and struggles.

I met Phil when he traveled from the United Kingdom to interview me at the Perimeter Institute for Theoretical Physics in Waterloo, Canada. Phil has a burning passion for understanding the Big Bang, and the list of cosmologists he has interviewed for his YouTube channel is remarkable. We instantly hit it off and decided to adapt his film series *Before the Big Bang* into a book. Then COVID-19 changed all our lives. I tried to use the tools of cosmology to model the trajectory of the disease. But as Phil could no longer travel, he decided it was time to put pen to paper. For much of the pandemic, when we were all trapped at home, Phil was carrying out extensive follow-up interviews with cosmology superstars, as he wanted to get a more personal insight into the minds and lives of those who strive to uncover the mysteries of the Big Bang. The quotes you will read in this book are mostly from those discussions.

We quickly agreed that we should not just explain the science as Phil did in his YouTube films, for—as the poet Muriel Rukeyser once said—"the universe is made of stories, not of atoms." So instead, here we will tell the story of the Big Bang and of the personal journeys of the scientists who struggle over it. By telling stories, we hope to be your tour guides on this remarkable journey. And as with any excursion, we anticipate guiding a variety of curious minds with a diversity of backgrounds and levels of knowledge. Many of the ideas we will encounter are truly mind-bending, but how fully to grasp them is up to each reader. For example, several models assert that a cold, diffuse universe can be equivalent to a hot, dense one. If you want to know how such a thing could be possible, we will provide the explanation (still without equations!). But there is no harm in skipping over the fine detail to get to the big picture if that's your preference. Particularly after the first three chapters, the foundation of our story should be well set and chapters after that can be considered mostly self-contained. So, for example, if string theory is tying you down, cutting yourself free and moving onto the next big idea is always an option. The grand picture of our scientific speculations should be visible to all. The modern competing narratives of the Big Bang stir controversies over whether our world is unique or one of many, whether it has always existed or had a definite beginning, and whether it undergoes cycles of destruction and rebirth. As we will discuss, these debates are not new but have antecedents that stretch back thousands of years. And despite what you may have read elsewhere, they have not been settled.

Just like our childhood memories, our view of the Big Bang is a window into the way we see ourselves and our place in the universe. But if we do not find ways to test our ideas against the data, there is a genuine danger they will become nothing more than secular creation myths. While some of my colleagues assert that their speculative models are already verified, others say it is impossible even in principle to know what happened at or before the Big Bang. Sabine Hossenfelder, my friend and former colleague at the Perimeter Institute, now a famous blogger, YouTuber, and research fellow at the Frankfurt Institute for Advanced Studies, made this exact argument in a video released in 2022.[1]

For her and fellow travelers, I'm reminded of the case of the philosopher Auguste Comte. In 1835, the German astronomer Friedrich Bessel finally measured the distance to the star 61 Cygni and found it was inconceivably vast. Comte, not unreasonably, concluded: "All investigations which are not ultimately reducible to simple visual observations are . . . necessarily denied to us. We shall never be able by any means to study their chemical composition." But by 1864 the astronomer William Higgins did the impossible and measured the chemical compositions of a star using a new technology known as spectroscopy. Similarly, Wolfgang Pauli, in 1930, admitted shame when he suggested the existence of elusive particles known as neutrinos, saying, "I have done a terrible thing, I have postulated a particle that cannot be detected." But, yet again, only a few decades later, experiments detected the impossible. Future generations can often learn from our mistakes, striving beyond our wildest dreams. What seems impossible today may be possible tomorrow. In our final chapter, we will show how we can put new speculative ideas that go beyond the Big Bang to the test.

As you consider the stories we are about to reveal and become your own cosmological genealogist, here are some pro tips. Theories must make predictions that can be tested against observational data. But even before the data are in, we can examine other criteria. These include *simplicity* (the fewer assumptions a theory makes, the better), *consistency* (both with fundamental physics and known data), and *continuity* (the ability to explain why the universe is the way it is, in order to answer the riddles that lie at the origins of our expanding universe). But only nature can decide which idea is the real Big Bang theory and which theories are, like imaginative children, merely pretending.

We shall see that the story of the Big Bang is not just that of a scientific endeavor but also one of humanity. It is a story of underdogs, as fringe ideas fly against the hegemony of the mainstream. It is politics, as power and prestige influence our intellectual pursuits and celebrations. It is tragedy, as some of us succumb to our most ancient demons. And finally, it is community, as scientists share their cosmic curiosities to reveal what was lost or never known.

✷ 1 ✷

THE BEGINNING: COSMOLOGY FROM THALES TO HAWKING

I was born during a revolution. The political winds that had changed the landscape of the twentieth century had eroded the dynasties that ruled the ancient land of Persia for millennia before sweeping them away to leave the Islamic Republic of Iran. The Iran–Iraq war, one of the longest-running armed conflicts of the twentieth century, started when I was a toddler and only ended when I was about to turn ten and finish primary school. It was strange then to grow accustomed to living without the state of constant warfare, without a perpetual enemy (spoiler: our leaders never did). But life goes on, and as a middle schooler growing up in a bustling metropolitan area with some of the worst air pollution in the world, I decided to pick a new challenge: amateur astronomy! It was not easy to get a telescope in Iran at that time, but there was something addictive about trying to make out ghostly celestial figures in the dead of the night, looking through an eyepiece on a rooftop in northern Tehran.

This era turned out to be a defining one in my life, one that later took me as a teenager to Oslo in 1996 to compete in the International Physics Olympiad. I was great at solving problems with pen and paper, but I was terrible at doing experiments. I later learned that I was embracing an illustrious tradition among theorists, a tradition that even has its own name: the Pauli effect, after the luminary theoretical physicist and founding father of quantum mechanics whose mere presence apparently broke any expensive experimental apparatus. Having tanked the experimental portion of the Olympiad, I was thus relieved to still win a silver medal.

This experience helped propel me to North America and to follow an academic career in theoretical astrophysics and cosmology. This book's co-author, Phil, a recovering derivatives trader, has also pursued amateur astronomy, in the heart of light-polluted London, another bustling and rebuilt metropolis. Thus, we both have in common the pursuit of our passions for astronomy, in some of the worst places in the world to study the sky.

Once you start to look at one celestial object—the rings of Saturn, the Ring Nebula, or the Globular Cluster M13, all visible from both Tehran and London—you can't help but wonder where they all come from. That curiosity led us both from astronomy to cosmology, albeit in very different ways: I did it professionally, while Phil traveled the globe interviewing the world's leading experts.

As we've now discussed, the current conventional picture of the Big Bang tells us that the universe began about 13.8 billion years ago in a *singularity*, a state of infinite density, pressure, and temperature. From this singularity, the universe began to expand and behaved like a giant nuclear-fusion reactor, cooking basic elements like hydrogen and helium in the hellishly hot conditions of the primordial cosmos. As the universe expanded, it cooled; galaxies started to form, stars shone into the darkness, and planets took shape, including our own beautiful blue marble. This account is the standard story of our universe, built over thousands of years of inquiry. To some extent, the story begins in my homeland of Iran, arguably the oldest country in the world and a nation that has been the favorite battleground for the likes of Alexander the Great and Genghis Khan and for the Roman, Russian, Ottoman, British, and American empires. But before the mighty Persian kings ruled the land, there were their predecessors, the Medians, who were at war with their neighbor, the ancient kingdom of Lydia (now in modern Turkey).

Our story begins during this war. The year is 585 BCE, the date the 28th of May, and the time 7:29 p.m. In the height of battle, something wondrous happened. It started in Nicaragua, traveled at supersonic speed across the Atlantic, and finally reached Anatolia, the scene of the fighting. Day suddenly turned into night, the stars came out, and a vast black disc engulfed the sun. Phil and I have both been lucky enough to witness total solar eclipses and were blown away by their majesty, even though we knew what was coming. One can only imagine the astonishment the combatants must have felt at the heavens above them. Their reaction, as the story goes, was to lay down arms and sign a truce. According to Herodotus, a philosopher called Thales had predicted the eclipse. The science fiction novelist Isaac Asimov called this amazing intellectual feat "the birth of science." I'm not convinced that Thales really did make this prediction, but we do know that he was a brilliant mathematician, and Aristotle credits him as the father of Greek philosophy. What made Thales and his school in nearby Miletus so revolutionary, the reason why I want to start our journey here, is that he was one of the first to assert that the cosmos's origin did not lie in the creative act of a deity.[1]

Take, for example, the Enūma Eliš, the Babylonian creation story; in this tale, the world was made from primordial oceans. The idea of elemental water is pervasive in antiquity. In Egyptian creation narratives, the world emerges from a sea of chaos called Nu. In the Hindu Vedas, Danu is the primordial water and Mother Goddess. In the Bible, God moves over the face of the waters before creating land, and, in the Islamic Hadith, Allah fashions a throne above the waters before making the Earth. The waters are often separated by a solid, transparent dome of the sky known as the firmament and assumed to be chaotic. The very word *cosmos* is perhaps better translated as *order* rather than as *the universe*. Cosmology for the ancients represented the birth of order from chaos, a mission that reverberates across modern cosmology to this day. It is possible that Thales's travels in the Near East influenced him. He reasoned that there must be some substance that persists through time from which all things originate; he called it the *arche*, from the ancient Greek word for "beginning." Arche can be thought of as the first cause. Noticing how water could so easily change its form, he concluded that it must be the arche and that the Earth rests on a sea of primordial liquid.[2]

Thales's best-known student was Anaximander, who was something of a rebel, refusing to follow his mentor blindly. Instead, he did what all good scientists should do and questioned the views of his teacher (though, from experience, I suggest that you do so diplomatically). Anaximander rejected the claim that the Earth needs something to support it, saying that it "is held up by nothing, but remains stationary owing to the fact that it is equally distant from all other things." For Anaximander, the world came not from water but from the *aperion*, or the infinite, and "all things originate from one another and vanish into one another according to necessity." Sixth-century philosopher Simplicius tells us that Anaximander was the first to suppose an unlimited *kosmoi*, with each individual *kosmos* coming from "that which has no boundaries." Anaximander proposed that there wasn't simply one world but an infinite number of them (a notion that arguably continues in physics today as the idea of the multiverse). Humans, he claimed, evolved from other creatures, and earthquakes and lightning have natural causes rather than representing the anger of the gods. Some have argued that this stance justifies calling Anaximander, Thales, and their compatriots in Miletus the world's first scientists.[3]

Regardless of who was first, many people today agree that science evolved gradually. By one account, we could say the first step was taken deep in antiquity, especially in Egypt and Babylon, when people recorded the position of the stars; science, after all, needs data. The philosophers of Miletus took a next step by looking for natural explanations for the world around them. Pythagoras, of the geometry theorem fame, took a third step by teaching that the language of nature is mathematics. Some say that Pythagoras was the first to teach that the Earth was a sphere.[4] Thales had thought it a flat disc, Anaximander a cylinder. Scholar of Greco-Roman philosophy David Furley claims, however, that it was Parmenides who had the idea first.[5] Parmenides said that "nothing comes from nothing," claiming the universe arose from mixing the elements of darkness and light. The sun, he said, is "an exhalation fire," as is the Milky Way. The idea that something cannot come from nothing inspired Empedocles to speculate that the universe must be cyclic, going through phases of destruction and creation. It is love that brings things together, but if love were the only force in the universe, everything would collapse in on itself. To balance the attractive force of love, strife tears things apart, and this pattern continues back and forth in equal duration. As unscientific as this narrative may sound, and despite our dramatic progress in understanding the forces of nature, we shall see that these themes continue throughout the history of cosmology.

The Ionian League of Greek cities that included Miletus eventually found itself in revolt against the Persian emperor Darius the Great (داریوش کبیر). It was brutally crushed. The older men were killed, the women and children sold into slavery, and the young men castrated. The war between Greece and Persia would last for fifty years, and, when peace finally returned, the Ionian tradition would be revived, most notably by Democritus. He claimed that the world is not made of earth, wind, fire, and air as was then commonly supposed. Instead, he held, it consists of tiny, invisible materials that do not decay. These cannot be cut into smaller pieces, nor can they be created or destroyed; he named them atoms. Democritus tells us that only two things exist: atoms and the void. Everything is made from atoms rearranging themselves, there is no purpose to the world, and our universe is created by "cutting off from the infinite" sea of atoms. Many other *kosmoi*, he argued, will be created in the same way, and the process is eternal. This huge sea of infinite possibility is the reason we have order in the world. As the Roman atomist Lucretius tells us, over infinite time the atoms "have made trial of all things their union could produce; it is hardly surprising if they come into arrangements and patterns of motion like those repeated by this world."[6] That is to say, "anything that can happen will happen." More than two thousand years later, this principle would become part of the foundation of quantum theory and a key piece of many proposals for the Big Bang, creating our universe out of a multiverse of madness.

Alas, the preceding story—attributing the theory of atoms to one man—seems a simplification of Westernized history books. Democritus, after all, had journeyed to India; his travels may have led him to the teachings of the Vedic sage Uddalaka Aruni, who hundreds of years before also suggested that "particles too small to be seen mass together into the substances and objects of experience."[7] As the atoms mass together, we have cosmos and as they come apart, chaos. This description of the transient nature of the universe was echoed by a rival Greek school known as the Stoics, who believed in Ekpyrosis, a cyclic universe that would have its birth and death in a cleansing fire. My former colleagues from Princeton and Cambridge revisited the idea in the twenty-first century. Even earlier, the belief in a cyclic universe was not unique to the Stoics but also permeated Hindu and Buddhist thought.

Plato hated the purposeless universe of the atomists and vowed to burn their writings. For him, God created the world. Order existed because it was good and so God fashioned the universe from chaos to cosmos. Plato's most brilliant student was Aristotle, tutor to Alexander the Great, a peerless general who was never defeated in battle. The enormous scope of Alexander's empire had huge impacts on all philosophical matters for more than a thousand years. To Aristotle, the heavens were the realm of perfection associated with God and so, contrary to the atomists' view, would not follow the same principles as the imperfect materials on Earth. The sun, the moon, and the five planets moved inside "the seven heavens," or crystal spheres, around the spherical Earth, traveling in circular orbits that stretched back in time to eternity.

As Greek civilization eventually went into decline and Christianity began to dominate the Western world, a new concept for cosmology took hold: creation ex nihilo, from nothing. God, philosophers claimed, did not need existing materials to make the universe. The idea slowly became popular, finally becoming official church doctrine in 1215 CE. But Jewish philosophers like Philo argued that the concept was absent from Genesis and that God generated the world without a beginning. It was a view shared by Aristotle himself, who thought that God produces the universe like an object casts a shadow: both have existed forever. One creation story that especially intrigues me, as it really does support creation ex nihilo, is that of North America's Kiowa Apache: "In the beginning nothing existed—no earth, no sky, no sun, no moon, only darkness was everywhere."[8] There is also a Hindu hymn (the Nāsadīya Sūkta) which states, "Then even non-existence was not there, nor existence. There was no air then, nor the space beyond it . . . At first there was only darkness wrapped in darkness . . . who can say Whence it all came, and how creation happened? the gods themselves are later than creation, so who knows truly whence it has arisen?"[9] I'm immensely impressed that our ancestors would entertain skepticism about their creation narratives. Such doubts about one's own lore would be less welcome in years to come as organized religions came to dominate the world. Through the Islamic Golden age that advanced science and astronomy during the European dark ages, theologians continued to debate whether God created the world from nothing (creation ex nihilo) or from preexisting materials (creation ex materia). And whether a deity created an infinite or finite cosmos.

For millennia, these competing models existed—of the atomists who preached an infinite random universe of multiple *kosmoi*, of the cyclic cosmology of the Hindus, Stoics, and Buddhists, or of creation from nothing. But the scientific revolution of the seventeenth century that so radically transformed the world would appear to settle the debate. It wouldn't be the last time a premature conclusion would be embraced by the establishment.

In his great 1687 work, *Philosophiæ Naturalis Principia Mathematica*, known as the *Principia*, Isaac Newton demonstrated that the same laws governed motion here on Earth and across the heavens.[10] But Newton soon realized that his universal force of gravitation would cause all objects to attract one another. He was faced with the same problem that occurred to fifth-century Greek philosopher Empedocles, who claimed that love brought all objects together. If that were so, why wouldn't the universe simply collapse in on itself? Empedocles appealed to strife, Newton to the infinite universe of the Greek atomists. In such a scheme, there is no center and so matter would not collapse but could exist eternally in a stable state—as unseen bodies balanced each other, tugging across infinity. The universe was not only enormous beyond measure, but also it had always existed. God may still have created the universe but did not do so at any point in time. The idea of a static, eternally old universe became the new dogma, and few in that era would question the wisdom of Newton, who had become the new giant of natural philosophy (what we now call science). But in a paradox attributed to the eighteenth-century German astronomer Heinrich Wilhelm Olbers, it was shown that if there were an infinite number of stars and the universe was infinitely old, every point in the sky should show a star's light. The heavens should be permanently illuminated, and the darkness of night would be impossible. Thus, the universe cannot be both static and infinite, a clue that a seismic shift in humanity's view of the cosmos was coming.

THE PRIMEVAL ATOM

As our story of science began with an eclipse, so did the birth of modern cosmology. In 1919, newspapers around the world heralded a triumphant set of observations made by a team led by British astronomer Arthur Eddington: the *Times of London* read "Revolution in Science: New Theory of the Universe. Newtonian Ideas Overthrown," and the *New York Times* said scientists were "agog over results of eclipse observations." As with the eclipse of Thales, a dazzling astronomical event had heralded a revolution in our understanding of the cosmos. What had been made visible by the eclipse was the bending of a star's light by the gravitational field of the sun—a bend of precisely the amount predicted by a new theory of gravitation. Over a decade earlier, Albert Einstein had puzzled over the strange behavior of light. In the nineteenth century, Michael Faraday and James Clerk Maxwell had proposed one of the most radical shifts in physical theory in history. They claimed that electricity and magnetism were linked together in a single entity, the electromagnetic field. In this theory, light was the propagating waves of this field, which traveled at a constant speed relative to an imagined medium known as the ether. Yet, in 1887, physicists Albert A. Michelson and Edward W. Morley made a surprising discovery that even though Earth travels around the sun through this hypothetical ether, the speed of light appears unchanged. The two used an interferometer, an experimental device that splits a light beam into two separate paths and then recombines them to create an interference pattern. The motion of the Earth through the ether should cause a distortion to the interference pattern, as according to our best understanding, one of the beams (depending on its direction) would be traveling more slowly than the other, alternating from day to night. But no effect was detected. This finding represented a deep conundrum, as it appeared that light was fundamentally different from other types of waves we knew; light's speed was rigid, bizarrely indifferent to the velocity of the observer.

Einstein's resolution to this problem was bold and daring. It started by acknowledging that speed is simply distance over time. Thus, if the speed of light is a universal constant, then it must be distance and time that are relative. Time slows down and lengths contract for a speeding observer. This effect is negligible at the pedestrian speeds of our humdrum existence, but if one could travel at light speed, the universe's maximum allowed velocity, time would stop ticking altogether. These amorphous roles of space and time suggested a fusion of the two into a single creature: space-time. Thus, as a consequence of Einstein's new principles, the ether was vanquished from modern physics (only to return, in a different form, a century later, with a vengeance; read on). Einstein's new theory became known as special relativity, "special" as it was limited to worldviews of observers at constant velocities. With ether's demise, all that was left were the fields, as the fundamental building blocks of nature. Particles are just waves that propagate when you excite a field, and the quantum revolution, initiated by Einstein amongst others, was the realization that these excitations happen in discrete units, or "quanta." Today, many scientists consider the quantum theory of fields as the most precise theory of nature. But there was one thing missing from that theory: gravity.

In 1915, Einstein unveiled the theory that would be confirmed by the solar eclipse in 1919. General relativity unified gravitation with the experiences of accelerating observers. But the new theory was far more profound yet, for it described nothing less than the dynamics of the very fabric of space-time. Every object in the universe bends the space-time around it, creating depressions that we feel as gravity. The deeper into a gravitational well one falls, the slower time ticks. Space-time is not just an empty nothingness but an elastic fabric that can deform and expand. And it wasn't just the local space around massive objects that the new equations could examine, but the entire fabric of the cosmos. Einstein soon found himself confronted with the puzzle that Empedocles and Newton had struggled with; in general relativity the universe was unstable: it would either expand or collapse, depending on how much matter and energy it contained. That brings us back to the "blunder" mentioned in this book's introduction. So, in 1917, Einstein modified his equations, adding a "cosmological constant" that represents the energy of empty space-time itself and could be set to match observations of a static universe. But new astronomical discoveries would show this addition to be superfluous.

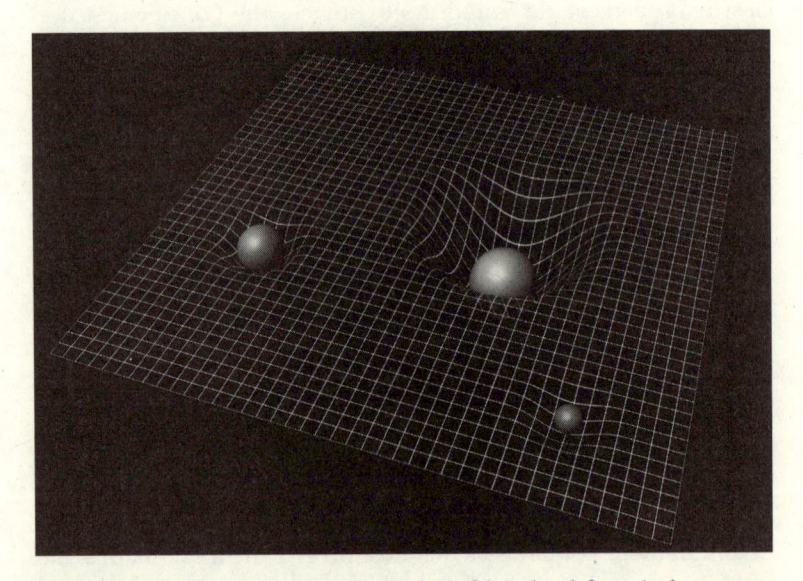

In Einstein's relativity, space-time is like an elastic fabric that deforms in the presence of matter. Planets, like Earth, follow the longest path (not the shortest, as is often said) through warped space-time. Gravity becomes geometry. Since it is hard to visualize curvature in four dimensions, this graphic only depicts the curvature of space. Copyright: ESA—C. Carreau.

For centuries, astronomers had observed nebulae, a term that comes from the Latin for "mist" or "cloud." These phenomena appeared as fuzzy patches of light, perhaps the most notable example being the spiral nebula in the constellation Andromeda, but their true nature remained elusive. The philosopher Immanuel Kant described these spiral structures as "island universes," but, with no way to measure their distance from Earth, they could easily have been stars and gas within our Milky Way. In 1920, the American National Academy of Sciences hosted "the Great Debate" between Harlow Shapley, who argued that these objects all existed within our Milky Way, and Herbert Curtis, who argued that some nebulae were island universes.[11] Neither man could sway the other; it seemed that the existence of other island universes was a controversy impossible to resolve. But Henrietta Leavitt, part of a small band of poorly paid female astronomers known as the Harvard Computers, had already devised a technique that would settle the issue. She studied Cepheid variables, relatively bright stars that pulse regularly over periods that range from days to months. Her analysis showed a clear relationship between the period of their variability and their average intrinsic brightness. With that knowledge, one could deduce how intrinsically bright a Cepheid was and thus infer its distance from its apparent brightness, as faraway objects appear fainter. Leavitt's variable stars would become "standard candles" to determine vast distances across deep space, still used to this day by astronomers. This revolutionary discovery enabled Edwin Hubble to finally measure the distance to the Andromeda nebula. His answer was a mind-boggling nine hundred thousand light-years, putting Andromeda far beyond the confines of our own galaxy, the Milky Way. The great debate was over. Thanks to Hubble's distance measurements using Leavitt's technique, we now believe that island universes do exist; today we call them galaxies.[12] A century later, I would end up in the same institute as Leavitt and revisit what she went through, but that's another story for later in this book.

The same year that Leavitt published her groundbreaking law for Cepheids, Vesto Slipher had begun to analyze the light coming from spiral nebulae. He found that most of them were shifted to the red end of the spectrum, indicating that they were moving away from us at mind-boggling speeds of hundreds of kilometers per second. This "redshift" is often said to be caused by the Doppler effect, described using the analogy of a person hearing the pitch of an ambulance siren change as it moves. When the ambulance is approaching, the sound waves get bunched up, making the pitch higher as the waves shift to shorter frequencies. When the ambulance recedes, the reverse happens: the sound waves get spread out, the pitch declines, and the waves are shifted to longer frequencies. In the visible part of the electromagnetic spectrum, red has the longest wavelength and blue the shortest. So, objects moving away from us are said to be "redshifted"; and those toward us, "blueshifted." What is true for sound waves is true for light waves. We can also think of the universe stretching light to longer wavelengths without any reference to objects moving past one another.

Whichever way we think of it, Slipher's data showed that most galaxies were moving away from the Earth. Building on this discovery, Hubble (with much help from Milton Humason, a former janitor at the Mount Wilson observatory, who without any formal education became one of its leading astronomers) showed that redshift measurements were no coincidence. The farther away from us galaxies were, on average, the faster they were continuing to move away from us: exactly what we'd expect to see if the universe were expanding. The centuries-old belief in a static cosmos was shattered. Hubble's observations had confirmed the thinking of scientists like Alexander Friedmann and Georges Lemaître, who had used general relativity to predict a dynamic cosmos, overturning the Aristotelian vision of the world that had ruled for millennia. Friedmann and Lemaître found that while galaxies could move apart from each other in an expanding universe, the story could be reversed such that the entire cosmos could originate from a superdense state, what Lemaître called "the primeval atom." In explaining this new vision of our origins, he made an analogy with "atomic" physics, describing the expansion of the universe as akin to a sort of radioactive decay. "We live," he said, "in the smoke and ashes of bright but very rapid fireworks."[13] As space and time are fused together in Einstein's relativity, the beginning of space was also the beginning of time. The primeval atom would be "a day without a yesterday." Lemaître became known as the father of the new theory, a pun on the fact that he was also a Catholic priest. Whilst some scholars debate who should be given credit for proposing an expanding universe, others wonder if both scientists weren't perhaps influenced from a rather unlikely source.[14]

In 1848, American poet and writer Edgar Allan Poe wrote a prose poem entitled "Eureka," describing a world coming from "unity" with an origin point as "primordial particle" with "absolute extreme of *Simplicity*. Here the Reason flies at once to Imparticularity—to a particle—to one particle—a particle of one kind—of one character—of one nature—of one size—of one form—a particle, therefore, without form and void, there would be diversity out of sameness—heterogeneity out of homogeneity—complexity out of simplicity."

STEADY STATE VS. BIG BANG

In 1931, Einstein visited Hubble, a meeting that symbolized the triumph of the expanding universe. This event is often painted as a fait accompli, and indeed Einstein admitted that the new results "smashed my old construction like a hammer blow."[15] But that's a caricature of a more nuanced story, for lurking in Hubble's data was a serious problem. It could be used to determine the time since all the galaxies were squeezed together; the answer was an embarrassing 1.8 billion years. I say embarrassing, because, by the 1940s, geologists determined the Earth was 4.5 billion years old. The universe, it seemed, was younger than the Earth.

A version of this problem, known as the Hubble tension, is still with us, even a century later, despite tremendous advances in astronomical measurements of the expansion of the universe. Similarly to how geologists use radiometric dating to determine the age of old rocks, modern cosmologists can use the afterglow of the primordial fireball to determine the time since the hot Big Bang, and they still do not agree with astronomers who use dying stars as standard candles (brighter versions of Leavitt's Cepheid variables) to make the same measurement. Hubble himself refused to endorse any cosmological interpretation of his data. Maybe the reason for his refusal was this tension in measurements of the Hubble constant, or perhaps he thought that as an observer it was not his place to endorse a specific model.

But that was not the only problem. As light redshifts and loses energy in an expanding universe, where does the energy go? Could it be that the sacred law of conservation of energy, like Newtonian mechanics, had to be let go in our brave new expanding world? The answer came thanks to Emmy Noether, a brilliant mathematician struggling to be heard in a male-dominated world. The academic senate of her university had, only two years before admitting her, declared that mixed-sex education would lead to the overthrow of all academic order. At the invitation of David Hilbert and Felix Klein, two of the most highly respected mathematicians of the age, she attended Göttingen University, the Mecca of the subject. But even here others objected, saying, "What will our soldiers think when they return to the university and find that they are required to learn at the feet of a woman?"[16]

Noether, for the first time, understood the deep connection between the conservation laws and the symmetries of the laws of nature. In Newton's world, the laws are the same, no matter which origin you choose for your clock (be it, say, standard or daylight saving time). In 1918, Noether proved that this consistency implied a conserved energy, no matter whose laws (Newton or Einstein) you follow. However, in Einstein's theory every observer is free to choose their own clock, independent of others, multiplying the number of these conserved quantities by an infinite amount, but only if you include cosmic expansion and curvature. In other words, conservation of energy is a reasonable assertion for local observers, but in the vast distances of the cosmos, there is really no good definition of conserved energy. In reading Noether's paper, Einstein wrote to Hilbert: "Yesterday I received from Miss Noether a very interesting paper on invariants. I'm impressed that such things can be understood in such a general way. The old guard at Göttingen should take some lessons from Miss Noether! She seems to know her stuff."

Nevertheless, Einstein was uncomfortable with the new picture of cosmology and secretly worked on a new theory that would resemble the classical picture of a static universe. He realized that "for the density to remain constant, new particles of matter must be continually formed within that volume from space."[17] Einstein imagined a universe expanding without an origin; new particles could be created, arising from the cosmological constant that he hypothesized gives energy to empty space. The universe's expansion diluted new matter, so the total density of matter stays constant. To understand this difference, cosmologists often imagine a cake baking with raisins. But let's imagine it contains chocolate chips, as Phil hates raisins. As the cake bakes, it expands, but the number of chocolate chips does not change. This means that the density (the number of chips divided by the space they take up) declines. However, let's imagine I am a baker with a sweet tooth and good oven mitts, and I poke new chocolate chips into the batter as the cake expands. Now the density could stay the same. Replace the chocolate chips with galaxies and the rising cake with expanding space, and we can see the contrast between Lemaître's primeval atom and Einstein's secret model. However, Einstein couldn't get the idea to work, as it violated Noether's conservation laws, and so he let it sit on the shelf, only to be discovered in 2014 in his archives, held at the Hebrew University of Jerusalem.[18]

Einstein was not the only one to ponder this alternative to the primeval atom. In Britain, three physicists who worked on top-secret radar research during World War II had a similar thought. One night in 1945, Fred Hoyle, Thomas Gold, and Hermann Bondi went to the cinema to see a new horror movie called *Dead of Night*, which ends exactly as it begins. This unusual narrative inspired Gold to wonder if the universe might behave similarly. Not that it would literally cycle back to its beginning. Rather, there would be no beginning, and any starting point would be as good as any other. As Hoyle described, "What the ghost-story film did sharply for all three of us was to remove this wrong notion. One can have unchanging situations that are dynamic, as for instance a smoothly flowing river."[19] Personally, I think more cosmology groups need to watch horror movies together.

Hoyle, Gold, and Bondi had independently stumbled upon the same model that Einstein had abandoned: a model in which, like ghosts appearing out of nowhere, new matter is created in the space between expanding galaxies, so the density of the universe never changes. The universe has no beginning and will have no end; they called the idea "the steady state." Unlike Einstein, they did not appeal to new matter arising from the cosmological constant. Instead, they added a "creation field" that would imbue space with the ability to make one hydrogen atom per cubic meter per hundred billion years—all that would be needed to create new galaxies that would later drift apart, matching Hubble's observations of the expanding universe. The trio argued this model was more parsimonious than one in which all the matter in the cosmos was created in a sudden explosion.[20]

Hoyle would soon become the public face of popular science in Britain. He introduced his radio show listeners to two rival cosmological theories, his own steady state and Lemaître's primeval atom. During one such broadcast monologue, he described the latter as saying that "all the matter in the universe was created in one Big Bang at a particular time in the remote past."[21] It has become legend that Hoyle invented the term *Big Bang* to ridicule the theory; this legend may have come about because some of its proponents recalled him using the term this way in a debate on BBC radio. As historian Helge Kragh pointed out, there never was such a radio debate, and in a later interview, Hoyle denied the phrase was meant pejoratively; he was simply explaining, on his own radio show, the rival models in layman's terms.[22] Nevertheless the name stuck. As Hoyle said, "Words are like harpoons. Once they go in, they are very hard to pull out."[23] The phrase *the primeval atom* faded into obscurity, replaced by *the Big Bang theory*.

The Big Bang proponents who misremembered Hoyle's comments were students of George Gamow, who had risked his life to defect from the Soviet Union to the West. In the 1940s and '50s, Gamow pondered what would happen if he traced our expanding universe back in time toward the Big Bang. As the universe contracted, it would get hotter and denser, and matter would break down into a soup of fundamental elements. The universe would become a giant nuclear reactor. Working with his student Ralph Alpher, Gamow demonstrated that the primeval universe could rapidly create basic elements such as hydrogen, helium, deuterium, and lithium—and in just the right proportions to match astronomical observations.[24] The press soon picked up on these calculations, the *Washington Post* explaining them with the headline, "World Began in Five Minutes."[25]

The story became so popular that hundreds of people showed up to witness Alpher's PhD defense. But that breakthrough was only the beginning of the profound discoveries Alpher would make. In collaboration with Robert Herman, the two soon theorized that, shortly after the Big Bang, the universe would be in an excited state of matter known as plasma, in which electrons are stripped from atoms. After 380,000 years of expansion, the cosmos would have cooled enough for the electrons to attach to protons and neutrons to form atoms, releasing primeval light as the fog of plasma disappeared. If the Big Bang theory was correct, the entire universe should be bathed in this relic radiation. Further, their theory predicted that the frequency of this radiation would have shifted from red light into cold microwaves as the universe continued to stretch. Hence this fossil of the Big Bang became known as the cosmic microwave background, or CMB.

Alpher and Herman gave multiple talks about the search for the CMB, but few other scientists appeared interested. Many were unconvinced by the ludicrous results that seemed to show the universe was younger than the Earth. Others thought the whole idea of studying cosmology was pointless. "Don't let me catch anyone talking about the universe in my department!" said leading atomic theorist Ernest Rutherford.[26] Herbert Dingle, former secretary of the Royal Astronomical Society, described cosmology as a "paralysis of the reason with intoxication of fancy" and bemoaned great scientists falling for "universe mania."[27] Russian physicist and Nobel laureate Lev Landau said, "Cosmologists are often wrong, but never in doubt."[28]

You may have read elsewhere that the CMB was first discovered in the United States in the mid-1960s. But, like the tale that Hoyle invented the term *Big Bang* as an insult, this story seems like yet another myth. The CMB was actually *detected* in 1940 in my adopted home country of Canada and was even published in a textbook printed as early as 1950.[29] The cold darkness of interstellar space seems to be empty but is filled with hundreds of different molecules; there are even giant clouds of alcohol in outer space. Andrew McKellar was an astronomer who pioneered techniques to look for these interstellar chemicals. His studies of cyanogen (a toxic carbon gas) and methylidyne (another carbon-based compound common in the vast regions between the stars) enabled him to probe the temperature of the surrounding space, which he found to be 2.3 degrees above absolute zero, and so he became the first person in history to see the afterglow of the Big Bang. But seeing it and recognizing it are not the same thing. Nobody at the time realized the significance of what McKellar had found. Ironically, the first person to use McKellar's data in a cosmological setting was Hoyle, who claimed it refuted the Big Bang as the CMB was expected (incorrectly) to be several degrees hotter than McKellar's measurement. As Hoyle said in an article in the *New Scientist*, "I recall my telling George [Gamow] that it was impossible for the Universe to have a microwave background with a temperature as high as he was claiming, because observations of the Methylidyne and Cyanogen radicals by Andrew McKellar had set an upper limit of 3K for any such background."[30]

Discovering the CMB was not the only way to explore the early universe. Astronomers also surveyed galaxies to see if their distribution changed with time. A positive result would support the Big Bang; a negative result would favor the steady-state model. Another British wartime radar veteran, Martin Ryle, began to improve his instruments' sensitivity to provide just such a probe. But before he could survey galaxies en masse, he found a strangely bright radio signal from an unknown source. At a conference at University College London in 1951, Gold and Hoyle asserted that the bright radio waves must be galaxies, but Ryle claimed they were nearby stars and publicly mocked what he thought was the ignorance of theorists. Follow-up observations from the 200-foot Hale telescope (the largest optical telescope in the world for more than twenty years in the postwar period) showed without doubt that the mysterious radio waves were indeed from a distant galaxy, just as Gold and Hoyle had claimed. When the results were shown at an astronomy meeting in Rome, Ryle was humiliated and allegedly burst into tears.[31] This incident was the start of a poisonous rivalry between Ryle and his fellow Cambridge astronomers Gold, Bondi, and Hoyle. Oxford cosmologist Pedro Ferreira, in his history of relativity, described Ryle as a "volatile, irascible character, competitive and suspicious."[32] Woodruff T. Sullivan III, a historian of radio astronomy, concurred, saying that Ryle and Hoyle would "go after each other like a mongoose and cobra."[33]

In 1955, Ryle published a new catalog of radio sources. It was a moment he thought would mark his triumph over Gold, Bondi, and Hoyle as the data seemed to favor the Big Bang, showing that the distribution of galaxies did change over time. But a follow-up study showed clear problems: 80 percent of their sources were nonexistent, but rather a blended jumble of fainter sources and noise. This finding was another blow for Ryle, and his desire for revenge snowballed. As the decade ended, the radio astronomers were learning from their mistakes, and, by the early 1960s, they had far more sources with fewer defects. When Ryle was finally confident in the new data, he called a press conference, inviting Hoyle to attend with his wife. Hoyle later recalled that he thought the invitation "must mean that Ryle was about to announce results in consonance with the steady state, ending with a handsome apology for his previous misleading reports."[34] As Ryle presented his newer and far more accurate survey, he showed they were incompatible with the steady-state model; the density of galaxies did change with time. He then pointed out that Hoyle just happened to be in the audience and asked if he would like to comment. The newspapers were filled with headlines showing how Ryle had turned the tables on his rival, demolishing the steady state in the process. Hoyle could appeal to previous mistakes from his Cambridge rivals, but other results were stacking up against his model.

Commentary on Ryle versus Hoyle by Barbara Gamow, George Gamow's wife:

"Your years of toil,"
Said Ryle to Hoyle,
"Are wasted years, believe me.
The steady state
Is out of date.
Unless my eyes deceive me,

My telescope
Has dashed your hope;
Your tenets are refuted.
Let me be terse:
Our universe
Grows daily more diluted!"

Said Hoyle, "You quote
Lemaître, I note,
And Gamow. Well, forget them!
That errant gang
And their Big Bang—
Why aid them and abet them?

You see, my friend,
It has no end
And there was no beginning,
As Bondi, Gold,
And I will hold
Until our hair is thinning!"

"Not so!" cried Ryle
With rising bile
And straining at the tether;
"Far galaxies

Are, as one sees,
More tightly packed together!"

"You make me boil!"
Exploded Hoyle,
His statement rearranging;
"New matter's born
Each night and morn.
The picture is unchanging!"

"Come off it, Hoyle!
I aim to foil
You yet" (The fun commences)
"And in a while"
Continued Ryle,
"I'll bring you to your senses!"[35]

Recall that one of the prime motivations for believing in the steady-state model was that the Big Bang faced an "age crisis" predicting that the universe was younger than the Earth. But new data showed that the age of the universe was based on incorrect measurements of the distance to galaxies like Andromeda (it turns out, there is more than one kind of Cepheid variable). Revisions implied a more sensible time since the Big Bang of around ten to twenty billion years, which now made it comfortably older than the Earth's 4.5-billion-year age. But the age crisis was not the only motivation for rejecting the Big Bang. Hoyle, a passionate atheist, claimed the theory smacked of divine intervention, saying, "The reason why scientists like the 'Big Bang' is because they are overshadowed by the Book of Genesis. It is deep within the psyche of most scientists to believe in the first page of Genesis."[36] This assertion was not necessarily true, as Gamow, the main proponent of the Big Bang, was also an atheist. And many supporters of the steady state were Christians like Bernard Lovell, the first director of the great Jodrell Bank telescope, and Philip Quinn, a Catholic philosopher—both of whom argued that the steady state backed the notion of God creating the universe continuously (as did Mulla Sadra, the seventeenth-century Islamic philosopher with his perpetual-creation doctrine). Quinn supported his assertions with passages in the Bible that talked of God as the sustainer of the universe. Nevertheless, Pope Pius XII declared in 1951 that the Big Bang "has succeeded in bearing witness to that primordial *Fiat lux* [Let there be light]." Lemaître, a Catholic priest, was furious with what he considered to be the pope's misguided pronouncements. One of his students recalled him "storming into class on his return from the Academy meeting in Rome, his usual jocularity entirely missing. He was emphatic in his insistence that the Big Bang model was still very tentative, and further that one could not exclude the possibility of a previous cosmic stage of construction."[37] It is also likely he knew that the pope's endorsement would play into the hands of opponents like Hoyle, and he asked for a private meeting with Pius; there is no record of what was said in the meeting, but in a subsequent speech, Pius dropped all mention of the Big Bang. Nevertheless, atheists and theists continue to argue over the Big Bang's theological implications. We will come back to this argument in our penultimate chapter.

Lemaître was not alone in wondering if some prior cosmic evolution might precede the Big Bang. One scientist who took this idea seriously was Bob Dicke at Princeton. Dicke was a man of many talents, being a brilliant engineer and superb theorist. He backed the Big Bang theory but saw it in the context of a longer evolution of the universe that cycled between expanding and contracting branches, a view shared by Gamow, who called the model the Big Squeeze. In the early 1960s, Dicke began to calculate what the universe would look like as it emerged from contraction into an expanding epoch. Like Alpher and Herman before him, he realized that there would be a hot fireball which would gradually cool down over billions of years and appear as a faint glow of microwaves in every direction of the sky. Unlike Gamow's students, Dicke possessed the skills to search for the elusive microwaves himself.

In 1964, Dicke's team were in position to make the discovery of the century, their expertise was second to none, and few scientists anywhere in the world seemed even interested in looking for the

CMB. But during one of their weekly meetings, Dicke shocked his colleagues, announcing that they had been scooped. Two scientists working for Bell Labs, just an hour's drive away from Princeton, had detected a strange hiss.

A noise source had plagued radio astronomers Arno Penzias and Robert Wilson—it was impossible to remove no matter what they tried. They had no idea of the significance of what they were observing. Eventually, Penzias discussed the problem with Bernard Burke, a physicist at MIT who had heard about Dicke's CMB project. Suddenly, the enormous implications of their annoying noise became clear. In that moment, the afterglow of creation had finally been, not just detected (as McKellar had), but also recognized, its temperature being 2.7° Kelvin: a chilly −270.3° Celsius, −454.5° Fahrenheit, or—as we like to call it here in Canada—spring. That is the conventional story; the real story has more nuance. In 1961, E. A. Ohm, an engineer at the very same laboratory where Penzias and Wilson "discovered" the CMB, had also reported an excess of radiation of between 1° and 3° Kelvin.[38] Dicke's colleague Jim Peebles later described this excess radiation as the "dirty little secret" of Bell Labs.[39] But the secret did not go entirely unnoticed by outsiders; two Soviet scientists, Doroshkevich and Novikov, had commented on it and speculated it might be linked to Gamow's theory; alas, few noticed their commentary from behind the iron curtain. So, Penzias and Wilson were not the first to discover the CMB, nor even the first to see it at Bell Labs, but they were the first to make a detection that got the attention of the world.

The discovery of the CMB seemed to seal the victory of the Big Bang over the steady state. Penzias and Wilson would win the Nobel Prize. The debate was over; the universe had evolved from a hot, dense state. At almost the same time, just across the pond, two innovative mathematicians were about to argue that this Bang marked the beginning of time itself.

SINGULARITY

As early as the 1920s, Friedmann had considered models where the universe contracted all the way to a moment with zero distance between any two points. He found that such a contraction would cause the temperature, pressure, density, and curvature of the universe to become infinite, marking what we call a singularity, where time stops and the laws of physics break down. But it is not clear that Friedmann ever took this possibility seriously. Even Lemaître thought it was possible that something physical existed before the primeval atom, and Einstein warned against trusting his theory at the impossibly high densities close to the Big Bang. Theoretical physicist and "father of the atomic bomb" Robert Oppenheimer, working with Hartland Snyder, had found similar results when considering collapsing stars so compact that even light couldn't escape their gravity—what we now call black holes. However, their conclusions were also limited to perfectly symmetrical balls of dust. So, for decades, physicists knew about the possibility of singularities but treated them mostly as novelties, with bemusement. Two Soviet physicists, Vladimir Belinski and Evgeny Lifshitz, for example, defended the view that singularities were an artifact of the unrealistic assumption of perfect symmetry.[40] This stance remained representative until the 1960s, when a brilliant English mathematician decided to take an interest in the subject.

Fred Hoyle's radio broadcasts got that mathematician, Roger Penrose, hooked on cosmology. But Penrose had questions. His brother Oliver, a Cambridge student, directed him to the person who could answer them: cosmologist Dennis Sciama. Penrose soon fell under Sciama's spell and began to develop a physical picture of black holes by focusing on the mysterious singularity. As Penrose explained to Phil during a 2020 interview, he had pondered the problem while walking with a colleague. Suddenly the conversation stopped, and Penrose remembers a "strange feeling of elation." He elaborated: "I went through everything that happened in the day, starting with getting up in the morning, and what I had for breakfast and all that . . . until it came to this crossing of the road." Penrose's idea, inspired by the traffic junction, was that of a "closed trapped surface," a space where photons (particles of light) would be captured and focused into a point at the center of a black hole, their path through time coming to an abrupt end: a singularity. Using this new insight, he showed that the unrealistic assumption of perfect symmetry in a star had no effect on Oppenheimer and Snyder's conclusion. No matter how uneven it was, a sufficiently compact star would always collapse into a singularity.

In the film *The Theory of Everything*, a young Stephen Hawking is shown listening to a talk by Penrose on his new singularity theorem at King's College London. "Stephen Hawking was not there," Penrose said, "despite the film showing sparks coming out of his head." Still, Sciama heard of the talk and encouraged Penrose to give a repeat performance in 1965 while visiting with Sciama's latest protégés, Brandon Carter, George Ellis, and Stephen Hawking. Later, he had a private meeting with Hawking, who had been working with Ellis to see if a singularity was present at the Big Bang. It was at this meeting that Penrose met Hawking. "He picked up things very quickly," Penrose said, explaining that Hawking saw the "trapped surface argument"—that traffic intersection inspiration—could be used on a cosmological scale if you considered time moving backward. "I thought that was really a clever step; this indicated that he was someone out of the ordinary." Nevertheless, Hawking's powerful intuition and quick calculations sometimes led to errors, but fellow theorist Brandon Carter helped to correct them. The important piece, Penrose explained, was that Hawking's underlying arguments were correct. Having met Carter early in the first decade of the 2000s during graduate school, I can confirm that it was impossible to get anything that was not super-rigorous past him, like the precise definition of a rotation in relativity, or, as he once complained to me, that the advertised prices in Starbucks in North America did not include taxes. But he was also a gracious man, and he would not have wanted to steal any credit from his colleague. Hawking's breakthrough was truly ingenious, turning Penrose's proof on its head. If the future of collapsing matter is a singularity, then the past of expanding matter must also have been a singularity. Hawking could use Penrose's techniques and assumptions to show that there really had been "a day without a yesterday," as Lemaître had claimed decades before.

A year after Hawking graduated, a major gravity conference was being held in Seattle. Penrose had been invited and decided to take one of Sciama's brilliant new protégés to accompany him. He considered Hawking but chose Brandon Carter, as he thought Carter would be a far better communicator. Penrose recalls feeling guilty, as "what Stephen was doing was really important." He changed his mind and invited Hawking to join him in Seattle. However, Hawking ended up leaving before he could present his own findings, forcing Penrose into a last-minute attempt to squeeze all of Hawking's ideas into the final three of twelve lectures. By his own admission, he wasted time explaining abstract ideas that nobody followed, but by the time he got around to the singularity, he had found a way to simplify the arguments drastically. On returning to London, Penrose called Hawking, excited it was possible to bring such clarity to the issues. As Penrose recalls, Hawking replied: "Yes, I've got that too."

Penrose and Hawking collaborated on several papers strengthening the argument, so that by 1970 it seemed clear that the beginning of the universe was unavoidable with ordinary matter in Einstein's theory of gravity. Time itself would stop ticking at the Big Bang. There simply was no before, so asking what caused the universe was therefore meaningless. The debate that had begun thousands of years prior, concerning the evolution of the cosmos, seemed finally settled. The universe, it was said, had a definite beginning, born of infinity, from a fiery event we call the Big Bang.

A STORY, UNFINISHED

You know by now that this story isn't finished. In this book, the Big Bang singularity is literally only the first chapter. It's now time to unpack some of the Big Bang's puzzles so we can see why cosmologists need a deep rethink.

The theory of the hot Big Bang has been enormously successful in explaining cosmological observations, from cosmic expansion to distribution of galaxies, the abundance of light elements (namely hydrogen, deuterium, and helium), and the patterns of tiny fluctuations in the intensity of the microwave light emitted by scorching-hot primeval plasma. Pushing the physics of the Big Bang to its limits, the Penrose-Hawking singularity theorems that supposedly proved an absolute beginning of time are powerful and rigorous. But any piece of mathematical machinery is only as good as the assumptions it rests upon. Hawking and Penrose made four assumptions:

1. Gravity is always attractive.
2. The universe has three space dimensions and one time dimension that we can measure with rulers and clocks.
3. Time travel into the past is impossible.
4. Einstein's theory of general relativity describes the evolution of the universe at all scales.

The reason the old picture of the Big Bang is unraveling is that we cosmologists have come to doubt every one of these assumptions. The first assumption, that gravity always pulls and never pushes, is already ruled out by cosmological observations. This finding may seem surprising, as the only gravity we ever experience is that of the attractive kind. Our feet are pulled down to the center of the Earth; the moon and sun tug at our oceans, manifesting the tides; and what goes up usually comes down. But nature has a way of pulling the rug out from under even our most basic presumptions. That was exactly what happened in 1998, when we discovered that the expansion of the universe is speeding up, driven by an enigmatic gravitational repulsion called dark energy—a repulsion that resurrected Einstein's "biggest blunder"—the cosmological constant he had originally invented to make the universe static.

Meanwhile, Penrose and Hawking have both disavowed the second assumption, admitting that the very scaffolding of the universe at the Big Bang may have been substantially different from our familiar world. Penrose argues that at the scorching heat of the Big Bang, it was too hot for nature to build a clock or ruler and so there is no meaning to the notions of distance or time that his theorems require. Meanwhile, Hawking would suggest that time morphs into a new dimension of space as we trace the evolution of the universe back to the Big Bang. Thus, the very authors of the theorem of the beginning have abandoned it.

Scientists often dismiss time travel into the past because it would violate the belief that causes must always precede their effects, so much so that Hawking conjectured it as a principle of nature. But conjectures don't have to be true. My old Princeton professor Rich Gott and I have both challenged Hawking's thesis, leading us to propose that the universe might have created itself. We'll talk about these ideas later in the book.

The last assumption, that general relativity works at all scales, is almost certainly wrong, as it contradicts the probabilistic principles of quantum mechanics, our well-established theory of the subatomic world. Quantum theory tells us that particles are not little balls, as you might have imagined in school, but rather wave packets that cannot be pinned down in a point in space; the world is probabilistic and almost anything is possible. Yet, general relativity only works when particles have definite locations that deform the space-time in their local vicinity. To imagine the Big Bang singularity is to ignore the contradictions of quantum mechanics and general relativity, the pillars of the physical world. Resolving this conflict by creating a theory that unifies the physics of the subatomic and the celestial, a quantum theory of gravity, remains the backdrop for much of this book, as it does for modern cosmology.

But the singularity is not the only enigma that plagues cosmologists. In the late 1970s, another set of mysteries directed cosmologists to rethink the Big Bang. Particle physicists highlighted the first problem, arguing that at the high energies of the primordial cosmos, the universe should produce huge numbers of strange particles called magnetic monopoles. Yet not only has none ever been found, but these particles would also be so heavy they would prevent the universe from expanding into the grandeur of the cosmos we see today.

Our second problem can be understood if we think of firewalkers walking on hot coals. This act might appear a supernatural feat of mind over matter, but the truth is that the coals are poor heat conductors, so it takes too long to transfer enough warmth to the walker's feet to cause any burns. If you replace the burning coals with a hot metal plate, a far more efficient conductor, the walker's mental powers will quickly fade. When we examine the oldest light in the universe, the CMB (cosmic microwave background), we see that the universe also appears to be performing a supernatural trick. Just as firewalkers' feet don't have enough time to thermalize (come to equal temperature) with the hot coals, calculations show that neither have CMB particles had enough time since the Big Bang to thermalize with each other. Yet they all manage to equilibrate at nearly the same temperature of 2.7° Celsius above absolute zero (or 2.7° Kelvin), far beyond their horizons (distance that is accessible at light speed). This horizon problem is the second clue that a radical revision of cosmology is needed. One might imagine it could be solved by assuming the universe was perfectly uniform to start with, but such a featureless cosmos could never have formed anything like the galaxies that adorn the heavens in their trillions. This origin-of-structure problem is our third clue that something is amiss.

The last difficulty is known as the flatness problem, and it was first pointed out in a lecture by our old Princeton friend Bob Dicke in 1978. The universe can have different degrees of curvature at large scales. This curvature is denoted by a number called omega: if its value is precisely one, then space is perfectly flat, like a sheet of paper. If omega is bigger than one, then space is positively curved, like a ball. If omega is less than one, the universe has negative curvature, like a saddle. Cosmologists of the 1970s could not precisely measure omega, but most estimates had it close to one.

According to general relativity, setting omega equal to one would require a fine balance between cosmic expansion rate and cosmic density. Think of galaxies as projectiles flying away from each other: if they are moving too slowly, they will stop and fall back, which happens when omega has a value bigger than one. If they move too fast, then they can happily escape each other's gravity and coast freely; that's what happens when omega's value is less than one. Omega = 1 is the precise twilight zone between these possibilities.[41]

What Dicke wanted his audience to understand was that a flat, expanding space was inherently unstable. If omega is close to one, but not exactly one, it will rapidly diverge away from that value.

To be close to one today, it would have had to have been extremely close to one in the past. The very shortest duration physicists contemplate in the presence of quantum mechanics and gravity is called the Planck time (more on this point later). Start one Planck time after the Big Bang, and omega would have had to have been set to one by an unbelievable fifty-eight decimal places. The consequences of omega not being one could be catastrophic for life in the cosmos, as omega's value controls the expansion rate of the early universe. If the universe expanded too rapidly, it could be torn apart; too slowly, its expansion could stop and collapse. One scientist who was in the audience at Dicke's talk that day was Alan Guth, who took note of Dicke's "fantastic" and "outrageous" claims.

Guth's fascination with science began when he saw the popular 1950s and '60s TV show *Watch Mr. Wizard*, which formed the basis for the Professor Proton character on the more recent television sitcom *The Big Bang Theory*. Guth's aptitude for mathematics enabled him to skip his senior year in high school and enroll at MIT. But as the saying goes, in academia one must "publish or perish," and three years after graduating, his sum total of published papers was a paltry one.[42] But everything was about to change. Guth and his colleagues would stumble upon an idea that not only had the potential to answer all these problems but would revolutionize cosmology and revive the debate that began in ancient Greece over whether there existed multiple kosmoi or a single, cyclic universe. Even Penrose and Hawking would publicly denounce the Big Bang singularity and embrace the notion of other universes, although they would not agree as to whether such universes were separated in space (a multiverse) or in time (a cyclic cosmology).

☆ 2 ☆

THE MULTIVERSE AND THE ULTIMATE FREE LUNCH

In early 2020, the entire world was reminded of an important mathematics concept: exponential growth. What had seemed like a problem that existed only on news screens and in China was beginning to threaten all our lives. To demonstrate the staggering effect of exponential growth, let's imagine that on December 1, 2019, there were only two cases of COVID-19, and the disease grew linearly, by two cases per day. After a month, there would be sixty cases. Now let's assume instead that the number of infections doubles every day; after just over a month, the entire population of the Earth would be infected. This is the difference between a minor nuisance and a global pandemic, the difference between linear and exponential growths. The latter is at the heart of cosmologists' prime suspect in the quest to understand what really happened at the Big Bang. Inflationary cosmology posits that the early universe underwent a fantastic exponential growth spurt. The doubling times of COVID at its worst was around every two days, but compare that to inflation—where it's believed the universe doubled in size every ten-trillionth of a trillionth of a trillionth of a second. Today, inflation is mainstream cosmology, but all is not what it seems.

A SPECTACULAR REALIZATION

Cosmic inflation was born at the end of 1970s, when particle physicists, high on their triumphs in unlocking the secrets of the subatomic world, turned their attention to the discipline of cosmology. The extreme temperatures of the Big Bang promised to unlock a physics far beyond the energy reach of any imaginable particle collider on Earth. But it soon became apparent that this fusion of particle physics with cosmology was not a happy union. Normally we think of nature as being governed by four fundamental forces: two of which we are familiar with and two which are beyond our normal experience. The first is electromagnetism, which describes the behavior of light waves, electricity, and magnets; the second is gravity, which Einstein showed arises from the curving of spacetime by massive objects. Then we have the two nuclear forces: the weak, which is responsible for radioactive decay; and the strong, which binds atomic nuclei together. A spectacular triumph of physics was to show that at high temperatures, two of these forces (the electromagnetic and weak forces), which at first glance appear to have nothing to do with each other, merge into a single force. In 1979, Steven Weinberg, Sheldon Lee Glashow, and Abdus Salam were awarded the Nobel Prize for this remarkable achievement. The idea occurred to physicists to repeat this trick and imagine that at higher temperatures still (as we get very close to the hot Big Bang), the strong force would also be incorporated into a single force. This incorporation is known as grand unification.

Unfortunately, theories that accomplished this merger seemed to predict the production of a large number of strange particles known as monopoles. You may know that all normal magnets have two poles that attract opposite poles and repel similar ones. But cosmic evolution within these Grand Unified Theories was bound to produce magnets with one pole, permeating the entire cosmos; yet none has ever been seen. Nor should they have been, as their enormous density would have halted the expansion of the universe, long ago. This is known as the monopole problem, courtesy of the careful calculations of Alan Guth and Sze-Hoi Henry Tye, and independently, the quantum information theorist John Preskill, who narrowly beat the duo to publication.[1] Tye is in many ways the unsung hero of this story. It was he who had persuaded the initially reluctant Guth to work on such an esoteric topic. As Guth and Tye had been scooped in discovering the monopole problem, the two were determined to be the first to find its solution. Either particle theories were wrong, the Big Bang was a mirage, or something else was missing. The answer would arise from the seemingly mundane physics of phase transitions. An ice cube that is heated and turns into a liquid has undergone a phase transition. Heat it more and it will turn into steam, another phase transition. These transitions are part of our common experience, but physicists have discovered other, more exotic phase transitions. One in particular would be key to resolving the mysteries of the Big Bang.

When liquid water cools, it turns into ice, but if the water is pure enough something strange can happen. The phase transition can be delayed; the water can stay liquid even though it is well below its normal freezing point. This phenomenon is known as supercooling. The standard Big Bang has the universe start at an infinite temperature that then rapidly cools as it expands. Particle physics models predicted that this cooling would produce too many monopoles to be consistent with observations. However, Guth and Tye noticed that if the transition is delayed, or supercooled, that delay could suppress monopole production. It was December 1979, and Tye was about to leave for a trip to Hong Kong, but something was nagging at his mind, something that would prove to be momentous, so he asked Guth to check whether supercooling would affect the rate of expansion of the universe.[2] We might wonder what would have happened had Tye not taken his trip. Perhaps he would have done the calculation himself; conceivably he would have realized its implications as Guth did. Maybe he would be one of the biggest names in cosmology instead. We'll never know. With travel tickets bought, the die was cast.

Supercooling means that space is trapped in a false vacuum. You might think of the vacuum of space as devoid of all content, but quantum physics describes it as a sea of chaotic activity, where species of particles come with partner species that have the same properties but opposite charges. These partner species are known as antiparticles. Quantum theory describes these particles and antiparticles continually popping into existence, moving apart, and then reuniting to disappear back into the vacuum in a tiny fraction of second. This phenomenon, along with other exotic quantum effects, means the vacuum has some energy to it. In fact, there could be different types or phases of vacuum (or vacua) with different energy densities (akin to different phases of water). Only one of those vacua, the one with the lowest energy, is the true vacuum, while all the other ones are false and can have very strange properties. In particular, unlike particles that have attractive gravity, a false vacuum with positive energy density has repulsive gravity in general relativity, as Einstein had discovered during his "biggest blunder" six decades earlier. Late that December evening, Guth started to calculate how quickly the universe would stretch if it were driven by the negative gravity of the false vacuum. The answer was stupendous: the universe would, in a tiny fraction of second, grow at least by a hundred trillion trillion times.

Guth immediately realized that this fantastic growth spurt might impact Dicke's flatness problem. Recall that the universe can have different degrees of curvature, a quality that cosmologists denote by a number called omega. A universe with positive curvature like a ball has an omega value greater than one; a negative, curved space like a saddle has an omega of less than one; and a perfectly flat universe has an omega of exactly one. But the flat solution was unstable, like a pencil balancing on its tip, and so it seemed like the universe had to be delicately fine-tuned to keep it from deviating from its flat geometry. By taking account of the repulsive gravity state, Guth found the equations were turned upside down. Instead of omega being rapidly driven away from one as the universe expanded, it was swiftly pushed toward it. Now, no matter where omega started, it would be forced to one with "exquisite precision." Guth realized he was onto something big and wrote in his diary, "Spectacular realization. This kind of supercooling can explain why the universe is so incredibly flat—and therefore resolve the fine-tuning paradox pointed out by Dicke." So confident of the potential significance of his discovery, he drew not just one box around his words, but two. However, Guth admitted to Phil that the night was "a mixture of extreme excitement and worry." Because he was new to cosmology, he feared "that as soon as I told other people about it, they would tell me what was wrong. . . . I was very nervous that it would all go away." I share this impostor syndrome, and it's one of the hardest obstacles to overcome for any theorist, especially as they embark upon understanding uncharted territories. Guth's idea did not go away, and his double-boxed note now hangs in the Adler Planetarium in Chicago, a marker to a revolutionary finding.

One way to think about how the false vacuum solves the flatness problem is to say that Dicke's calculations assumed that gravity was attractive, but that if the sign of gravity is reversed, then so is the conclusion. But an easier mental aid is to consider standing on the surface of a ball. If you manage the feat of standing upright, give yourself a round of applause. Now imagine the ball is inflated to the size of a planet. While the object you are standing on is still a curved ball, the enormous change in scale means it appears perfectly flat to you. Unless you are eight months old, you're not likely

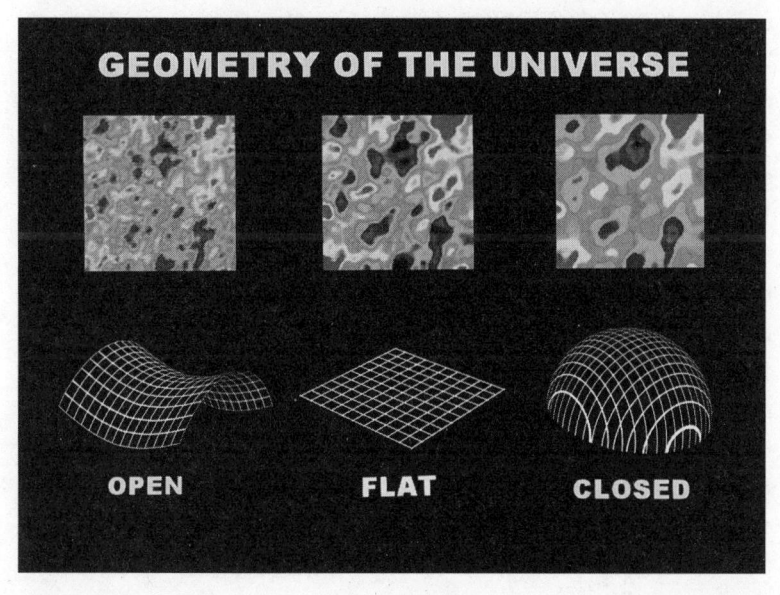

GEOMETRY OF THE UNIVERSE

OPEN   FLAT   CLOSED

What the universe would look like under different possible values of omega, a number that measures its (spatial) curvature. In an open universe, omega is less than one. In a flat universe, omega equals one. In a closed universe, omega is greater than one. Each potential geometry distorts the patterns of hot and cold spots in the cosmic microwave background. By measuring these spots, cosmologists determined the value of omega and thus the shape of the universe. Credit: NASA / WMAP Science Team.

to get applause for standing on this enlarged ball, as essentially that's what most of us do on Earth. Similarly, by blowing up the universe, the false vacuum would become the ultimate ironing machine, removing any cosmic curvature from view. Little did Guth know that its resolution of the monopole and flatness problems was just the beginning of the "miracles" of repulsive gravity. Clearly influenced by the economic worries at the time, he coined a name for this fanatic growth spurt: inflation. If the universe could grow so enormously in a tiny fraction of a second, then the initial nugget of false vacuum might have been smaller than a subatomic particle. He concluded that economists and ancient philosophers were wrong to say you can't get something for nothing. The universe, Guth claimed, was "the ultimate free lunch."

A few weeks later, while eating at his university cafeteria (where the lunches were not free), Guth learned from a colleague of another puzzle of the Big Bang, the horizon problem: Why is the universe so uniform, when there hasn't been time for it to reach equilibrium? He quickly bicycled home to see if his new idea could shed light on the question. Inflation's rapid stretching implies that any given patch of the universe was much smaller in the pre-inflationary era than would be predicted by the standard Big Bang cosmology. So, two points in the universe that were too far apart to be able to communicate in the old model, even at the speed of light, would have started out much closer together, allowing their photons to go back and forth (creating the uniform 2.7° Kelvin temperature of the CMB today) and thus solving the horizon problem. It was one thing for inflation to have solved the puzzles Guth was familiar with, but now it was fixing issues of which he had never previously heard. Inflation seemed like magic; it had solved three of the most severe difficulties in cosmology—the monopole problem, the flatness (or omega = 1) problem, and the horizon problem—but its biggest triumph was still to come.

Guth presented his model of inflation in a lecture series, touring American universities for months before sending his research off for publication. Many in the audience at Guth's lectures were awestruck. Fellow particle theorist Paul Steinhardt described it to Phil as "the most exciting and depressing talk I've ever been to." Steinhardt's depression came from the fact that inflation had to end and bring the universe to a more sober rate of expansion. The process was often modeled as a pot of water (representing the false vacuum) coming to boil. For a smooth universe like ours to emerge out of this chaos, the bubbles (of true vacuum) had to merge, but inflation's violent expansion would prevent that merging. Thus, inflation would never end and could not possibly result in the universe we see today. This conundrum became known as the graceful-exit problem.

Steinhardt's new PhD student, Andreas "Andy" Albrecht, had been in training to be a violinist before he realized his true calling was physics. He had no interest in cosmology, which was the reason he had picked a particle theorist as a supervisor. So, he was shocked to find that Steinhardt was now obsessing over inflation. As Albrecht recalled to Phil, cosmology was "this field with big questions and no way to answer them and no data . . . and so my heart totally sank." Steinhardt insisted they try to fix inflation's problems anyway. They studied other forms of exotic phase transitions and found that the energy in the false vacuum could decline much more gradually (in a process known as a slow roll) than was previously thought possible. A gradual transition from false to true vacuum, modeled as a uniform field slowly declining in strength, could allow inflation to end smoothly. But Steinhardt was worried that the model might need to be jury-rigged to work (physicists call such jury-rigging "fine-tuning," requiring very special parameters or initial conditions). The same doubts didn't bother a rival from the other side of the Iron Curtain and soon, as Steinhardt recalls, they would find themselves in a race to publish.

In the 1970s, Soviet scientists like Andrei Linde and Alexei Starobinsky had toyed with ideas like Guth's for an entirely different reason: resolving the Big Bang singularity (more on this later). None had appreciated that an exponential expansion would also solve the monopole, horizon, and flatness problems. Looking back, Linde now calls his early work "garbage."[3] But as news of Guth's research filtered through to Russia, he saw the promise of inflation and vowed to find a solution to its problems. One night in 1981, the same thought that occurred to Steinhardt and Albrecht—that a slow roll could fix inflation—came to Linde. He called up his friend, the late Russian physicist Valeria Rubakov. In the era before cell phones, Linde had to stretch the phone's long cord to hide in the bathroom so as not to wake his family. He then explained the simplicity and effectiveness of the slow-roll process and how it could rescue inflation from the graceful-exit problem. Finally, he couldn't control his excitement and decided to wake his wife, fellow physicist Renata Kallosh, to tell her: "I think I know how the universe was born."

Linde had been destined for a bold career in science. Both his parents were physicists, and his mother had also piloted night bombing runs for the Soviet air force during World War II. The Nazis called her unit the Night Witches. But back in North America, forty years later, Guth was not ready to accept Linde as the savior of inflation, as he and his colleague Eric Weinberg had tried a similar fix to no avail. Reading on though, Guth realized that Linde's slow-roll process would put inflation back on the map. Inflation had scored another major victory; it would soon be the hottest topic in cosmology. Meanwhile, in Japan, entirely independently of Guth and colleagues, Katsuhiko Sato also published a paper implying the universe could expand exponentially. He doesn't get much credit in other histories of inflation as his paper was released in 1981, but it was submitted in February 1980, just two months after Guth's "spectacular realization."[4] Inflation was an idea whose time had come.

Not everyone was impressed with Guth's fashionable new proposal. Roger Penrose admitted to Phil that he "didn't think it would last a week." Stephen Hawking was also skeptical, explaining in a Moscow lecture why inflation might not succeed. Lacking the speech synthesizer that would later become iconic, his speech was difficult to understand. A graduate student, Nick Warner, interpreted in English for the group; that interpretation then had to be translated into Russian. Linde was chosen for the latter job, and he was delighted to find himself translating praise for his own work to the elite of Soviet cosmology. But as the talk progressed, Hawking would argue that Linde's solution to inflation's woes was a failure.

Linde had to continue through this painful seminar until he could explain to those who held the keys to his future why he thought Hawking was wrong. Afterward, Linde and Hawking adjourned in private to discuss further the matter, a dangerous move as Soviet citizens were not allowed to meet in private with foreign nationals, and as Linde recalled to Phil, "at that time all [the] institute was in panic because the famous genius disappeared." Eventually Hawking became convinced that Linde was right. Inflation had been endorsed from on high.

From the uniformity of the CMB, cosmologists already knew by the 1970s that on the largest scales, the universe is incredibly uniform but with small deviations that allow galaxies to form. Prior to the introduction of the inflation theory, there was no explanation for these deviations. In the early 1960s, Leonard Parker began to analyze quantum fluctuations in an expanding space. As discussed, in normal quantum theory, pairs of particles constantly pop out of the vacuum, but they only last a very short time before recombining back into empty space. These pairs are known as virtual particles. Building on Parker's work, Stephen Hawking and Gary Gibbons discovered that the violent cosmic expansion during inflation can tear these pairs apart, so much so that they cannot recombine as they would normally do, and thus that they can last forever. Just like Pinocchio, who could thank a fairy for turning him into a real boy, quantum virtual particles can thank inflation for turning them into real particles. Empty space could generate matter and radiation; inflation would ramp up this metamorphosis exponentially. But as the process was a random, quantum one, it would not be perfectly uniform. Some regions would have more matter than others; these regions would be stretched by inflation, creating the huge swathes of galaxies we see adorning the heavens today. In other words, inflation would smooth the universe but not perfectly; rather, it would leave just enough imperfections—quiet ripples—to form us and our cosmic structures. Incredibly, the largest structures of the universe were thought to have been seeded by tiny quantum fluctuations stretched by inflation. If the idea could be made to work, it would be the theory's greatest triumph, explaining the entire distribution of matter in the cosmos. In 1982, Gibbons and Hawking invited the world's leading cosmologists to the Nuffield Workshop in Cambridge to find out if galaxies really were born this way.

The fact that galaxies exist now suggests that there must have been density fluctuations in the early universe after inflation. But whether quantum fluctuations during inflation could really generate these fluctuations was an open question. The calculations were subtle and tricky. As Guth recalled in his memoir, he spent so much time working on his sums that he forgot the conference banquet. If you've ever tried to check your addition of a long series of numbers only to get a different result each time, something analogous happened at the Nuffield Workshop. Every team that attempted the calculations found different answers. But after a week, they finally converged on what is now considered the right method to compute density fluctuations, only to find out that two Soviet scientists, Slava Mukhanov and Gennady Chibisov, had already beaten them all to it. Years later, when I asked Mukahnov about those early days, he was keen to point out that he had been the first in 1981 to compute the quantum fluctuations. He also sent us the cover page of his 1980 preprint at the Academy of Sciences of the USSR, where they showed very similar physics to inflation using quantum fluctuations to germinate the seeds of galaxy formation.

Back at the Nuffield, the attendees concluded that inflation would generate a fireball of matter and radiation, so hot that it was a plasma, much like the surface of the sun. It was almost the same everywhere but with small deviations from perfect uniformity caused by quantum fluctuations, which would propagate as sound waves through the hot, dense plasma after inflation ends. You might think space has no sound, but that's only true in the cold, sparse universe we see today. The scorching plasma of the early universe was dense enough for sound waves to travel easily, and this primordial cosmic soundtrack was nearly scale-invariant; that is, the fluctuations of density would have nearly the same amplitude on all scales. You can think of a constant hum in the primordial plasma, which sounded nearly the same, independently of time or frequency. Fortunately, the predicted small deviation from scale-invariance was not in conflict with observations at the time, but few scientists held out hope for experimental confirmation.

This picture of the infant cosmos reminds me of the Japanese creation story, Tenchi-kaibyaku, which describes the beginning of the universe as "immersed in a beaten and shapeless kind of matter (chaos), sunk in silence. Later there were sounds indicating the movement of particles. With this movement, the light and the lightest particles rose but the particles were not as fast as the light and could not go higher."[5]

Similarly, Hindus believe the universe was born with a certain hum that permeated the cosmos; for this reason, many yoga practitioners begin their classes with a chant of this sound they call Om. Maybe we have emerged from the deep Om of a primordial yogi.

The modern paradigm of inflation that emerged in the aftermath of the Nuffield conference is the life story of a hypothetical field known as the inflaton (rhymes with proton). Like the magnetic field that permeates Earth today, the inflaton permeated the universe, with two main differences. First, the inflaton is a scalar field, which means it can be simply described by one number at any point in space-time, unlike the magnetic field, a vector, which is described by three numbers and has both strength and direction.[6] The second difference is that while the magnetic field of Earth varies a lot with geographical location, but slowly over time, the inflaton was almost uniformly evolving everywhere in space.

Today, there are more than a hundred different inflationary models with one or many inflaton fields. And just as hikers can descend mountains through alternate routes and landscapes, so a variety of hypothetical inflaton fields can explore different paths as they transition from the false to the true vacuum. Each version of inflation leads to varying predictions for observations. Inflationary cosmology has evolved from one theory into a myriad of different theories. As the inflationary pioneer Albrecht describes the competing theories, "People just started saying, 'Well, what if you have this field? What if you have that field?' . . . [These fields were] totally disconnected from a complete theory of particle physics. That was distressing to me." Personally, I find this criticism of inflation unfair, as physicists do not really have a more "complete theory of particle physics." I think inflation is just at the edge of our understanding, and whatever more fundamental theory lies beyond inflation would be even more speculative. However, I agree with Albrecht that having so many different variations of inflation, with little connections with particles that we know and love, makes it a theory that is hard to test. But if the plethora of inflationary models was disturbing for some, it was nothing compared to the shock of what was coming next.

THE ETERNALLY INFLATING MULTIVERSE

The evidence for the Big Bang was essentially that the universe evolved from a hot, dense state. But during inflation the universe is cold, so when it ends, the energy of the false vacuum should decay into matter and radiation igniting the hot primordial inferno. Even though this phase might have been the first time the universe became hot, cosmologists (confusingly) refer to it as *reheating*. So, one could say that inflation did not happen after the Big Bang, as Guth had originally supposed; rather, it happened before it. He now calls inflation a "prequel" to the standard cosmological narrative. Guth had originally marketed inflation as happening just *after* the Big Bang, and textbooks today still label inflation as a post–Big Bang event. But in our modern understanding, it's not the Big Bang that generates inflation but inflation that generates the Big Bang. Scientists began to wonder: if inflation could generate one Big Bang, why not more? First to publish this idea was my old friend and Princeton professor Rich Gott, who had been thinking about cosmological phase transitions as bubbles of steam popping out of hot water in a boiling pot.[7] What if, he wondered, our universe was simply one of those bubbles? Soon, others—including Steinhardt and Linde—were converging on the same thought. Initially, these solutions were treated as curiosities, but Alex Vilenkin argued that in fact they were a generic property of almost all inflationary models. His reasoning can be illustrated by thinking again of water turning into steam. Normally, only so many bubbles can be made, as eventually the water runs out. But if we kept pouring in water at a rate faster than the rate at which bubbles were produced, then the number of bubbles formed would be unlimited. In inflation, space is expanding exponentially, but this state is unstable and decays, creating bubbles of Big Bang universes. Like a radioactive substance, the inflaton decays with a half-life, but the remaining portion of inflating space that has yet to decay continues its relentless exponential expansion, always outcompeting its progeny universes, which expand at comparatively paltry rates. It would be like eating a cake that always doubled in size every time you took a bite. Inflation wasn't just a free lunch; it was cake you could have and eat, repeatedly. Normally, more water (or cake) would have to come from somewhere, but vacuum energy is different; there is simply no known limit to how much vacuum there can be in the universe. This process would repeat endlessly. Vilenkin called it "eternal inflation," creating an infinite ensemble of bubble universes, or a "multiverse." To make it work, he claimed, one only must assume that the inflating space expands faster than it decays, which happens in almost all inflationary models.

If the multiverse wasn't shocking enough, Guth, working with Edward Farhi and Jemal Guven, later wondered whether it might be possible to create a bubble or "pocket universe" in a lab. While the technology required to do such a thing would be far beyond anything our civilization could muster, calculations showed that creating a pocket universe might just be possible by heating some material to a fantastic temperature and then supercooling it rapidly. This process could create a patch of false vacuum, which, with some quantum luck, would begin to inflate, just as Guth had originally envisaged. A scientist in such an advanced civilization might imagine they were akin to Mangala, the creator god of the Mandé people of Southern Mali, who tell of the world being born of experiments with tiny seeds that grow to form the world. Aspiring gods might be disappointed though, as the new bubble would be completely disconnected from its source. No prayers could ever be heard or acted upon, and there would be no reason to think the creator god would be a perfect being.

Just as the Earth, the sun, and—later—the Milky Way galaxy had all been demoted from their unique status, Vilenkin was suggesting that even our universe was not special. Having checked his calculations and now sure a multiverse resulted from inflation, he went to see the originator of the theory. Vilenkin says he couldn't contain his enthusiasm when he called Guth and planned a visit to MIT "to discuss some physics with him." But as Guth encountered the idea for the first time, his reaction was not the shock and awe Vilenkin expected. Instead, Vilenkin says, Guth simply fell asleep. Later, Vilenkin discovered this was one of Guth's superpowers. He often dozes off during seminars only to awake at the end and ask the most penetrating question of the speaker. In this case, Guth followed Vilenkin's logic and thought it sound but simply didn't care about hypothetical other universes, dismissing them as irrelevant to physics. It's an attitude many cosmologists still share. But this picture of an eternally inflating multiverse revives a view of cosmology that was thought to be dead and buried.

Almost all accounts of the history of the Big Bang tell us it defeated the steady-state model of Bondi, Hoyle, and Gold. But in some way, I think this account is yet another myth. The version of the steady state proposed by the British trio is no more. Recall, though, that their model had to postulate new matter being created in the expanding space between galaxies, to continually replenish the universe, keeping the density of the cosmos constant over time, with no need for a Big Bang origin (think of Phil's favorite chocolate chip cake from the last chapter). But eternal inflation implies a return of the steady state, although at vastly different scales. The creation field of the former model has now been replaced with an inflaton field, and instead of new galaxies popping up in the eternal expanding space, it is bubble universes that do so. The details are different, but conceptually both models are remarkably similar. Alan Guth explained the irony well that the evolution of an inflationary universe "will strongly resemble the old steady state."[8] In other words, inflation fixed the Big Bang's problems by (almost) resurrecting its old rival.

These sorts of issues were far too esoteric for most scientists to worry about in the 1980s, and for a while cosmologists followed Guth's indifferent attitude. But at the close of the twentieth century, increasingly accurate measurements from space telescopes—and developments in fundamental physics—would set the scene for another cosmological civil war.

COBE SEES GOD

In 1982, the same year as the Nuffield Workshop, NASA (the National Aeronautics and Space Administration) approved the development of a spacecraft that would revolutionize cosmology. The COBE (Cosmic Microwave Background Explorer) satellite was poised to measure in fine detail the properties of the CMB. The Big Bang theory implied that the CMB should be what is called a black body. This term was coined by the physicist Gustav Robert Kirchhoff, co-inventor of the spectroscope, which enables astronomers to break up the light falling on their telescopes and reveal the secrets of the stars. An idealized object that absorbs all light is called a black body. But Kirchoff noted that, in thermal equilibrium, all the energy absorbed by the black body should be re-emitted in a way that only depends on the temperature of the object. The coup de grace happened thanks to Max Planck, who explained black-body radiation using two fundamental tenets of physics: thermodynamics and quantum mechanics, leading to a simple formula that realized Kirchoff's idea.

The early universe was a hot and dense fog of photons repeatedly scattering electrons (like countless balls stuck in a pinball machine), unable to form the atoms that make up the world of our experience. In the 1940s, Alpher and Herman theorized that this scattering process would generate the perfect black-body spectrum described by Planck, as particles reach thermal equilibrium. And so, if the Big Bang really happened, not only should we find its relic radiation, but this ancient light must have the signature of an idealized black body.

46 CHAPTER 2

In 1990, NASA scientist John Mather showed COBE's first results: the CMB was the most perfect black body ever seen. There was simply no discernible difference between the predicted spectrum (or its intensity as a function of frequency) and the satellite observations. A standing ovation greeted Mather, who along with George Smoot would later receive the Nobel Prize for this stunningly profound measurement.

While the groundbreaking CMB black-body spectrum was obtained by a detector called FIRAS (Far Infrared Absolute Spectrophotometer) on the COBE satellite, another instrument on the spacecraft, called DMR (Differential Microwave Radiometers), was slowly mapping the CMB temperature in the sky. By 1992, COBE

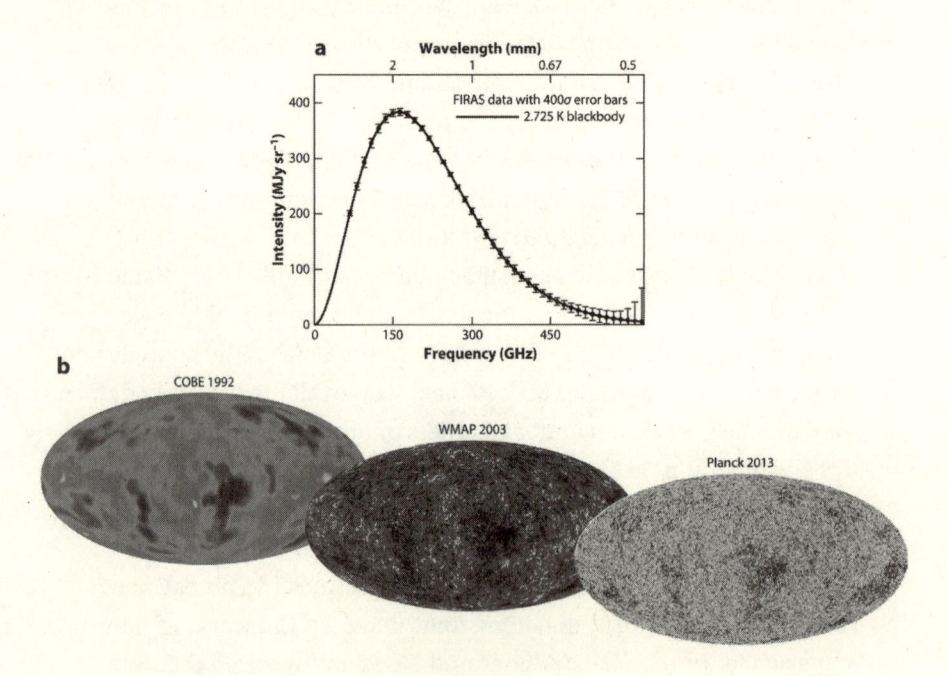

The top panel shows the intensity of the cosmic microwave light as a function of its frequency, as measured by NASA's Cosmic Background Explorer (COBE) probe. Operating until 1993, COBE was also the first to map the fluctuations of cosmic microwave background temperature in the sky. Over the following two decades, the resolution of this map improved a hundredfold, thanks to new probes, NASA's WMAP, and ESA's Planck satellites (*bottom panel*). Credit: Ned Wright; NASA (public-domain data); Michael S. Turner, *Annual Review of Nuclear and Particle Science* 72, no. 1 (September 26, 2022): 1–35.

DMR had collected enough data and saw small fluctuations (at a level of one part in a hundred thousand) in the temperature maps of the CMB. These tiny yet profound deviations from a uniform temperature were a gold mine of information, as they provided a picture of a most ancient era and would turn "early universe" cosmology into a precision science. Stephen Hawking declared the results "the scientific discovery of the century if not of all time."[9] George Smoot said at the NASA announcement, "If you are religious, it's like seeing God."[10] The press had a field day with this comment. Smoot was on every major broadcasting outlet and years later received perhaps (to me) as great an accolade as his Nobel: an appearance on the sitcom *The Big Bang Theory*.

As described in Michael Lemonick's history of the project, behind the scenes apparently many COBE scientists were unhappy with Smoot and wanted him to resign for, among other offenses, having his home institution produce the results in a press release that should have come from the team.[11] Lemonick's book even describes some of the team members as "ready to murder George Smoot."[12] David Wilkinson, another COBE scientist, had been approached by Alan Guth for hints of the data before release but had, in my opinion, done the right thing and refused. He did, however, say that Guth would consider the results "good news."[13]

The data from COBE enabled scientists to compare inflation against a competing theory: topological defects. The idea here was that, rather than inflation stretching quantum fluctuations, the CMB temperature fluctuations we see came from cracks and imperfections formed at primeval phase transitions, similar to those that form in ice when it freezes. Some of these defects might form stretched-out lines of deformed space-time known as cosmic strings and could be detectable by astronomers. These ideas were plausible, but the data simply didn't support them. Princeton astronomer David Spergel, who had worked on the framework, lamented, "We're dead," at the conference where COBE results were first announced.[14]

Spergel was my PhD supervisor. When I first started my graduate studies in the United States in 2000, I was working on inflation and how it might have ended. However, the subject was all a bit too abstract for me. Cosmology focuses on the universe as a whole, whereas astrophysics is more generally concerned with the science of heavenly bodies. Remember that I started as a teenage amateur astronomer in Iran. Astrophysics seemed like a good compromise, where fantastic ideas were often met with hard physics and tested against detailed observations. I ended up working on clusters of galaxies and accretion into black holes, before going back to cosmology. But this transition was only part of the story. Going from the suburbs of bustling Tehran into a quaint college town in New Jersey in your early twenties is probably just as cataclysmic as switching from cosmology to astrophysics (possibly more so). I came from a deeply religious country with a tortured relationship between mosque and state, and boom, 9/11 happened within a year of my arrival at Princeton. My wife, Ghazal, and I (both graduate students at this point) did not leave the United States for over four years, as we were afraid that we wouldn't be allowed back into the country. With Spergel, though, we were in safe hands. He was extremely supportive, and he knew everyone. In 2021, I was at his sixtieth birthday party at Princeton, where Senator Chuck Schumer sent him a video message thanking him for his service to setting US science policy. During the meeting, Spergel recalled the tense moments at the 1992 American Astronomical Society meeting, where he was supposed to speak following the talk on COBE results, but at that point it was clear to him that his five years of work on "topological defects" had turned out to be a wild-goose chase. Others, notably Neil Turok, whom we shall meet later in the book, were still holding out. But by 1994, Smoot declared, "The inflationary Big Bang is the standard model in cosmology"—although he conceded that about 10 to 20 percent of cosmologists didn't accept Guth's theory.[15]

However, as the twentieth century was ending, a serious problem was becoming apparent for inflation. Omega, which measures the geometry of the universe, should have a value extremely close to one according to most inflationary models, as the exponential expansion forces it into a flat state; but astronomers, by adding up the energy contents of the universe, were homing in on a value of

0.3. Those were awkward times for inflationary advocates. Guth described dinner with observational astronomers as uncomfortable, while Vilenkin recalled the same astronomers laughing at desperate inflationary cosmologists. Guth's beautiful theory seemed disproved by the ugly fact that 0.3 does not equal 1.

In 1998, Guth published a popular account of his idea and commented on the problem of omega being 0.3.[16] One solution was to modify the theory in what became known as open inflation, which described a universe undergoing considerably less exponential expansion than was normally assumed, thus allowing omega to be lower than one. However, Guth described this solution as "unattractive," as the model looked as though it had been contrived to give an answer to fit observations. He then turned to another possibility, resurrecting Einstein's "biggest blunder," the cosmological constant, as a possible way to salvage their model. Recall that the cosmological constant can be thought of as the energy of empty space, which has the peculiar property that as space expands, its density (which is its amount of energy divided by its volume) does not decline. Contrast this constant density with molecules of gas in a box: if we double the size of the box but keep the number of molecules the same, the density goes down. But for a cosmological constant, the density doesn't go down, because the cosmological constant's energy is a feature of space itself. In order to know the true value of omega, cosmologists have to sum up all the energy that exists in the universe. If the cosmological constant exists, then its energy could be just enough to push omega from 0.3 to 1.

Indeed, in 1995, Lawrence Krauss and Michael S. Turner argued that "the cosmological constant is back" and described several problems—from the age of the universe to the issue of omega and inflation—that all pointed to the same conclusion.[17] However, this line of argument had a huge cost. As discussed earlier, in quantum mechanics, the vacuum is a bustling sea of virtual particles that spontaneously pop in and out of existence, giving it an energy density. This energy has both positive and negative contributions, so adding them all up was a difficult task. Those who tried found an outrageous number; the vacuum had so much energy it would instantly tear the universe apart. The excess energy was one hundred and twenty orders of magnitude larger than what was observed. Some called this forecast the worst prediction in all of physics, one otherwise known as the cosmological constant problem. Others assumed that a hypothetical new symmetry principle would force the calculations to cancel out, giving the cosmological constant a value of zero. But if inflation was to be rescued with a small cosmological constant, any hope of such a symmetry principle would be lost. Cosmologists were faced with a choice to modify or abandon inflation or to revise the energy content of the entire universe. Lucky for the inflationary proponents, a series of extragalactic explosions would soon come to their rescue.

In the late 1990s, observations of white dwarf supernovae were being used by astronomers to determine how fast the universe's expansion was slowing down due to gravitational attraction. A white dwarf is the remnant of a once-bright star; dense and cool, it glows with a faint glimmer, a mere shadow of its former glory. Billions of years from now, our sun will puff up and shed its outer layer, leaving a white dwarf in its place. If this stellar corpse finds a partner star, it can steal its mass and start to grow anew. But there is a limit to this highway robbery, as the great astrophysicist Subrahmanyan Chandrasekhar showed: a white dwarf that has more than 144 percent of the mass of our sun will be destroyed in a catastrophic explosion that can outshine even an entire galaxy. As astronomers know the intrinsic brightness of these supernovas, they can be used as "standard candles," just as Hubble and Humanson used Leavitt's Cepheid variables to measure vast distances across space. With this method, astronomers could measure distances far deeper into the past than had ever been done before. These distance measurements could be then used to determine how fast some of the oldest galaxies are moving away from each other and thus to learn how the expansion of the universe changes over time. But much to their surprise, astronomers found that the expansion was speeding up, implying a mysterious repulsive force that counteracts the gravity of all of the matter in the universe. There had to be some energy in empty space after all; Einstein's "biggest blunder" was not that he introduced the cosmological constant but that he gave up on it. The findings meant the energy budget of the universe had to be revised; lo and behold, omega was one (to the best of our measurements)—just as simple inflationary models predicted. Whether this vacuum energy was really a constant is still debated today, so Turner dubbed it dark energy. Whatever its explanation, it was a pill many found hard to swallow, but within two years new data would leave little room for doubt, when my colleagues managed to measure the shape of the entire cosmos.

Amazingly, determining the geometry of the universe—and thus determining whether omega is equal to one—is no more complicated than summing the angles of a giant triangle. On a flat piece of paper, the angles of a triangle add up to 180 degrees. On a positively curved surface, like a ball, the angles sum to more than that; on a negatively curved saddle, they sum to less. So, all one needs for an independent check on our cosmic geometry is to find a triangle the size of the universe and measure its angles. Since the CMB has been traveling through almost the entire history of the expansion, astronomers could use it as a proxy for our universe-sized triangle to determine our cosmic geometry. The shape of space directly influences the patterns of the fluctuations in the CMB, so by measuring these fluctuations, astronomers could tell whether space was flat or not. A balloon experiment over the South Pole called BOOMERanG (Balloon Observations of Millimetric Extragalactic Radiation and Geophysics), led by Andrew Lange, did precisely this. Like other cosmological probes, it flew instruments called bolometers that are cooled to almost absolute zero. When microwave light hits the detector, its temperature changes, allowing astronomers to measure the tiny difference in heat between CMB photons, a treasure trove of cosmological data. At the beginning of the new millennium, the mission had made sufficient measurements to reveal the geometry of the heavens. News outlets around the world ran the headline that BOOMERanG had discovered the universe was flat, implying that dark energy was real and omega was one. Inflationary cosmologists no longer had to fear astronomers at dinner.

A follow-up satellite experiment, called MAP (Microwave Anisotropy Probe), launched by NASA and co-led by Spergel and experimentalists out of Princeton's physics department in 2001, confirmed the flat shape of the universe and more evidence for dark energy. The mission was a breakthrough in "precision cosmology"; no longer could anyone accuse the field of being light on data. Year after year, the satellite's mission was extended to collect increasingly more precise measurements of the CMB. When MAP had collected sufficient data, NASA announced they had seen a signature of inflation saying they found "compelling evidence that the large-scale fluctuations [in cosmic density] are slightly more intense than the small-scale ones, a subtle prediction of many inflation models."[18] This phenomenon is known as a red tilt. *Red* because red light has the longest wavelength in the visible spectrum, so for red tilt there is more power at longer wavelengths. Incidentally, it's for this same reason that light being stretched to longer wavelengths is called *redshift*. Two types of tilt are essential for probing the Big Bang. One, called scalar tilt, is concerned with sound waves; the other, known as tensor tilt, refers to gravitational waves. WMAP measured scalar tilt, but we shall tell the fascinating saga of tensor tilt later on in the book. It had been almost ten years since David Wilkinson had told Alan Guth that COBE had good news. Tragically, during this time, Wilkinson was losing a struggle with cancer. Prioritizing work on the MAP satellite amid doctor's appointments, he died in 2002. NASA agreed to change the spacecraft's name to WMAP, the Wilkinson Microwave Anisotropy Probe.

I recall meeting Wilkinson during the first week that I arrived at Princeton in 2000. I was keen to rekindle my passion for amateur astronomy, and he was in charge of the small rooftop observatory at the department. He welcomed the idea, but unfortunately the stress of starting graduate school distracted me from following up with him. And then it was too late. Wilkinson was not the only cosmologist to lose his life at an early stage: Friedmann died aged thirty-seven from typhoid; Andrew Lange, who led the BOOMERanG experiment, committed suicide at fifty-three; and Andrew McKellar, the Canadian astronomer who first detected the signatures of the CMB, died at fifty of cancer. It may seem that fate does not look kindly upon those who dare to look billions of years back into the heart of the Big Bang. Yet I tend to think that the demons that haunt all our earthly mortal lives will not spare even the connoisseurs of our most fantastic and eternal memories.

One of the most influential parts of my own PhD work was discovering independent physical evidence for dark energy.[19] This discovery came through finding large-angle correlations between maps of galaxies and those of CMB photons, something that could only be caused by late-time cosmic acceleration. The day after our first paper appeared on *arXiv* (pronounced *archive*, an electronic repository for yet-to-be-peer-reviewed papers, which most physicists aspire to check daily), the *New York Times* reported: "Astronomers Report Evidence of 'Dark Energy' Splitting the Universe."[20] This headline apparently panicked the host of the *Late Show with David Letterman*, and one of our collaborators had to go on air to reassure the viewers that, as mysterious as this new cosmic dark force might be, human demise at its hands was not imminent.

The discovery of dark energy, confirmed by WMAP, meant that particle theorists could no longer appeal to some exact symmetry principle to explain why the vacuum energy was so low today. In fact, Nobel laureate Steven Weinberg had suggested that there was no such principle, and thus that vacuum energy cannot vanish.[21] Rather, he argued that, within the grand multiverse, vacuum density varied from bubble universe to bubble universe. If it were too positive, the bubble would be torn apart; if it were too negative, the bubble would implode. We must live in a bubble with just the right value or else we wouldn't be here to observe it. This type of reasoning is often called the anthropic principle; it was proposed in the 1950s by Princeton's Bob Dicke, the same Bob Dicke who almost discovered the CMB and who pointed out the flatness problem.[22] But its origins can be traced all the way back to the atomist school of ancient Greece. Recall that Lucretius had said that atoms "have made trial of all things their union could produce; it is hardly surprising if they come into arrangements and patterns of motion like those repeated by this world." Suddenly, the idea of eternal inflation was deployed to solve what some were calling the biggest mystery in physics. Why was the vacuum energy so close to zero but not actually zero? Guth sought this kind of relevance for eternal inflation and the multiverse. Now it could actually do something, namely solve the cosmological constant problem. He quickly became a convert to anthropic reasoning, and many physicists followed suit. But others would not share their enthusiasm.

REBELLION

When Alexander the Great learned of the atomist ideas of a multiple kosmoi, it is said he wept, knowing that his empire, which had seemed so vast, represented only a tiny aspect of reality. He wouldn't be the last to react with horror to such a notion. Talk of an infinite number of universes was for many scientists more like philosophy or even fantasy. Just as the Platonists despised atomist ideas, scientists would fiercely criticize multiverse theorists, several of whom, like the ancient Stoics, favored a cyclic cosmology over a multiverse. Anthropic reasoning can also be used to explain why we live on a planet that has just the right properties for life. There are lots of planets, so we must be on one that has life supporting properties. But for this line of argument to work, there must be a lot of other planets. Here is a crucial difference, though: we can observe other planets, yet we can't observe other universes. It was hardly surprising then to find pushback against such radical ideas, but few could have predicted that a key rebel would be one of inflation's founding fathers, Paul Steinhardt.

When I first met Paul as a prospective graduate student, visiting Princeton in winter 2000, it was clear that he was starting to turn sour on mainstream inflationary cosmology. His doubts about finetuning in his original paper had undergone their own exponential expansion. In a popular book written in 2007, Steinhardt and Turok claimed that inflation made six empirical predictions, the first five of which had been confirmed by experiment.[23] But they argued it would fall from grace at the final hurdle, being replaced with their own cyclic alternative. These predictions were (a tick indicates they have been confirmed by the data):

1. The geometry of the universe should be very close to flat. ✓
2. Much like a fractal image, the fluctuations that seeded the temperature difference in the CMB should be the same magnitude no matter what scale you look at. This pattern is known as a scale-invariant spectrum. ✓
3. The scale-invariance is not perfect, as longer wavelengths have slightly more power. This power increase is the red scalar tilt WMAP detected. ✓
4. CMB photons all have almost the same temperature, with tiny differences at one part in a hundred thousand. However, if we add up the spots that have a slightly colder temperature, we should find just as many spots with a slightly warmer temperature. This phenomenon is known as a nearly Gaussian distribution. ✓
5. The patterns of fluctuation are the same for all the different types of matter, a consistency known as adiabaticity. ✓
6. There should be primordial gravitational waves, ripples in the geometry of space and time coming from the Big Bang. ?

At the turn of the century, then, Steinhardt was still of the opinion that inflation made definite predictions that could be checked by experiment, the first five of which had been confirmed but the last of which was still in doubt. He even rebuked some cosmologists who claimed that inflation was too flexible to be tested, saying their claim was "a myth" and arguing that "if you are cautious to discriminate between what inflation naturally predicts versus what happens when you add extra ingredients, then you can rightfully claim triumphant agreement between the simplest theory and observation."[24] However, in more recent years his criticisms became far more damning, saying that inflation's claims to definite predictions were untrue. When Phil asked him about his previous, more accommodating stance, he replied: "To be honest, I probably didn't recognize the seriousness of the problems with inflation at the time. . . . I think I was being overly optimistic that we could resolve them. It's as simple as that."

The ultimate judge of any scientific theory is nature. Theories must make precise predictions which can then be tested against the data. Inflationary cosmologists claim they followed this scientific procedure by predicting the five properties of flatness, adiabaticity, Gaussianity, near-scale-invariance, and red scalar tilt. But when a theory predicts a multiverse where an infinite number of things can happen, such comparisons are problematic to say the least. For the rebels, probabilities no longer make sense—a development that makes inflation unscientific. If the chance of drawing an ace from a deck of cards is four in fifty-two, the odds will stay the same with one, two, or even an infinite number of decks, so probabilities can still be meaningful. But in infinite set theory, they crucially depend on the order of counting, known as a measure. So, for example, if we keep the order of the cards the same as in a normal deck, we will keep the ratio of four aces per fifty-two cards for our infinite deck. But we could reorder them so there were four aces, then four picture cards, then four aces, then four number cards, then four aces again and so on. Such a reordering cannot be done for a finite deck, as we will eventually run out of aces—but in an infinite deck, however many aces we count, there are always infinitely many left. So, the ratio of aces to other cards will depend on the ordering. In the case of cosmology, the order or measure corresponds to the frequency of events in the multiverse, as experienced by a typical observer, and one problem is that rare events could be ordered such that they are common. Put more simply, we cannot agree on what a "typical" observer in the multiverse is. Inflationary theorists have succeeded in finding measures that avoid these problems and give sensible predictions. But they cannot agree on one definite measure: this puzzle is known as the measure problem. This same problem may also apply to some of the very difficulties inflation is thought to solve. For example, if we think it's unlikely that the universe starts off flat, we have to ask: from what probability distribution? If we assume that the flatness of the universe can take on any value, then we have an infinite range of possibilities and we are stuck with a measure problem.

In my mind there is a middle ground between the two extreme views: admitting that probabilities can make sense in an infinite multiverse but recognizing that the measure problem is a challenge not yet met, one that poses a serious obstacle to the theory's capacity to be predictive. In other words, any time you think something is a generic prediction of inflation, say a flat universe, then you wonder whether it is a feature of the theory or rather an outcome of how you define your probabilities. The only way out of this dilemma is to identify a physical principle that would fix the measure. When Phil asked Guth to explain what the measure problem was and how to solve it, he laughed before replying that he could do the former but not the latter. Add to that the reality that we have a plethora of inflationary models that can predict essentially any observation (that I can think of), given the right set of parameters. What I also have a problem with is that most astronomy textbooks talk about inflation but fail to mention these difficulties. If we are to make progress in science, we need to be honest about the strengths and weaknesses of our models, not simply sing their praises to the public.

Another difficulty I think we should highlight for inflation is that the theory does not address the issue of what came before the exponential expansion began. Guth often replies by saying that inflation is analogous to Darwinian evolution, which is a theory of how life evolves and has nothing to say about how life started. It's just a different question. That's fair enough, but if an alternative theory can probe deeper into the quantum gravity era, then such a theory might look more attractive. This possibility is what motivated me to work on alternatives to inflation. A quantum gravity model has the potential to resolve the singularity; in fact, doing so was Starobinsky's initial motivation for inflation. Recall that the Penrose-Hawking theorems that supposedly proved an ultimate beginning assumed that gravity is always attractive, but that during inflation gravity is repulsive. Dark energy provides another independent confirmation that gravity does not always pull. So, a key assumption of the singularity theorem had been undermined, and cosmologists could no longer justify the view that the Big Bang singularity was the beginning of time. But if you want to take the opposite opinion, your best bet may be to appeal to a theorem proved by Borde, Guth, and Vilenkin that showed that while inflation is eternal into the future, it cannot be eternal into the past. Vilenkin did once argue that this theorem proved the universe had a beginning, an opinion that remains controversial amongst many of his peers. Linde admitted to Phil that he was the peer reviewer for the paper and said of the theorem: "The mathematics is good; the interpretation [that the universe had a beginning] is iffy." I think Linde is right here; even if there was some ultimate beginning, it would not be one we could ever observe. In order to travel from one pocket universe to another, you would have to travel faster than light, so there is no meaning to the question of which universe came first according to Einstein's theory of relativity. Even if there was, we would have no way of telling if it were ours.

Today, enthusiasm for the multiverse varies amongst my colleagues. Linde said he would bet his life on it. Martin Rees, the British Astronomer Royal, was also a fan, but would only wager his dog's life on the proposal. When Weinberg was asked the same question, he replied, "I have just enough confidence about the multiverse to bet the lives of both Andrei Linde and Martin Rees's dog." But many scientists were alarmed at the thought of embracing unobservable realms. Sabine Hossenfelder has said that believing in the multiverse is "logically equivalent to believing in God" as they are both "unobservable by assumption."[25] Hawking's old colleague George Ellis, along with Oxford astrophysicist Joe Silk, wrote a scathing critique of the idea with the headline "Scientific Method: Defend the Integrity of Physics," arguing that physics must banish "imperceptible domains" such as other universes from the literature or face a fate where "theoretical physics risks becoming a no-man's-land between mathematics, physics and philosophy that does not truly meet the requirements of any."[26]

The fact that we can't look through a telescope and see another universe is a serious problem in my opinion, as science is supposed to be empirical. To be fair, if you have a concrete theory with predictions that are empirically verified, and the same theory predicts a multiverse, then I would take it more seriously. However, even our most successful theories to date, relativity and particle physics, can and (arguably) do fail well beyond the regime in which they are tested, so why would inflation be an exception? Again, I think we should be walking a difficult tightrope here. The multiverse critics have valid points, but there is a possible escape from Ellis's no-man's land, if we directly probe the existence of the multiverse. Given that other universes are necessarily beyond our observable horizon, this kind of probing might appear impossible even in principle. But the secrets of nature are only hidden for the unimaginative. In the early years of the twenty-first century, one cosmologist would find a method that had the potential to do the impossible: to see another universe imprinting itself into the primordial light of the CMB.

BUBBLE, BUBBLE, TOIL AND TROUBLE

In eternal inflation, bubble universes are continually arising from the sea of false vacuum. They then spread apart from each other as space expands exponentially. What theorist Anthony Aguirre realized was that not only was it possible for bubbles to occasionally collide, but the remains of such cosmic concussions could also be visible to us today. While there was no guarantee that inflation wouldn't have erased any sign of impact, there was a chance that a bruise in the skin of the universe could show something like soap rings intersecting, a circle in the CMB. Aguirre recalled in an interview on Phil's YouTube channel that his former student and collaborators searched the WMAP data for signatures of bubble collisions and found four candidates. "This led to my favorite newspaper science headline of all time," Aguirre said, "which was 'Scientists Look for Other Universes, Find Four.'" On the eve of publication, Stephen Feeney, the lead author and advisor to Phil on his astronomy dissertation, told him that he almost fell off his chair when newspaper headlines appeared describing circles in the sky revealing "violent pre–Big Bang activity." It looked like they had been scooped for one of the greatest discoveries of all time. This apparent scoop, to Feeney and his collaborators' relief, turned out to be a false alarm.

Roger Penrose had been claiming evidence for his own cyclic alternative to inflation that gave superficially similar patterns. Like the stars in the night sky that appeared as mythical human or animal constellations to our ancestors, the CMB is a vast dataset of hot and cold spots whose distribution can easily fool scientists' imaginations. Neither Penrose's circles nor the potential imprints of bubble collisions had enough statistical significance to mark them out as a discovery. Those who hoped for a clear signature of the multiverse would be disappointed. But the circle hunting cosmologists claimed that a higher-resolution CMB satellite would make it easier to sieve through the data to reveal what was signal and what was noise. As they wrote their paper, a spacecraft more than a million kilometers away was quietly collecting the information they needed.

PLANCK AND "THE PROOF" OF INFLATION

In 2009, almost fifty years after Penzias and Wilson made their historic detection, the European Space Agency's (ESA's) Planck probe was launched into deep space. Planck was by far the most advanced detector ever constructed to study the CMB from the heavens, and riding with it were the hopes of many cosmologists for a resolution to the inflation debate. Alas, the high-definition CMB maps showed no signs of cosmic bubble collisions. While this lack of evidence was a missed opportunity for multiverse proponents, Planck had other results that kept the idea afloat. Firstly, the official paper from the collaboration stated that the Planck data was in very good agreement with simple inflationary models. Some scientists, like string cosmologist Gia Dvali and science communicator Brian Cox, would take it further and tell the media that Planck had proven inflation. Here, I'm reminded of a T-shirt I once saw that simply said, "I think you'll find it's more complicated than that." Planck also found anomalies that might or might not be bubble collisions, like an unusually large cold spot, already observed by WMAP. Most scientists had written these off as noise but when they turned up in Planck's map, eyebrows were raised. More importantly, Planck found that types of inflation known as plateau models were favored by the data, and these models seemed especially suited to make inflation eternal. George Efstathiou, the scientist chosen by ESA to present the cosmology results, later said to Phil in an interview, "The Planck data say that inflation is eternal; the simplest models would have an eternally inflating universe."[27] When Phil asked if Planck therefore had provided evidence of a multiverse, Efstathiou replied, "If inflation is eternal, then you have a multiverse, and that's why I say I think this is the most important result from Planck. We're sort of being pushed towards . . . the direction of a multiverse."

With strong words from Efstathiou and others backing the inflationary multiverse, one might imagine the critics would be planning a retreat. Instead, they came out fighting. Steinhardt, working with Anna Ijjas and Avi Loeb, both from Harvard, argued that the fact that the multiverse seemed to stem from Planck's results meant that "all cosmological possibilities (flat or curved, scale-invariant or not, Gaussian or not) and any combination thereof are equally possible, potentially rendering inflationary theory totally un-predictive."[28] If right, inflation could never be proven wrong! Loeb went further, telling Phil that indulging in fantastic ideas like the multiverse was like "taking recreational drugs . . . you will feel good. I'm sure Alan Guth feels really good about the idea of the multiverse, but doing physics is not about feeling good."

After this paper, Guth collaborated with Yasunori Nomura and David Kaiser, Guth's former student who is both historian and physicist. The three responded by saying that the data confirmed the "generic predictions" of most inflationary models. Yes, there were free parameters that needed to be fixed by observations but that was normal in physics; the standard model of particle physics had, they argued, the same problem. Nomura later said to Phil that Ijjas, Loeb, and Steinhardt were thinking too classically and that their critique of the multiverse was "garbage." Infinity, he said, "doesn't mean relative probability is one; that does not make any sense." This claim is really a restatement of the measure problem, and Nomura is right to say that infinities do not make probabilities automatically meaningless. But as I argued before, without a definite measure, that meaning still eludes us. We are waiting for a principle by which to choose the one true measure of how to define probabilities in an infinite multiverse, and until we get it, inflation remains at best, an incomplete hypothesis—not the proven fact it is often marketed to be. The debate seemed to have no resolution in sight and was becoming increasingly bitter. Steinhardt, who had once described Guth as a lifelong friend, now looked more like an enemy. And then came a signal, detected at the South Pole, that promised to change everything.

ASHES TO ASHES, DUST TO DUST

In 2014, John Kovac, an experimental CMB physicist, had been holding discreet meetings with Guth. For years Kovac had been working on an experiment called BICEP (Background Imaging of Cosmic Extragalactic Polarization) situated under the desert dry skies of Antarctica. It had been searching the heavens for a swirling pattern in the polarization (the direction of the electric fields in electromagnetic waves that make the microwave light) of the CMB known as B-modes. These are imprints of gravitational waves, quantum ripples in the fabric of space-time amplified by the violent expansion during the inflationary epoch. Such imprints were the last prediction of inflation, a prediction that Steinhardt and Turok had said would not be found as their cyclic alternatives to inflation would produce no such ripples. Cyclic evolution is typically a more gradual and gentle process than the violent expansion of space seen in inflation. When BICEP announced their results, headlines around the world heralded "the smoking gun" of inflation and there was talk of Nobel Prizes for Guth and Linde. Arguably Steinhardt, despite his passionate anti-inflation stance, should have been considered—all three men had been awarded the prestigious Dirac medal twelve years earlier, before the inflationary civil war tore them apart. Today, Steinhardt says he would refuse a Nobel for inflation. BICEP claimed that the power in the Big Bang's gravitational waves was 20 percent of the power of the sound waves in the primordial plasma, far stronger than many had thought plausible. Such a large signal should have been seen by the Planck spacecraft, but they had reported nothing. Perhaps they were keeping the data to themselves ready for their own Earth-shattering announcement, or maybe the signal was a mirage. There was only one way to find out: publish or perish.

I had done my PhD on analyzing WMAP CMB observations with David Spergel, and I knew that the only way to know whether what you see is from the Big Bang, as opposed to being junk in our galaxy or others, is by getting the same result in different frequencies. Radio emission from electron cosmic rays is brighter at low frequencies; hot dust (tiny particles floating in space, typically the size of molecules or even atoms) at high frequencies, yet the Big Bang signal should stay the same. BICEP had seen something unusual but as it observed the sky at a single frequency (150 Gigahertz; thirty times the frequency of a 5G Wi-Fi modem), we couldn't tell which one it was. Spergel was the first to point out this problem and other inconsistencies in BICEP claims, spoiling everyone's celebration. To be sure, the signal was coming from inflation; BICEP astronomers had to determine emissions at higher frequencies, which are more sensitive to the contaminating dust. The problem was that the keepers of such precious information were their rivals working on the Planck satellite. Of course any rival wouldn't want to give competitors their results, lest they be scooped. But, as luck would have it, a BICEP researcher saw a talk given by Planck astronomer Jean-Philippe Bernard. It contained a revelation, a slide showing Planck's dust estimate at 353 Gigahertz. That was their golden ticket to declare the discovery of the century.

In a viral video, BICEP's Chao-Lin Kuo is seen arriving at Andrei Linde's house in Stanford, California. As the door opens, Linde and Renata Kallosh greet Kuo who says nothing but "5 sigma; 0.2." The couple instantly understand. *Five sigma* refers to a high enough level of statistical confidence to declare a detection (corresponding to a false-alarm probability of less than one in three-and-a-half million); 0.2 represented the detected power of the Big Bang's gravitational waves, relative to its sound waves. B-modes had been discovered; inflation in the eyes of the press was now a fact. Kallosh hugs her guest, almost crying, and Linde asks him to repeat the incredible news as if he can't believe it. The couple seemed genuinely shocked, claiming they had no idea Kuo was coming. Kallosh thought it was the post and had asked Linde if he had been expecting a delivery. In the video Linde replies, "Yeah, I ordered something thirty years ago. Finally, it arrived." But even as the celebration was in full swing, Linde could be heard worrying that the results might be a trick.[29]

Within a few weeks, the blogosphere was alight with rumors that indeed, something was not quite right. The slide that the team had captured from Planck noted a caveat: "not CIB subtracted." The CIB is the cosmic infrared background, emitted by warm interstellar dust, and is another source that must be accounted for in determining the primordial component of any potential signal. The rumor was that BICEP had simply missed this fine print, creating a spurious result. The criticisms were not warmly received; John Kovac denied any errors, and the (in)famous Harvard string theorist–turned-blogger Luboš Motl dismissed BICEP 2 skeptics as conspiracy theorists, comparing them to geocentrists forcing Galileo to recant his sun-centered worldview.[30] But the majority opinion was increasingly with critics like Spergel at Princeton, who along with his student Colin Hill and a postdoc, Raphael Flauger, thoroughly debunked the BICEP claims. When Matias Zaldarriaga, another influential cosmologist at Princeton, told Guth about the trouble with dust, Guth's heart sank; "That immediately had me worried," he recalled to us later in an interview, "because I trust Matias a lot." BICEP was trying to put out these fires in the community by responding to the critics. But like eternal inflation itself, this fire was one that could not be extinguished. When Planck finally released its analysis, the results were devastating: BICEP had not shown evidence of primordial B-modes. Dreams of a Nobel Prize had bitten the proverbial dust. By the time BICEP's paper was published in the prestigious *Physical Review Letters*, it had been thoroughly refuted.

With BICEP's claims in ruins, some inflationary advocates felt betrayed. As they saw it, the story created the false impression that there had been no evidence for inflation before the "discovery." It was the perfect time for the rebels to launch their counteroffensive, claiming the BICEP signal was right where inflationary cosmologists had expected it. The theory, critics like Ijjas, Steinhardt, and Loeb suggested, should suffer the same fate as the bogus signal and be consigned to the scrap heap of forgotten science. In my view, both sides probably held a double standard. The non-discovery of gravitational waves no more disproves inflation than the non-discovery of life on Mars disproves the hypothesis that alien life might exist somewhere in the cosmos. Nor can one claim that inflation makes no predictions and then declare triumph when its "predictions" aren't found. On the other hand, claims that B-modes (had they been discovered) would have constituted definitive proof of inflation are also false, as other competing ideas might also produce these swirling patterns in the primordial light. There will be subtle differences we can look for in the observed patterns to determine who is right and who is wrong—a point we shall come to in later chapters.

Circling back to the notion of trying to reconstruct your oldest collective memories, say your family tree, it may take more than an old, grainy picture of your grandmother with a handsome gentleman to tell who your real grandfather might have been. Sometimes

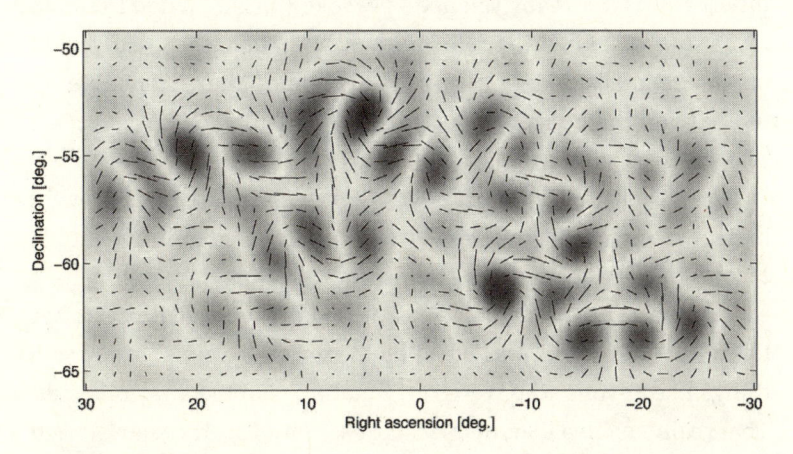

The Background Imaging of Cosmic Extragalactic Polarization (BICEP) experiment's infamous false detection. The line segments show the direction of polarization of cosmic background light. The swirls, known as B-modes, could have been the smoking gun for the Big Bang gravitational waves or, as it turned out, dust in our galaxy. Credit: BICEP2 Collaboration.

it's best to avoid rash judgments and wait for more data. I prefer to wait for more data.

THE INSURGENCY GOES PUBLIC

Despite BICEP's failure, most cosmologists still saw inflation as the only game in town and largely ignored alternative paradigms for explaining the origin of cosmos. The cyclic theorists needed some better publicity, and the popular magazine *Scientific American* delivered by publishing a strong critique of inflation by Ijjas, Steinhardt, and Loeb in 2017.[31] I remember feeling déjà vu while reading their article, realizing it was eerily similar to one I had read six years earlier entitled "Inflation Debate"; I'd assumed that one would feature a proponent of inflation confronting a critic, but instead Steinhardt had played both roles.[32] The outcome (in both articles) was unsurprising; inflation was considered a failed paradigm. Alan Guth decided enough was enough. He contacted Nomura and Kaiser, and the three decided that they needed to change tactics. They discussed strategy with fellow theorist Linde. In particular, the claim made by Ijjas, Steinhardt, and Loeb—that inflation was not really science—provoked Guth. "When I read it, I was somewhat horrified," he told us. "I thought they really went off the deep end . . . To assert that I claimed that science needs to abandon the concept of empirical testability, in my view, is a blatant lie."

One idea that Guth, Nomura, and Kaiser considered was to adapt their response in the peer-reviewed literature in a more digestible form for *Scientific American*, but Nomura recalls thinking, "That's exactly what they wanted . . . there is no real controversy." Kaiser told Phil that he, too, was upset by the article's claims: "They said things like anyone who works on inflation is in the sway of a cult. That really seemed like rhetorical excess." And so, the team started to ask some of the world's leading CMB experts to sign a letter to the editor defending inflation as a valid scientific enterprise. They recruited senior Planck team members and both Nobel Prize–winning COBE scientists as well as the principal investigator from

NASA's WMAP. Finally, helping to seal the deal, Stephen Hawking, Steven Weinberg, and Martin Rees all accepted the group's invitation to sign the letter as well.

Inflation, they said, was not only testable, but it had been tested. "Empirical science is alive and well," the letter concluded.[33] Nomura admitted to us that he knew the long list of names was overkill, but they felt that including it was necessary to reinforce the fact that the debate wasn't three versus three but a community versus a small minority.

Loeb, however, told Phil that he felt the letter was not a legitimate response but an attempt to shut down the debate, claiming that "the originators of inflation wanted the Nobel Prize . . . if you have good arguments, you should make them, rather than bully the criticism." Loeb went on to compare their ordeal to the time when a hundred critics condemned relativity shortly after its publication; Einstein, with some snark, retorted, "To defeat relativity one did not need the word of 100 scientists, just one fact."

Loeb then compared inflationary cosmologists to religious zealots who "feel offended if someone doubts their idea."

Tensions between the two camps remain high. When Phil asked the critics to debate the inflationary multiverse with a proponent, Ijjas and Steinhardt both refused, Steinhardt claiming the idea had done too much damage to the credibility of science and was too trivial to be worth discussing. Perhaps being the key partisans in a scientific rebellion had taken its toll and the two had lost the energy or desire for perennial infighting. Loeb, though, seemed to us to have no such qualms and was used to being in the forefront of controversy, having given (literally) thousands of interviews promoting his view that comets and meteorites coming from outside the solar system might be extraterrestrial technology. After being labeled "Harvard Astrophysics *enfant terrible*," he showed no hesitation and quickly accepted Phil's offer of a debate.[34] But Guth also declined, saying he couldn't discuss the issue with people who lied and misrepresented his views. I believe that most cosmologists side with inflation, but many are uneasy with its accompanying multiverse. And while I think Nomura is right to say the controversy isn't three versus three, the list of well-respected inflation critics is certainly larger than just Ijjas, Steinhardt, and Loeb. Phil eventually hosted a showdown between Guth and another well-known cyclic-universe proponent, Penrose, and months after the broadcast the two were still parrying claims and counterclaims by email, with Phil copied on the exchanges.

At this point, you may be left wondering if there will ever be a way to resolve this debate. The approach I have pursued is to try and develop alternatives to inflation and test those; if they can be more predictive and are borne out by the data, then we don't have to worry about the measure problem and other metaphysical speculations. We shall examine these alternatives later in the book. Those in the inflation camp take different approaches. Debika Chowdhury and colleagues recently claimed that the multiverse is only a worry if it comes with a definite measure; since it doesn't, we can follow Guth's original strategy and ignore it.[35] But multiverse champions like Guth and Nomura argue that the inside of an inflationary bubble should have a tiny amount of negative curvature (like the geometry on a horse's saddle), so we should try and more accurately measure the shape of space. So far, it's close to flat, but detecting any positive curvature could, it is claimed, rule out the inflationary multiverse.[36] Yet, I am not convinced that there won't be ways around this conclusion, by messing with the measure. Vilenkin, and colleagues are working on a scenario in which bubble universes could form in our patch of space and appear as supermassive black holes. It might just be possible to predict black-hole properties from this scenario.[37] I wish them luck, but so far I don't think anything concrete enough has arisen from this line of thinking. A final option would be to find the primordial gravitational waves, that sixth prediction of inflation we left unticked from our list. Detecting them would rule out most but not all alternatives to inflation, and examining their more detailed properties would pin things down even further. That possibility is something I'm really excited about and something we'll come back to in our final chapter.

We've met our first candidate for the origin of the hot Big Bang; at first glance it seems like the case is compelling. Inflation explains cosmological puzzles with a very simple hypothesis—space can expand exponentially—and relativity certainly allows this possibility. Inflation also makes predictions that have been confirmed by observations. But things aren't quite as they seem. My biggest qualm is that all the so-called predictions predated inflation and were later adopted by it. One instance is the adiabaticity of fluctuations (that is, that fluctuations are the same across all the different types of matter), which has been so spectacularly confirmed by CMB data. This adiabaticity merely requires that the early universe had only radiation and that everything else was born from it later on, independently of whether any inflationary era preceded it. Another is the scale-invariant power spectrum (that is, that primordial sound waves had nearly the same strength on all scales), invented independently by Edward Harrison and Yakub Zel'dovich in 1970 and 1972, on opposite sides of the iron curtain—long before Tye convinced Guth to work on cosmology and make his "spectacular realization."[38] Others like Ijjas, Steinhardt, and Loeb complain that inflation has too many different versions, leading us to question whether it really is as predictive as is claimed. Finally, the measure problem casts a shadow of doubt on any claim involving an inflationary model.

But even if inflation is right, we can still ask, "What happened before the exponential expansion began?" That is the puzzle we shall turn to next. One scientist who thought he might have an answer was Stephen Hawking. It was an answer that would completely undermine his earlier work with Penrose on singularities, yet that undermining—and the attention his U-turn generated—didn't bother Hawking.

✴ 3 ✴

HAWKING'S U-TURN AND A UNIVERSE FROM NOTHING

In 2012, the Olympic and Paralympic Games came to London. Phil and his wife Monica attended the Paralympic opening ceremony, a dazzling son et lumière with a prelude featuring disabled ex-serviceman and pilot David Rawlins flying over the stadium in a glowing aircraft with pyrotechnics bursting from the wings like fireflies scattering from a forest. As the audience chanted "Five, four, three, two, one, zero!" a voice emerged—one that was instantly recognizable. Stephen Hawking welcomed the public to the Games. The ceremony, called "Enlightenment," was a love letter to science, complete with floating apples to represent Newton's theory of gravity and a transformation of the stadium into an artistic vision of the Large Hadron Collider. But the whole show began with a recreation of the Big Bang itself, with a flaming explosion and performers fanning out to represent the expansion of the universe. Whether the countless bubbles that filled the sky that night were meant to represent the multiverse of eternal inflation remains a matter of speculation. But Hawking did not shy away from promoting his own model for the birth of the cosmos during the ceremony. "There ought to be something very special about the boundary conditions of the universe," he said to the crowd, "and what could be more special than it has no boundary? And there should be no boundary to human endeavor." Witnessing Hawking narrating (and even performing a song with techno duo Orbital) was something Phil says he will never forget. Little did he know, a few years later he would be meeting with Hawking in a more private setting: inside the physicist's office in Cambridge. Hawking's words seemed fitting to the Paralympians; as some commenters said, "In Sydney, Paralympians were treated as equals. In London, they were treated as heroes."[1] But few in the audience were likely to appreciate what Hawking was really talking about when he said that the universe had no boundary.

Two years later, another enigmatic reference to Hawking's theory entered popular culture as Eddie Redmayne won an Oscar for portraying the scientist in *The Theory of Everything*. One scene that provided some small insight into Hawking's work began with Jonathan (the future lover of his first wife Jane) raising the idea over dinner:

JONATHAN: Jane tells me you have a beautiful theorem that proves that the universe has a beginning. Is that it?
STEPHEN: That was my PhD thesis; my new project disproves it.
JONATHAN: Disproves it?
STEPHEN: Yes.
JONATHAN: Oh, so you no longer believe in the creation?
STEPHEN: What one believes is irrelevant in physics.
JANE: Stephen's done a U-turn. The big new idea is that the universe has no boundaries at all. No boundaries, no beginning.
JONATHAN: And no God? Oh, oh I see, I thought you'd proved the universe had a beginning and thus a need for a creator. My mistake.
STEPHEN: No. Mine.

Hawking's U-turn was the culmination of a decades-long quest to understand the origin of the universe. After establishing the Big Bang singularity during his PhD, it was his supervisor Dennis Sciama who interpreted the state of infinite density, temperature, and pressure as revealing the breakdown of classical relativity. Thus, what Hawking had really done was to prove not that the universe had had a beginning but that a need existed for a quantum update to the Big Bang. For, while relativity has some strange, counterintuitive notions about the nature of time and space, its oddity is nothing compared to the bizarre world of subatomic physics that is governed by the rules of quantum mechanics. Hawking would become one of the pioneers of what is now known as quantum cosmology, erected upon the ruins of the singularities he had championed in classical cosmology. But hot on his heels was Alex Vilenkin, the multiverse champion whom we met in the last chapter. Vilenkin is a brilliant theoretical physicist from Ukraine who escaped from the Soviet Union to the United States at the height of the Cold War and would go on to challenge Hawking for the correct description of the quantum Big Bang. Both Hawking and Vilenkin claim that their models imply the universe came from "nothing"—but what this claim really means remains controversial, as do the proposals themselves. You might think the notion of a universe from nothing would appear to be absurd, but so is much in quantum physics. To understand these two schemes (known as [1] the no-boundary proposal and [2] tunneling from nothing), we need to appreciate the bizarre rules of engagement in the subatomic world. Having skirted around its notions for two chapters, it's now time to take a deep dive into the strange world of the quantum and how it is guaranteed to rewrite our story of the Big Bang.

THE STRANGE WORLD OF THE QUANTUM

When I was twelve, I read a lot of popular-science books on astronomy. They all had some description of the quantum world, particle-wave duality, and the uncertainty principle. Having read so much, I was then confident that I knew quantum physics, a belief I declared to one of the instructors, a physics graduate, in our amateur astronomy class. He then asked me whether I knew what the Schrödinger equation was, and I shut up. Quantum mechanics is the exotic mathematical framework needed to describe the behavior of the subatomic world and underpins much of the technology that powers our contemporary society, from computers to smartphones, nuclear power to lasers. I think most of my colleagues would agree that it is the most fundamental set of laws that exist in physics. But at its heart, the quantum world is a truly bizarre domain, nothing like our everyday experience. It operates on a few fundamental tenets, the consequences of which philosophers and physicists still grapple with to this day. The first is the principle of discreteness; just as the ancient Greek atomists said that the world is divided into objects that have a certain minimum size, so quantum physics says that energy comes in uncuttable packets. The very word *quantum* comes from Einstein's and Planck's supposition that light is emitted in discrete "quanta" of energy. Similarly, electrons orbit the nucleus of the atom in discrete energy levels. They can be excited to higher energies but again only in discrete units, as if an elevator were teleporting from one floor to the next without having to traverse the space between. Then there is the notorious Heisenberg uncertainty principle, which states that certain pairs of physical quantities cannot be measured to an arbitrarily high level of precision. For example, the more we know about a particle's energy, the less we can know about when it had that energy. In this way, energy and time are examples of complementary variables in quantum mechanics; another case is momentum and position. The more we know about the momentum of a particle, the less we can know about where it might be. An old joke to illustrate the uncertainty principle involves physicist Werner Heisenberg being pulled over by a traffic cop, who asks him if he knows how fast he was going. "No, but I know exactly where I am," Heisenberg replies. The cop tells him he was going 150 kilometers an hour. Heisenberg replies, "Great! Now I'm lost!"

This uncertainty has nothing to do with the lack of precision of our instruments. It's something deeper, something fundamental. Nature itself, according to the equations of quantum theory, seems intrinsically fuzzy. This extraordinary picture of the world posed a challenge to the regular clockwork-like laws of physics that had governed the Newtonian worldview for centuries. Heisenberg, one of the founders of this new physics, even said, "The law of causality is no longer applied in quantum theory."[2] Whether that's true or not, probability is central to the workings of quantum theory. Scientists can precisely determine the chances of an event in the quantum world, but they can't determine the outcome of any one trial. It's as if nature has rationed how much we can know about reality.

One of the implications of Heisenberg's uncertainty principle is particularly relevant for the Big Bang. If you try to pin down the position of a particle really, really well, it will have a very large momentum and thus a lot of energy. That huge energy, packed into a tiny space, will inevitably form a black hole. This observation suggests that any measurement of position will at best have a precision of 10^{-33} cm in a quantum theory with gravity—a limit known as the Planck length, sometimes described as the smallest length with any meaning: the realm where quantum gravity effects take hold. The time it takes light to travel across a Planck length is roughly 10^{-43} seconds, a duration known as a Planck time. That is how close it is possible to get to the Big Bang singularity without having to modify the laws of nature.

A startling consequence of quantum theory that lies at the heart of inflationary cosmology is that the vacuum of empty space can never be truly empty, as such emptiness would violate the uncertainty principle. Instead, as we discussed before, particle and antiparticle pairs pop in and out of existence in tiny fractions of a second, during which their energy is indistinguishable from zero.

A similarly bizarre feature of quantum mechanics is superposition. Normally we think of objects as being at particular locations or states in space. A baseball cap may be worn forward or backward, but it can't be worn both ways at the same time. Not so in the quantum world, where particles can be in two (or more) states simultaneously.

Another peculiar feature of uncertainty is that it implies that particles can penetrate barriers that would be impossible to cross in the world of classical physics. This phenomenon is known as quantum tunneling and is key to understanding how everything from the sun to flash memory drives work. One doesn't expect a ping-pong ball thrown at a brick wall to suddenly teleport to the other side, but something analogous can happen in the quantum realm. The only reason we don't see this teleportation happening in everyday life is that the probability that every particle in a ping-pong ball will simultaneously tunnel to the other side of a solid barrier is vanishingly small. But, luckily for us, vanishingly small is not zero! For example, protons in the core of the sun are positively charged and thus repel each other; according to Newton's laws, they will never meet. Nevertheless, the tiny probability of quantum tunneling—allowing some of the sun's gargantuan number of 10^{57} protons to stick together—produces enough nuclear fusion to allow the sun to shine. I think there is an important lesson here for understanding the Big Bang. If we use only classical physics, we will find that the sun will not shine, so instead we must apply quantum mechanics. The Big Bang is likely to require a quantum-mechanical understanding, too.

In 1963, Richard Feynman gave his famous lectures in physics and claimed that the outcome of one notorious experiment captured the essence of the quantum world, which is impossible to understand classically. The setup is known as the double-slit experiment and involves a device that shoots particles (usually electrons) onto a screen. In between the gun and the screen is an obstruction with two holes. Feynman considered the scenario a thought experiment that could not actually be performed. But theorists often underestimate the ingenuity of experimentalists. In fact, German physicist Claus Jönsson had already performed the experiment two years before Feynman considered it.[3] The results were every bit as bizarre as Feynman had said. As electrons are fired at the screen, one might expect them to pile up in two patches behind each of the two slits, as if one were pouring sand into a large bucket with two tiny holes. But that is not what happens. Instead, the screen will light up in a zebralike pattern with bright patches interlaced by dark fringes, an unmistakable signature of wave interference. The reason this pattern shows up is that, unlike classical particles that have positions and velocities, every quantum particle is described by a fuzzy wave function that can go through both slits at the same time, generating both constructive and destructive wave interference. What is even stranger is that if we try to cheat and place a detector at the slits to see which one the electron went through, the wave pattern disappears, and the outcome is precisely the classical picture with electrons concentrated in two piles behind each slit. It is no wonder that Niels Bohr, Heisenberg's mentor, famously quipped that those who aren't shocked by quantum mechanics can't possibly have understood it.

Just like ripples on the surface of water, the wave function of electrons creates interference patterns passing through two narrow slits. Credit: Niayesh Afshordi.

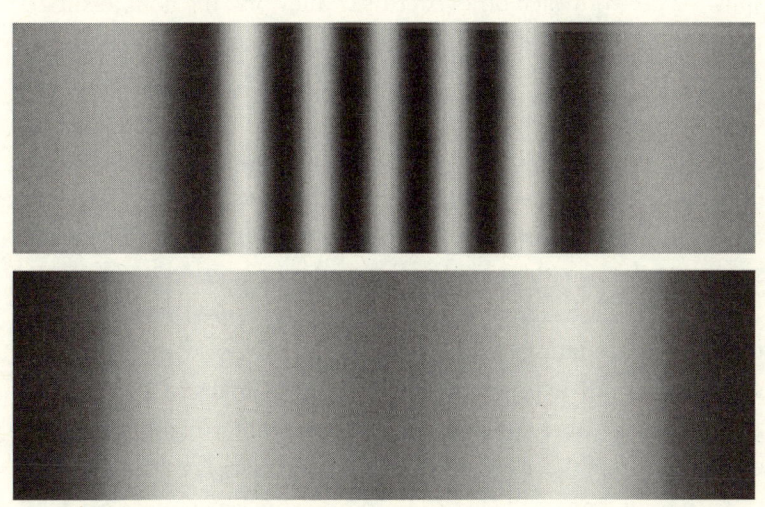

What a detector would see for quantum (*top*) and classical (*bottom*) electrons passing through the two slits in the previous figure. Credit: Niayesh Afshordi.

We physicists still struggle to make sense of these strange rules and so have various interpretations of quantum mechanics. Perhaps the two most notorious are the Copenhagen and many-worlds interpretations. The former has its origin in the musings of Bohr and Heisenberg in the 1920s, during the founding days of quantum theory. It says that the position of an electron has no reality until someone decides to look for it; that is, to make a measurement. The outcome of that measurement will be random, with probabilities that are predicted by the wave function. The act of measuring temporarily collapses the wave function into the position measured by the experiment, only for subsequent evolution to make it fuzzy again. In this framework, the interference pattern is seen because the electron is a probability wave with no definite location until it is measured. So, it can pass through both slits and interfere with itself. It is the act of measurement that collapses the wave function and forces the particle to appear in a definite location.

In contrast, the many-worlds interpretation was developed in the 1950s by Hugh Everett, who claimed that the wave function never collapses; we only see one branch (or world), but all those other possible ways the universe can exist, forever cut off from our view, are no less real. A myriad of other worlds exists, all continually splitting off from each other. In this framework, the double-slit experiment is explained because the electron passes through different slits in different worlds, which are in a superposition state and can interfere with each other. These quantum "many worlds" are generally considered to be a very different notion to the multiverse of eternal inflation, although theoretical physicists like Leonard Susskind, Raphael Bousso, and Yasunori Nomura have recently suggested the two ideas may be one and the same.[4]

Despite their disagreements, most quantum physicists agree that the world is fundamentally made of fields that spread out across the universe; a particle is just the excitation of a field in a localized region of space. There could also be particles that (roughly speaking) go back in time; these are known as antiparticles (such as the positron, the evil twin of an electron). In the same way that we can't tell where an electron is in quantum mechanics, we may not even be able to tell how many electrons there are, as electrons and positrons are constantly popping in and out of the vacuum. This quantum field theory forms the basis of almost all modern physics, but add gravity and all hell breaks loose. You may be tempted to apply the rules of quantum field theory to Einstein's theory of general relativity, but you shall soon realize that it stops making any sense for processes happening on the Planck length or time scales. But could there be a shortcut, a civil union of sorts between quantum and cosmology, without attacking the full problem of quantum gravity head-on? Finding such a shortcut was the task Hawking set himself in the 1970s. Having become the establishment in cosmology, it was time for him to become a revolutionary once more—but before he could challenge the Big Bang, he set his sights on the objects that had inspired him in the first place: black holes.

A CLUE FROM THE ABYSS

The cosmological singularity theorems had their origin in Roger Penrose's study of black holes, completed after he returned from Princeton collaborating with the American giant of relativity, John Wheeler. Most of my peers know Wheeler through his famous, voluminous textbook *Gravitation*, co-authored with his former students Charles Misner and Kip Thorne, who went on to win the 2017 Nobel Prize in physics. We often joked that the book was so massive that you can feel its gravity. In fact, Wheeler is almost as famous for his never-ending list of brilliant students as for his own work. One of these students, who would change the course of physics forever, was Jacob Bekenstein, the son of Jewish refugees from Poland who had been lucky enough to escape the Holocaust. Wheeler asked Bekenstein to think about what would happen if his cup of tea were to be poured into a black hole. The answer seemed obvious; once the drink and accompanying biscuits crossed the point of no return, known as the event horizon, they would be consumed by the black hole, never to be seen again. But every object in the universe, including a cup of tea, has some entropy associated with it. Entropy is, roughly speaking, a measure of disorder (counting the number of all microscopic possibilities). While you may think temperature is a measure of how hot your fever is, physicists understand temperature as the rate at which increasing disorder (entropy) raises the energy of a system. This understanding is called the first law of thermodynamics, and it suggests an intimate relationship between entropy and temperature. The second law of thermodynamics states that entropy almost always increases, because, as anyone with a small kid knows, there are always many more ways to make a mess than there are to make a tidy room. But to return from my son's bedroom to the cosmos: Bekenstein realized that if an object were to disappear from the universe as it was swallowed by a black hole, its entropy would vanish—a clear violation of the second law. Black holes are known to be uniquely bizarre objects, but the idea that they could really spell doom for such a fundamental principle doesn't sit right with most of my colleagues. Looking back, Bekenstein's insight was a clear signpost that a major new discovery was waiting in the wings.

Bekenstein reasoned that if an object falls into a black hole, the black hole will grow, and so will the area of its event horizon. He was building on an earlier discovery by Hawking that said that the area of the event horizon could only ever increase. That assertion sounds remarkably similar to a statement that entropy always rises. The area of the event horizon and the entropy of the black hole, Bekenstein argued, had to be the same thing. The entropy of an infalling cup of tea does not disappear but rather goes to increase the entropy of the black hole; the second law was saved. In 1973, Hawking, together with Brandon Carter and James Bardeen, published a paper suggesting the contrary.[5] If black holes have entropy, they reasoned, then they must give off heat, as in thermodynamics heat is associated with changes in entropy. But nothing can escape a black hole, not even light, and so they argued that black holes simply couldn't emit heat.

In fact, it was Bekenstein who was right. Ironically, Hawking himself would prove it. Hawking realized that when particles and antiparticles pop in and out of the quantum vacuum (such pairs are called virtual particles) near the edge of the black hole's event horizon, one may fall in while the other escapes. The particles that escape can then result in the emission of radiation by the black hole. Hawking's calculations showed that this radiation is like what you get from a hot rod with a temperature. So black holes do give off heat, as Bekenstein had proposed, and thus they obey the laws of thermodynamics just like everything else in the universe. Today, we call this emission Hawking radiation and the formula used to calculate the entropy of a black hole, the Bekenstein-Hawking formula. Understanding the meaning of this entropy remains the holy grail for any attempt at a quantum theory of gravity. At his sixtieth birthday celebration, Hawking famously stated that he wanted this formula on his gravestone. Strangely, it is instead the formula for black-hole temperature that can be found inscribed on Hawking's gravestone at Westminster Abbey, where he lies buried beside other giants of scientific discovery, Isaac Newton and Charles Darwin. Among those who knew Hawking, I have heard that there's some uncertainty about whether he had a change of heart or the gravestone formula was a blunder.[6]

As Hawking found isolated black holes emitting radiation, he realized that they would ever so slowly shrink and then suddenly blast and disappear into the void. Hawking was suggesting that there wasn't just a quantum loophole that allowed tiny particles to escape black holes; rather, the process would cause these gargantuan objects to vanish from existence. The idea was truly revolutionary although not immediately welcomed. The moderator of Hawking's first talk on the subject said, "Sorry Stephen, but this is absolute rubbish." Even the Soviet physicists Yakob Zel'dovich and Alexei Starobinsky, whose work he had built upon, found it hard to believe.[7] But the force of Hawking's logic turned out to be more inescapable than the gravity of the objects he was describing. Soon, physicists recognized Hawking's tremendous breakthrough. It wasn't just a surprising result. For the first time, a theory combined quantum mechanics and relativity to analyze a real physical object. The results were profound, reversing previously held wisdom; something that reverberates in the corridors of physics departments to this day. It was now time to do the same for the Big Bang.

In the 1970s, Hawking, collaborating with James Hartle, rederived his black-hole entropy formula using a new method called Feynman's path integral, a warm-up for their ultimate goal of tackling the Big Bang. Thirty years earlier, Feynman (another famous student of John Wheeler) had realized that in order to create the wavelike pattern of electrons moving through the two slits, he could model them by summing over every possible path to get from A to B. That's a lot of paths, an infinite number in fact. But waves have peaks and troughs, and in calculating the wave function most of these peaks and troughs cancel out. This cancellation enables physicists to get a finite answer even from an infinite series of numbers. This path integral (literally, sum of paths) technique allows physicists to calculate which factors make a dominant contribution to quantum probabilities, and with this knowledge, they can make exquisitely precise predictions for experiments, from the spectra of atoms to the underground energetic collisions of subatomic particles in the Large Hadron Collider.

These methods use a strange mathematics that may seem fantastical but is nonchalantly used by today's physicists. Think of the square root of one: what number times itself, gives one? There isn't one answer but two. Negative and positive one both square to positive one. But there is no conventional number that when multiplied by itself becomes negative one. However, mathematicians do not lack creativity and so they decided to invent a new number they labeled i, the "imaginary" number, that squares to negative one. It was Heron of Alexandria, a first-century follower of the atomist school and inventor of all kinds of contraptions including the world's first vending machine, who first used this bizarre mathematical concept. Heron's work was further refined in the sixteenth century by Rafael Bombelli, who suggested that numbers could have more than one axis, like a graph that has height and width rather than the simple one-dimensional number line we are familiar with. This picture creates a much larger space for new numbers, and the expanded area is known as the complex plane. Remarkably, mathematicians found ways to do calculations in the complex plane that could not be easily performed with conventional numbers. René Descartes hated the idea and named the new digits imaginary numbers as an insult to Bombelli. But insulting terms are often reclaimed as badges of honor, and scientists and engineers use imaginary numbers today without any sense of shame. In quantum mechanics, they are essential. Descartes would probably not approve.

THE NO-BOUNDARY PROPOSAL

Feynman's path integral approach to computing the quantum wave function involves carefully and laboriously summing over infinitely many complex numbers. Physicists then use a mathematical shortcut known as a Wick rotation, where the time variable is described by an imaginary number to enable infinite sums to be computed more easily. When Hawking and Hartle tried this trick on black holes, they found something remarkable: the singularity at the core simply disappeared. There was no region of infinite density; instead, there was a quantum domain where our everyday world with three space dimensions and one time dimension transforms into a place with no time dimension but four space dimensions. Soon Hawking would promote the idea that the Big Bang singularity could be similarly removed, in what became known as the no-boundary proposal.

In the no-boundary proposal, instead of a singular point where all light rays converge, there is a smooth surface; imagine a badminton shuttlecock rather than a cone. There is no point on this surface that can be marked out as the beginning, as there is no time, only space. Just as the Earth does not suddenly cease to exist at its poles, neither does the universe at the Big Bang. Unlike the singularity theorems, the laws of physics do not break down at some sudden juncture; instead, the proposal is a gradual transition from a purely quantum world where time does not really exist. As the cosmos grows in size, quantum randomness decays, correlations between cause and effect are established, and clocks start ticking where units of time turn from imaginary to real. This is the wonderland of the no-boundary proposal.

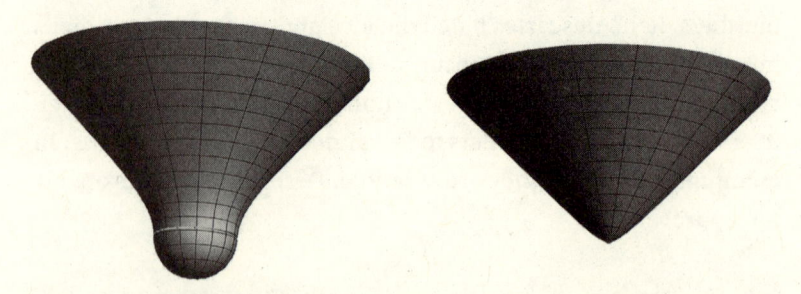

The shuttlecock-like, smooth space-time geometry of the no-boundary proposal (*left*) contrasts with a singular Big Bang geometry with its sharp tip (*right*). The gap in the shuttlecock geometry represents the switch from quantum to classical geometry. Credit: Niayesh Afshordi.

At the 1981 Vatican conference where Hawking presented his model, he recalled the pope telling the attendees they could inquire about the universe after the Big Bang but not earlier, as that was the realm of God. Hawking would have none of that, saying, "The no-boundary proposal is a model of the physical condition at the beginning. It describes how our familiar notions of space and time can emerge from a quantum state of the universe. In the no-boundary proposal, the same laws of nature hold at the beginning as in other places. This removes the need for an intelligent creator. The beginning of the universe would be governed by the laws of science."[8]

By 1983, Hawking and Hartle had published their first paper on the quantum origins of the cosmos, entitled "Wave-Function of the Universe."[9] We physicists are used to the idea of a wave function for a single particle (or even several of them) but not for the entire universe—that's nothing short of outrageous. Hartle's former advisor and Nobel Prize–winner Murray Gell-Man asked, "If you know the wave-function of the universe, why aren't you rich?" implying one might use such a function to predict whether the markets would rise or fall. This idea might explain Phil's eagerness as an ex-trader to interview quantum cosmologists, but to his dismay, Hartle's reply was that the markets' chances of rising in the no-boundary proposal were probably fifty/fifty. Yet the more immediate issue was not whether one could use the no-boundary proposal to get rich but rather whether it could predict features of the universe that astronomers observe. Before Hawking and Hartle could begin such a program, they would face stiff competition from another cosmologist keen to use quantum theory to describe the origin of the universe.

TUNNELING FROM NOTHING

Alex Vilenkin was born in Ukraine shortly after World War II. It had been a devastating conflict; the conflict killed seven and a half million of his fellow compatriots, and one and a half million Jewish people were murdered by the Nazi SS. When peace returned, conditions were harsh; he and his family lived in one room for four people, and four other families lived in the same apartments and had to share a single toilet. The sole bathtub was used to store whatever water they could acquire. In high school, Vilenkin and a friend pursued their passion for physics. That was when he fell in love with cosmology. He finished top of his class at university and was all set to do a PhD until the brutal nature of the Soviet regime intervened. "I was told my candidacy was not approved by authorities in Kyiv," he recalled to Phil. "I was told by people who had access that I was blacklisted by the KGB," he explained, saying it was retribution—the agency had earlier tried to recruit him as an informant, and he had refused. Instead of studying the universe, Vilenkin was drafted into a military building battalion, essentially a forced-labor unit, consisting of mostly hardened criminals whom the authorities would not trust with weapons. Despite long, harsh days, Vilenkin would occasionally sneak into the officer's mess to work on his physics. After the army, he found himself recruited for a list of other jobs: night watchman and then guard at the Kharkiv Zoo.

In the 1970s, Vilenkin eventually managed to emigrate to the USA to complete his PhD. When I was a postdoc at Harvard, I once took Vilenkin out to dinner following his colloquium at the Center for Astrophysics. He told me that he had always been interested in cosmology, and when he'd applied to graduate school, he had decided to go to Buffalo to work with Rainer Kurt "Ray" Sachs, the preeminent cosmologist who had predicted in the 1960s the fluctuations—later confirmed by COBE—of the cosmic microwave background. Except it had turned out that the Professor Sachs in Buffalo was an entirely different person, Mendel Sachs, a condensed-matter physicist. Therefore, Vilenkin had had to wait to do cosmology until after he became a professor in Boston. It was then that he learned of a now-legendary comment in a lecture given by Sciama, who you'll remember was the mentor of both Penrose and Hawking. The comment did not come from the speaker, though; instead it came from a scientist in the audience named Edward Tryon. Vilenkin recalls the story: "When the speaker stopped to collect his thoughts, [Tryon] just blurted out that maybe the universe is a quantum fluctuation. . . . Everybody laughed because they thought it was a funny joke, but he was serious."

Recall that the Heisenberg uncertainty principle implies that particles fluctuate in and out of existence. But the more massive these virtual particles are, the shorter their lifespans, typically lasting for around a billionth of a trillionth of a second. The universe is somewhat larger than a tiny particle, so there seems no possible way for it to fluctuate into existence without having to disappear in an even briefer moment. But Tryon pointed out that as gravity has negative energy, the positive energy of matter might balance it out, implying that the total net energy of the entire universe could be zero. If so, the length of time it could exist, as a vacuum fluctuation, would be infinite. Amazingly, the spontaneous creation of the universe would not violate any conservation law, a free lunch indeed.

The idea of the positive energy of matter and the negative energy of gravity balancing out was not new and goes back to the early days of quantum theory. When Big Bang pioneer George Gamow first mentioned it to Einstein, the second man supposedly stopped suddenly in the street and was nearly run over (recall that, in spite of Emmy Noether's insights, the concept of energy conservation in general relativity has always been treacherous). But while the idea may have almost killed Einstein, neither of the two giants of cosmology ever dared to make the leap that Tryon was proposing. The problem with Tryon's model was that although it was possible for a large universe to fluctuate into existence, a smaller one seemed far more probable. An initial quantum nugget of energy might be microscopic, but Vilenkin grasped that inflation could blow it up into the vast universe we see today. But still he wondered where the initial nugget of inflationary energy had come from. Even a subatomic-sized speck of inflating space is not nothing, and then he had his own flash-of-light moment, arguably one more profound than Tryon's.

In 1982, Vilenkin proposed that space-time itself could spontaneously fluctuate into existence from a state where there was no space or time, just as particles pop out of the vacuum.[10] More technically, he asked whether a minuscule universe, doomed to be crushed under its own gravity, can nonetheless quantum tunnel out to a nugget that would inflate to make our big universe. The answer was yes. He then wondered how small the initial radius could be and found that the answer was zero. Thus, the universe was born from literally nothing—space and time would quantum mechanically fluctuate into existence. One might wonder what caused the universe to tunnel in the first place. Just as Lemaître had envisaged the birth of the primeval atom as a radioactive decay, Vilenkin thought the universe began as a quantum process and so required no cause; cause and effect are classical concepts, arguably absent from the quantum realm (more on this in chapter 10). This understanding might suggest why Einstein did not consider this possibility; as he famously said, "God does not play dice with the universe," to which Bohr replied, "Einstein, stop telling God what to do!"

A universe that is spontaneously created from nothing and has no cause made even Tryon's model seem conservative. So, you might have expected that Vilenkin's idea would have been similarly ridiculed. But in the decade since Tryon was shot down in a fit of laughter, things had changed dramatically. Inflation had convinced physicists that talking about the initial conditions of the universe was not so absurd after all. Vilenkin recalled warmly to Phil that his reception was therefore very different: "After the talk, Lawrence Krauss came up to me and said it's amazing to give a talk like this and survive." The first discussion Vilenkin had regarding the idea was with Malcolm Perry (Hawking's former student and star of the

Netflix documentary *The Edge of All We Know*). After an hour at the blackboard, Perry was so impressed that he was surprised he hadn't thought of it. "What better compliment can you get from another physicist?" Vilenkin asked.

The reaction has not always been so positive. In 2012, conflict ensued when Krauss published his book *A Universe from Nothing*, promoting the models of Vilenkin, Hartle, and Hawking.[11] David Albert, a philosopher of physics, wrote in a *New York Times* review that Krauss "has an argument—or thinks he does—that the laws of relativistic quantum field theories entail that vacuum states are unstable. And that, in a nutshell, is the account he proposes of why there should be something rather than nothing . . . The fact that some arrangements of fields happen to correspond to the existence of particles, and some don't is not a whit more mysterious than the fact that some of the possible arrangements of my fingers happen to correspond to the existence of a fist and some don't. And the fact that particles can pop in and out of existence, over time, as those fields rearrange themselves, is not a whit more mysterious than the fact that fists can pop in and out of existence, over time, as my fingers rearrange themselves. And none of these poppings—if you look at them right—amount to anything even remotely in the neighborhood of a creation from nothing."[12]

I don't think Albert's is a fair criticism. The nothingness of quantum geometry is not the same as the nothingness of quantum particles. The latter do pop in and out of existence from the vacuum; that's right. But if space is quantum mechanical and it pops into existence, then it does so not from a vacuum but from a state without classical space or material objects as we know them. A universe from nothing, indeed. Well, kind of. There are no material objects, but there are the laws of physics—and where they come from is not explained by Vilenkin's model. We have to assume they exist somehow, even when there is no universe. This notion I find hard to contemplate, if the laws of physics are just descriptions of how the world behaves. But maybe they are more than that. Honestly, I don't think anyone can say with any confidence what the true nature is of the laws of physics. When Phil put this objection to Vilenkin, he replied, "It seems to me that the laws may well have some Platonic existence." Many Greek philosophers thought there was some primordial entity the universe arose from: for Thales, water; for Anaximander, the Aperion; for Anaximenes, air; and for theists, God. Vilenkin's radical suggestion was that the primordial entity was none of these. Instead, it was the laws of physics, for they, he claimed, are all that is required to conjure the universe into existence.

With the quantum creation of the universe being explored from two separate directions, it wasn't long before the rivals began to debate their proposals. Vilenkin recalls, "I was talking to Stephen, and I said something about my wave-function to which he said, 'You don't have a wave-function.'" Comparing Hawking to the Oracle of Delphi—given his short, probing pronouncements that have to be pondered to make sense of them—Vilenkin went on, "I realized immediately what he meant. Basically, at the time I only worked my wave function to the simple mini super-space model . . . not for the general case. . . . I spent the night rewriting my talk." A mini super-space model is one in which the evolution of the universe is described by a single number, its radius. Vilenkin described Hawking's reaction after the talk: "I thought, okay he will now destroy me with some critique; instead, he invited me to dinner. . . . His mother cooked dinner, and we discussed wormholes, and it was a very charming evening." I had a similar experience to Vilenkin when I was describing one of my models to Hawking. After waiting literally for fifteen minutes for his response, his assistant informed me that Hawking had accidentally erased it. So, I waited another fifteen minutes for him to tell me: "But this is too ugly!" Just like Vilenkin, I didn't have to work too hard to figure out what Hawking meant.

QUANTUM COSMOLOGY AND THE REAL WORLD

One of the difficulties I have with digesting the no-boundary proposal is that the classical universe starts with maximal initial size and minimal energy density. The tunneling proposal initiates the universe with the minimum size and maximum energy, thus seeming more in tune with an inflationary origin. But to be fair, the universe doesn't care what seems intuitive to me. If the evidence says it is compelling that it started in such a state, then so be it. But as both models claim to lead to eternal inflation, any trace of these beginning events will probably be erased from view. Vilenkin admits that nobody has found a way to test it against the data, and in my opinion, that absence of testability is the real problem. Yet both models have something to say about the very shape of the cosmos.

In these early quantum cosmology models, the universe should be closed like a sphere; this geometry allows the universe's negative and positive energy to balance. But after (a long enough) inflation, it would appear flat; hence, there seems little hope to confirm its closed nature. So, Hawking and Hartle began to train several graduate students to develop the no-boundary proposal to find another way for it to confront the real world. One of these students was Jonathan Halliwell, who was inspired to go into physics after watching a popular-science documentary. "It was the first time I saw Stephen Hawking on the telly," Halliwell says. "He's this kind of amazing figure in a wheelchair, this sort of sage-like figure.... And he's got all this incredible maths in his head." Soon Halliwell would sit the notoriously challenging Cambridge Tripos exam, which tests students in science and mathematics; he did well enough to have first choice of supervisor, and there was no question as to whom Halliwell would pick. Hawking's students helped with his nursing duties, which gave them more opportunities to talk to him. Halliwell describes him as charismatic: "He's incredibly funny and cheeky and rude and irreverent.... He'd say, 'Have a look at this calculation.' And then I'd come back two weeks later, and say, 'Oh, that's really interesting.' And he'd say, 'I don't believe in that anymore.' That would happen repeatedly.... Some of it was just plain wrong and some of it was just not very good, but then you know there was a big chunk of it which was pure genius."

If the no-boundary proposal was to have any chance of being a realistic model, it had to reproduce the conditions that cosmologists believed were present in the early universe. At first, Hawking and

Hartle had shown that this proposal might work in the idealized mini-superspace mentioned by Vilenkin above, but then Halliwell was asked to relax the simplifying assumptions. He found that this general case really does create the conditions cosmologists had imagined were the start of inflation, with all fields in their quantum mechanical ground or lowest-energy states. If correct, Halliwell's conclusion would have been a major triumph for the model and another impressive feat for the no-boundary proposal. In standard physics, we often split our modeling between setting the initial conditions of a system and specifying the dynamics that govern it. Think of the position of a set of balls on a pool table at the start of the game. These are the initial conditions. The dynamics are the laws of motion that describe what happens when the balls are hit. In the no-boundary proposal, the initial conditions and the dynamics are unified in one single equation. No wonder people call Hawking a genius. But all was not well; the model predicted too little inflation, and to get it to work, they had to use a rather, shall we say, innovative approach to probability theory, where instead of just calculating what the theory predicts, you calculate what the theory predicts given that you exist (recall the anthropic principle from chapter 2). That approach certainly gives me pause for thought, and many others think it's a severe flaw in the proposal, akin to the measure problem that, as we saw in the last chapter, has been haunting inflationary cosmology.

RETHINKING THE NO-BOUNDARY PROPOSAL

In Vilenkin's tunneling-wave function, the universe has an unambiguous beginning. But in the Hartle-Hawking model, things aren't so straightforward. In his book with Leonard Mlodinow, *The Grand Design*, Hawking famously said, "Because there is a law such as gravity, the universe can and will create itself from nothing. Spontaneous creation is the reason there is something rather than nothing, why the universe exists, why we exist." But when Phil asked James Hartle if he shared this view, Hawking's collaborator replied, "What do you mean by *nothing*? In quantum mechanics, there is always something." If the two authors of the model don't agree as to how to interpret *nothing*, then surely we should be a little cautious ourselves. Is this really a universe from nothing or not? In my view, most physicists think space-time geometry has a fundamentally quantum nature. Some states of this geometry might lack any classical description, resembling nothing in our everyday experience. If a classical universe were to emerge from this bizarro-land, you might call it a universe from nothing or from something. At the end of the day, it doesn't make a difference what you call it. However, we would still like to know if the universe had a beginning or not, and the answer to this question may depend on how we think about those strange imaginary numbers we discussed earlier in the chapter.

Just as the introduction of imaginary numbers led to a number space with two axes, so the introduction of imaginary time can be thought of as giving the universe two dimensions of time, one imaginary, the other real. What few people realize is that the no-boundary beginning may only be in imaginary time.

But what is imaginary time? Like all other humans, physicists like shortcuts, and in quantum mechanics, using imaginary time is one of those shortcuts we use to describe complicated processes. Case in point: recall that protons can travel through impenetrable barriers using quantum tunneling, allowing the Sun to shine. They can be thought of as taking shortcuts in imaginary time. While the classical description of events operates in real time, you can use imaginary time to quantum tunnel from one real-time story to another. Hartle and Hawking had used imaginary time in their initial work to avoid the singularity inside the black hole, and they performed the same trick to avoid it in the no-boundary proposal.

But Hartle and Hawking's story, told in real time, describes a universe undergoing contraction followed by a bounce to expansion. As a result, Aron Wall, Hartle's former colleague, wrote that those who claimed the proposal had a beginning were lying. His blog post continued: "OK, they weren't really lying, but rather oversimplifying. I'm sure their motives were as pure as the driven snow, but the end result in their readers' minds is the same . . . the

Hartle-Hawking model doesn't actually have a beginning!"[13] When challenged, another of Hawking's close collaborators, Thomas Hertog, replied that although the model shows a bounce in the real-time axis, the arrow of time points in the opposite direction, away from the bounce—as a point of minimum entropy exists in the middle, arising from the ground state enforced by the model. Those in the contracting branch of the universe would see the bounce in their past just as we do; time runs forward in both branches of the universe. Time behaves like duelists pacing away from each other, each thinking they are marching forward and away from their own Big Bang. Julian Barbour, who does physics and philosophy out of his English farmhouse, calls these models the Janus universe, after the Roman god of transitions, who is depicted as having two faces: one looking to the past, the other to the future.[14]

Two visions of the universe, with arrows representing the direction of increasing time. In the Janus universe (*left*), the arrow of time reverses, and observers in each half see their space-time as expanding. In the Big Bounce universe (*right*), the arrow of time always runs in the same direction, so observers in the bottom half see their universe contracting. Credit: Phil Halper and Nick Franco, 1185 Films.

Personally, I find these real-time descriptions somewhat misleading and removed from what we know from quantum mechanics. For example, imagine a movie of an unstable nucleus decaying into two smaller nuclei (as happens when an atomic bomb detonates). Now, play the movie backward to your physicist friends. They would see two positively charged nuclei approaching each other, and you would naturally expect the nuclei to bounce off each other due to electric repulsion. But, of course, you know better; the two nuclei didn't really exist at the start of the movie. There is a simple classical description for the bounce, but a quantum picture of the universe could be more complex and allow for either a bounce or a creation from nothing. We haven't seen the start of the Big Bang movie, so while the bounce scenario seems plausible (and we shall discuss it in the rest of the book), so is the creation-from-nothing scenario (that is, creation from nothing like anything we know today).

The real Achilles' heel of quantum cosmology, however, is whether the cartoon of a perfectly uniform, expanding (or shrinking) balloon is a faithful description of our complex cosmos at its inception. My colleagues at the Perimeter Institute, Neil Turok and Jean-Luc Lehners, had something to say about that.

BLOWING UP THE QUANTUM UNIVERSE

In the no-boundary proposal, to determine the most probable history of the universe, one has to sum over all possible histories, like interfering waves; some contributions reinforce each other, and others cancel each other out. An endless stream of constructive and destructive interference is mathematically equivalent to something that oscillates forever. It should be no surprise that physicists struggle to calculate in such a regime. But using a mathematical technique known as Picard-Lefschetz, Turok and Lehners found a sort of dictionary where one can translate these oscillations into a space where everything is finite and precise solutions can be found. Together with a graduate student, Job Feldbrugge, Lehners and Turok realized that this approach could be used to analyze what quantum cosmologies predicted for the universe's evolution. They found that the universe would not evolve from its timeless beginning into our smooth homogeneous cosmos; instead, the initial perturbations would tear the universe apart.[15] Even worse, they found that their results applied to Vilenkin's proposal as well. For the past two decades, Turok has been advocating for a view that the universe was endless in both directions of time and had won a bet with Hawking (against inflation) that the Planck probe would not find primordial gravitational waves (it did not). If this new work was right, it might show there never was a beginning, yet again snatching a victory from his famous former colleague at Cambridge.

However, the proponents of the no-boundary proposal were unimpressed with this latest challenge. The goal of quantum cosmology is to treat the universe quantum mechanically and therefore find a wave function for the entire universe, but Halliwell complained that they never did calculate a wave function—instead, they had calculated something called a Green's function, a tool that compares two different wave functions, describing how the universe evolves. Since a Green's function looks like a wave function, Halliwell says, they interpreted it as if it were one—and the universe blew itself to pieces. "It was, to be honest, all extremely frustrating," Halliwell explained, "because they strung together a number of questionable statements and concluded that the no-boundary proposal doesn't predict everything we wanted to predict. They were presenting the work as if they'd really found a massive flaw in the no-boundary proposal. So, I mean, me and Jim Hartle started writing papers to refute this and eventually, we just couldn't get them [Turok and collaborators] to agree with anything." In other words, they could not agree on a basic definition of the no-boundary proposal. To quote former US president Bill Clinton: "It depends on what the meaning of the word 'is' is." This confusion is somewhat inevitable because the truth is that the no-boundary proposal is not precisely described; it's a sketch of an idea. There are reasons to take it seriously; even Lehners agrees that it has many attractive features, especially forcing the universe to emerge from a quantum mechanical ground state, the point of minimum entropy that cosmologists believe may precede inflation. But even that point in its favor is a far cry from making it a rigorously formulated model of cosmology.

Cosmological debates can get heated when participants are trying to nail down precise definitions. In an interview with Phil, Halliwell called Turok "super smart" and "perfectly charming." Still, he described him as a "firebrand" who loves to challenge orthodoxy. Halliwell went on to say that the debate was "somewhere between a court case and an arm-wrestling match. I have to admit that it was at times hugely enjoyable but there was also a lot of frustration." The disagreement over how to define the no-boundary proposal led Hartle and Hertog to a manifesto describing what they think it should mean. Namely, they held that the universe should emerge from a state without time.[16] But nothing forces a cosmologist to accept that definition. Take Lehners, who described Turok as an excellent collaborator, with "a hundred ideas a day." Recalling this collaboration to Phil, he said, "We showed that it's not so easy to make a precise definition that actually works." According to these critics, the definition that should be used leads to wild, untamable perturbations. As Lehners put it, "I thought this is going to be a way of making it much more precise, much more solid, and then actually to my surprise it did not work at first, and this was to me actually quite shocking."

Ironically, Hartle and Hawking had claimed that Vilenkin's tunneling proposal also led to an unstable vacuum. Still, Vilenkin, working with his postdoc Masaki Yamada, recently argued that it did no such thing.[17] This particular claim of Vilenkin's is, at least, now accepted by the no-boundary proponents. At the time, I wondered if others might similarly be persuaded that both proposals could lead to a smooth universe. Much to my surprise, in 2019, Lehners, with his student Alice Di Tucci, claimed to have found a path that might just work.[18] Recall that in quantum mechanics, there is a fundamental uncertainty relationship between the position of an electron and its momentum. If we think of this uncertainty in a cosmological setting, the positions would be analogous to the size of the universe and the momentum analogous to its expansion rate. It's important to remind ourselves that this kind of uncertainty is not just about our knowledge but is a feature of reality itself. So, if the universe began with zero size, it would have unlimited uncertainty over its expansion rate. But why not fix the expansion rate and let the initial size of the universe become maximally uncertain? This condition can lead to a different formulation of the no-boundary proposal, but one that shares the same essential results—as the most likely size of the initial universe turns out, again, to be zero.

Whether this new proposal can resolve the instabilities of the original proposal remains to be seen. Such back-and-forth is part of the process of doing science—playing with ideas and seeing where they lead. Both the tunneling-from-nothing and no-boundary proposals are impressive first attempts at doing quantum cosmology. If the universe can come from "nothing," the idea is mind-blowing. But we have to admit that right now, we don't know whether or not it can. One of my biggest worries is the difficulty of verifying these models via experiments, a prospect that doesn't currently seem viable. But as defenders and critics examine and refine the models in more detail, such experiments might present themselves. Yet, there is an even more profound concern.

The models discussed in this chapter are considered to be semiclassical. In other words, elements of quantum mechanics and Einstein's gravity are combined in an uneasy alliance. As we've discussed, these two most fundamental theories of the universe are simply incompatible with one another. For the most part, this incompatibility can be ignored; physicists use quantum mechanics to study the very small and Einstein's gravity to explore the very large. But both frameworks become relevant when the entire observable universe is far smaller than an atom. As we discussed before, this scenario calls for a radical new theory that unifies the paradigms of large and small: a quantum theory of gravity. In the early 1980s, when the no-boundary proposal and tunneling-wave function were put forward, none of the leading candidates, such as string theory (chapter 4) or loop quantum gravity (chapter 5) had been developed. And while we do not know if these proposals are correct, scientists have already built models of the Big Bang using insights from these theories. As we are about to discover, in many of these scenarios, the universe did not come from a point of zero size but instead from something far more bizarre.

✹ 4 ✹

STRING THEORIES AND COLLIDING BRANES

When Einstein discovered his theory of gravity, he was so struck by its elegance, his reaction was lackadaisical when Eddington's eclipse expedition showed clear, dramatic evidence for it. If the data had not been so supportive, Einstein said, "I would have felt sorry for the dear Lord. The theory is correct."[1] Yet even as he said these words, Einstein knew they were not entirely true. Hidden at the heart of physics was a conflict. Relativity tells us that nothing can travel faster than the speed of light. However, if the fabric of the cosmos is quantum, distances would also have to obey Heisenberg's uncertainty principle, making it impossible to give space-time a hard speed limit. Doing so requires a definite fixed distance between any two points (at least, for any given observer) and a known position at any moment in time for a particular object—a requirement at odds with the fuzziness of the quantum. Similarly, when general relativity describes the distortion of space-time by a massive object, it assumes that the object has a definite location—an assumption that conflicts with the quantum description of a spread-out wave. So, quantum mechanics appears to defy the locality of relativity: we have a contradiction. Fortunately, the gravitational force generated by subatomic particles is minuscule, as are the quantum effects of large-scale structures like stars and galaxies, and we can mostly ignore this conflict. But such dismissal won't do for the Big Bang, where the extreme energy densities require both gravity and quantum theory to blend seamlessly together. As early as

1916, Einstein realized this necessity, saying, "It appears that quantum theory would have to modify . . . the new theory of gravitation."[2]

For almost forty years, string theory has been physicists' great hope for such a modification. But, like a sunset, it is beautiful, beguiling, and forever out of reach. Today physicists still don't agree as to what string theory means, whether it should even still be pursued as a theory of quantum gravity, or whether it really deserves its leading place in the hierarchy of speculative science. But we agree that we need it. Or something like it. We will never understand the mysteries of the Big Bang (or black holes) without some theory of quantum gravity. String theory is a search for unity in physics. Our current picture describes four forces of nature: electromagnetism, which governs the behavior of light; gravity, which dictates the motions of mass; the weak nuclear force, which describes radioactive decay; and the strong force, which binds atomic nuclei together. These forces are carried by a class of particles called bosons. Electromagnetism is mediated by the photon, the strong force by the gluon, and the weak force by the W and Z bosons. If gravity is a quantum force, like the others, then the hypothetical graviton mediates it. But string theory says that these myriad forces (and the matter particles that make up our material reality) are all manifestations of the vibrations of fundamental strings. In other words, an invisible reality exists in which all is one. Unfortunately, the physics community is deeply divided over this hypothesis. Like inflation, string theory seems to predict that our world is one of many; the others in string theory's "landscape" are forever hidden from sight. For the critics, this landscape reflects an abandonment of science. For the supporters, it reveals a deep truth about reality that we have to learn to live with.

In my view, cosmology provides a possible way to resolve the tension over string theory: while the hidden worlds that many string theorists believe exist may be unreachable by empiricism, the birth of the universe is not. It is, at least in principle, possible to create a model of the Big Bang using new lessons from string theory and then test that. We are about to meet the scientists who are attempting to create and test such a model; but alas, if unity is the hope, division is still the result. For there is not one model

Standard Model of Elementary Particles

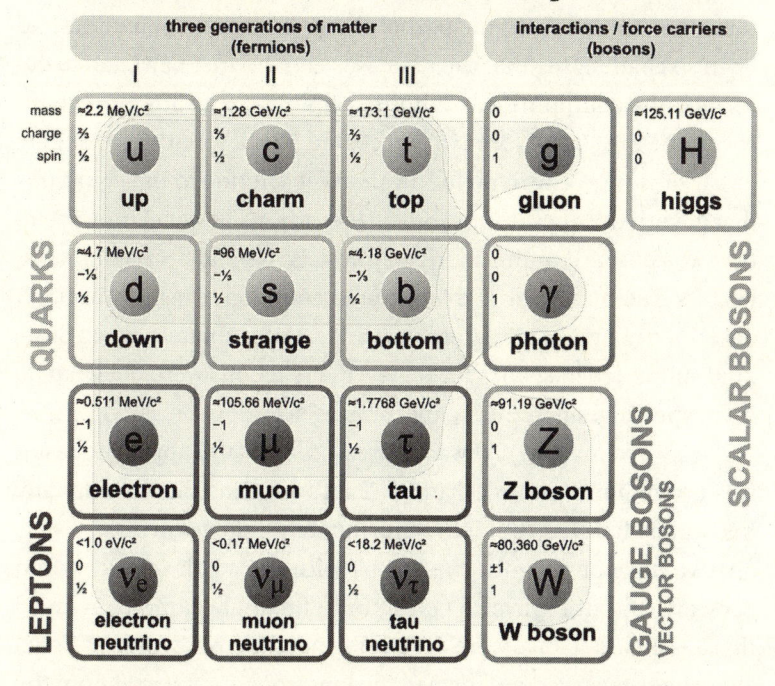

	three generations of matter (fermions)			interactions / force carriers (bosons)	
	I	II	III		

QUARKS

≈2.2 MeV/c² | ≈1.28 GeV/c² | ≈173.1 GeV/c² | 0 | ≈125.11 GeV/c²
⅔ | ⅔ | ⅔ | 0 | 0
½ | ½ | ½ | 1 | 0
u up | **c** charm | **t** top | **g** gluon | **H** higgs

≈4.7 MeV/c² | ≈96 MeV/c² | ≈4.18 GeV/c² | 0
−⅓ | −⅓ | −⅓ | 0
½ | ½ | ½ | 1
d down | **s** strange | **b** bottom | **γ** photon

LEPTONS

≈0.511 MeV/c² | ≈105.66 MeV/c² | ≈1.7768 GeV/c² | ≈91.19 GeV/c²
−1 | −1 | −1 | 0
½ | ½ | ½ | 1
e electron | **μ** muon | **τ** tau | **Z** Z boson

<1.0 eV/c² | <0.17 MeV/c² | <18.2 MeV/c² | ≈80.360 GeV/c²
0 | 0 | 0 | ±1
½ | ½ | ½ | 1
νₑ electron neutrino | **ν_μ** muon neutrino | **ν_τ** tau neutrino | **W** W boson

GAUGE BOSONS / VECTOR BOSONS · **SCALAR BOSONS**

The Standard Model chart of the fundamental particles, from which all known matter is built. Note that protons and neutrons are made of up and down quarks. However, the chart does not include constituents of dark matter and dark energy, as they have not been directly detected. Credit: MissMJ.

of string cosmology but many. Each has its own story to tell about cosmic origins. Some string theorists seek to find inflation in its complex mathematics; others look to replace inflation with models we shall explore, like string gas cosmology, the pre–Big Bang, and Ekpyrosis. And while we don't know which, if any, of these models is correct, they do make falsifiable predictions that can be tested against experiment. And yet again, they all seem to predict that the Big Bang was not the beginning of all things but only a transition event in a much richer narrative.

String theory was born in the 1960s, a period of enormous social upheaval, protest, and revolution. But the person who would lay its foundations was not looking to change the world. Gabriele

Veneziano was examining the equations that govern the strong force that holds subatomic particles together. These equations had been so challenging that the great Russian theorist Lev Landau described the community as "almost helpless" in its attempts to incorporate the strong force into accepted physics. But after roughly a year of work, Veneziano found an exact solution to the equations that govern this mysterious force. However, its physical interpretation was obscure until American physicist Leonard Susskind realized its true meaning. The Veneziano equation, Susskind argued, described not point particles but strings. Unfortunately, when Susskind submitted his paper for review the reactions were not what he was expecting. Susskind's paper was rejected. As he said in a PBS *NOVA* TV documentary, "I was depressed, I was unhappy. The result was I went home and got drunk."[3] That's a feeling that I personally find all too familiar (though, unfortunately, I do not drink).

It would not be until the 1980s, with the work of Green and Schwarz, that the physics community finally became convinced that Susskind's strings were describing not the strong force as Veneziano had envisaged but instead the long-sought-after graviton, the hypothetical particle that would carry the gravitational force in a quantum theory of gravity. The new theory replaced all particles with vibrating strings. Just as guitars can pluck different notes, so strings can create different particles. Fundamental particles like electrons and photons might seem distinct, but in string theory, they can just be particular manifestations of a single phenomenon. For this reason, physicists would eventually call string theory a "Theory of Everything." But, to make the theory work, theorists had to assume there were extra invisible dimensions of space. One way to think about this strange new higher-dimensional world is by picturing ants walking on a flat piece of paper. Their motion is limited to a two-dimensional surface, as ants don't have enough energy to jump up and down. However, ants with wings can discover an entirely new dimension of altitude. Similarly, we may have no knowledge of extra dimensions (six or seven of them, according to string theorists), if they are curled up and we don't have enough energy to explore them in our particle colliders. We need wings.

String theory also requires that every known particle have an unseen counterpart known as a superpartner. This mirroring is known as supersymmetry, hence the name superstring theory. Another difficulty that beset the new theory was that it seemed to come in five separate versions, which differed in how the strings were allowed to move around. This inconsistency was highly embarrassing for a so-called theory of everything, which should have a unique and consistent description of the world. But Ed Witten found that all five theories were simply different manifestations of a deeper underlying framework that he called M-theory. What the M stood for was not clear (mystery? Master? Membrane? An upside-down W for Witten?).[4] But Witten could not fully uncover M-theory, and its basic equations are still unknown. Instead, he gave mathematical reasons that convinced physicists it must exist. In M-theory, strings are just one type of higher-dimensional objects known as membranes. A one-dimensional membrane is a string, but there can be higher-dimensional "branes" as well. After Witten's breakthrough, physicists had renewed optimism that they were about to discover the final laws of physics. But, just as mountain climbers know the road to the summit is often treacherous and filled with obstacles, theorists discovered that M-theory contained a treacherous landscape that threatened to derail the entire program.

One hurdle to making sense of string theory is that as we don't see the extra dimensions, it needs to assume they are curled up from sight. The shape (or compactification) of the hidden dimensions determines the properties of the particles we see around us (for example, their mass and charge). In their youthful optimism, string theorists believed they might find the right shape for the extra dimensions corresponding to the standard model of particle physics. But the more physicists searched, the more shapes they found. It was as if a city dweller had looked up at the sky and—after seeing only a handful of stars—decided to count them all on a countryside camping trip. As they headed out to darker skies, the stars became increasingly numerous. While they might have marveled at the swelling number of luminous bodies, they could only despair at the prospect of counting them. The actual number of stars in the observable universe is estimated to be more than a billion trillion. That number has twenty-one zeroes. But even that figure is nothing compared to the possible ways to compactify the extra dimensions of string theory. In fact, this number is often said to have at least five hundred zeros, and some believe it is infinite. This vast space of possibilities is known as the string landscape. But with little hope of locating our world in this immense ocean, the prospect of coming ashore to real-world physics appears dim. I certainly don't see it happening in my lifetime, and I am not alone. Nobel laureate Sheldon Glashow has famously said that string theorists' work is so removed from the real world that it should be compared to medieval theologians pondering how many angels could fit on a pin.

Some have claimed that Glashow has it all wrong, though, and the landscape can make empirical predictions. Chief amongst these advocates is Laura Mersini-Houghton, who recently popularized the idea in her book *Before the Big Bang*.[5] There, she tells us that the presence of other universes in the landscape can be felt in our cosmos via a purely quantum process known as entanglement.

While the idea of entanglement dates back to the foundation of quantum theory, it has been recently resurrected as a key ingredient that gives the quantum computers of the future their superpowers to operate exponentially faster. It allows for the states of different particles, even ones from across the cosmos, to be strangely linked together—a phenomenon that troubled Einstein due to its "spooky action at a distance." But what if entanglement doesn't just happen for particles within our universe but also for whole universes across a vast multiverse? Mersini-Houghton and her colleagues calculated what impact this feature would have on the CMB sky, and amazingly, their predictions perfectly matched later observations. But some physicists question how perfectly. Inflationary cosmologist Will Kinney examined these assertions in a paper in 2016 and concluded that "the key claims do not survive even cursory scrutiny," saying they were ruled out by the Planck data.[6] In particular, the idea proposes corrections to inflation due to the landscape, but it turns out such corrections to simple models of inflation make their agreement with data worse (unless the corrections are very small). Needless to say, Mersini-Houghton did not agree. Still, what is disturbing to me are the rumors of attempts to stop Kinney from publishing the paper altogether.[7]

Personally, I also have always had a hard time wrapping my head around Mersini-Houghton's claims, and around how she derives things like a potentially large-scale flow of galaxies, known as dark flow, that some scientists claimed to have observed early in the first decade of the 2000s but that was later ruled out by researchers, including me. But as much as I would like to say that having one's beautiful theories tested and ruled out by data is a great part of the scientific process, I have to be honest and confess that it is horrible. As biologist Thomas Henry Huxley once said, "The great tragedy of Science—[is] the slaying of a beautiful hypothesis by an ugly fact."[8]

The string theory landscape may be poison to some, but it is an elixir for others. With it comes the possibility of explaining why our universe seems "finely tuned" for life, with all the constants of nature having just the right values to allow complex chemistry to form. One way to tackle this problem is by applying the anthropic principle, which appeals to an observer's selection effect; this effect biases the data in favor of life-friendly universes, even if such universes are rare in the multiverse of possible universes. Steven Weinberg famously used this principle to explain why the cosmological constant had a value close to what was observed. If it were too positive, the universe would be torn apart; too negative and it would collapse in on itself. We should expect to observe a value that is small enough to be compatible with our existence but not necessarily zero. Weinberg published this argument a decade before the nonzero value of the cosmological constant was discovered, and multiverse proponents use it as evidence they are on the right track.[9] However, I am not convinced they are right. Why? After all, didn't we say the most important criterion for assessing our possibilities is that they make predictions that are then confirmed by observations? Isn't that what Weinberg did? Well, yes, but as we shall see in chapter 7, Rafael Sorkin—at around the same time—used a different theory of quantum gravity to make the same prediction.

The anthropic argument only works if there are many other universes, each with different values for the constants of nature. While eternal inflation can provide us with a vast quantity of universes, it's the string landscape that allows them to explore every possible configuration, ensuring that at least one is hospitable to life. According to the landscape, the constants of nature, like the masses and charges of elementary particles, are malleable and have different values in other parts of the multiverse. According to Linde, the values of the constants are "like water, which can be solid and can be liquid." Disparaging the critics of the theory, whom he compares to the captain of the Titanic, he says: "If you do not think about it, then you crash your ship." But the problem is that we have very good reasons to think that water can be solid and liquid; after all, we've seen it happen, and theories of matter imply it. We clearly don't have the same level of confidence that the constants of nature can take on different values. Maybe they can and maybe they can't. With the controversy over anthropic reasoning dividing physicists into opposing factions, another way would have to be found if string theorists were to convince the rest of the community their elegant mathematics actually described reality. And there is no better place to probe extreme physics than in the fiery cauldron of the Big Bang itself.

STRING GAS COSMOLOGY

Robert Brandenberger did not set out to become a cosmologist, let alone a string cosmologist. He was born in Switzerland but studied for his PhD in mathematical physics at Harvard. Following a fellow student's suggestion, Brandenberger approached his supervisor for permission to do his PhD on inflationary cosmology. His mentor replied, "Sure, as long as I don't have to read it." Brandenberger was impressed with inflation, but the theory's lack of grounding in a more fundamental theory concerned him. He wasn't alone in his worries. Cumrun Vafa studied physics at MIT just before the revolution that swept the ayatollahs to power in our native Iran. Having completed his PhD under Witten, Vafa went on to show (with Andrew Strominger) that string theory can give a precise reproduction of the legendary black-hole entropy formula derived by Bekenstein and Hawking. But before tackling black holes, Vafa had his eyes on an account of the very birth of the universe. And so, he searched for a cosmologist to collaborate with. Robert Brandenberger was in Boston at the time and the two struck up a bond. However, Brandenberger admitted to being daunted by this superstar of string theory. "I get intimidated quite easily by people who are quick and sharp," Brandenberger told Phil. "Cumrun is an extremely impressive physicist, very very quick and very deep and very ambitious and very down to earth, so no nonsense."

Despite the attractive features of inflation, Brandenberger—as a good scientist—didn't want to put all his eggs in one basket. So, armed with new tools of string theory, the pair began to ponder an entirely different scenario, aiming directly at the Big Bang's Achilles' heel, the singularity. The history of the universe, in reverse, is often modeled as if it were a gas of particles inside a shrinking box; the gas gets compressed as we approach the Big Bang, heats up, and finally reaches an infinite temperature at the singularity. But what Brandenberger and Vafa discovered was that a gas of strings behaves differently: initially, temperatures will rise just as in classical physics, but eventually a maximum temperature will be reached, and strings just cannot be heated past this point. If you try, the temperature will decrease. This fact means that the universe cannot reach an infinite temperature as the Penrose-Hawking theorem implied. Nor can it shrink to zero size. The size of a string limits the minimum size of the universe.

The Big Bang in this framework was not the beginning of time, and there was no inflation either. Instead, the early universe featured a "quasi-static phase," an oven of fundamental strings in thermal equilibrium. What we would see if we could travel back in time before the quasi-static oven of thermal strings is unknown. One possibility is that eventually we would emerge in another expanding universe, implying a Big Bounce, although it is also conceivable that the static phase might remain in place indefinitely into the past (but thankfully, not into the future); this scenario is known as an emergent universe. Other physicists, like Hawking's old colleague George Ellis, have also advocated versions of an emergent universe—although controversy surrounds the notion that such a universe can really be static into the eternal past. Either way, the predictions of this model, which became known as string gas cosmology, are determined by the period of thermal equilibrium, so whatever happened before that period will not affect what happens afterward. The thermal bath of strings would, like inflation, erase all memory of its past. But if string gas cosmology were to succeed, it would have to address the problems that inflation seemed to miraculously solve.

Recall that the horizon problem is a quandary in the conventional Big Bang, because there hasn't been enough time for the universe to thermalize. In string gas cosmology, there could have been an indefinitely long period before the expansion began. So no horizon problem exists. The monopole problem was a feature of speculative particle theories popular in the 1970s, but there is no guarantee that strings produced monopoles. So, there may be no monopole problem to solve. But I think inflation theory's biggest triumph was to create a scale-invariant spectrum of perturbations in the early universe, causing a pattern of hot and cold spots in the primordial plasma that is remarkably well confirmed by measurements of the CMB.

I've always admired (and tried to emulate) Brandenberger's balanced take on these issues. We both agree that inflation is extremely successful in this regard, but more is needed to confirm the theory. If string gas cosmology could also generate a scale-invariant spectrum, it could become a genuine competitor to inflation. It would, however, take more than a decade before string gas theorists began to tackle this problem head-on. Just as the flames in a fire are not perfectly uniform, neither would a gas of heated strings be. It is these thermal fluctuations that string gas proponents hoped might generate the spectrum of perturbations that inflation was thought to predict so spectacularly. Ali Nayeri, who completed a postdoc with Alan Guth, was brought in to finally undertake the arduous calculations needed to check whether such a spectrum could really be generated. In fact, Nayeri, who is also from Iran, was like a mentor to me when I was starting to do cosmology. I remember a lecture he gave once, when visiting my high school, about the Coriolis force (which, contrary to popular belief, does not determine the way the water drains in your bathtub). Nayeri later finished his PhD in India, and after a few excursions, we both ended up as postdocs at Harvard. As Brandenberger recalls, "Ali had writer's block throughout his career, so I think Cumrun suggested that he work on this problem . . . Ali worked very hard. He did very good work." With the computations completed, Nayeri had shown that the hoped-for scale-invariant spectrum was indeed present.[10] Brandenberger described it as one of the high points of his career: they now had a possible scenario based on superstring theory that had no singularity and could be compared with observations.

At the time of Nayeri's perturbations work, inflation pioneer Andrei Linde was based at Stanford. As Brandenberger recalled the story, one of his former students told him that, after reading a pre-publication copy of the research, Linde said, "This is serious. We have to destroy it."[11] An alternative to inflation's pièce de résistance was apparently an affront to its pioneer.

When Phil asked Linde about the new model, he told Phil that the string gas papers were written in a "strange" way—essentially a series of conjectures in one paper, with a promise to prove them in the next. But what really irked the critics was the issue of what frame to use.[12] While in his theory of relativity, Einstein assumed that standard clocks and rulers exist that can be used to measure times and distances, in string theory, there can be different types of clocks that react differently to their environments. The choice of frame—that is, the right clock to use to model a gas of strings—was the point of contention amongst adversaries. The new fields associated with string theory produce new frames, and neither Linde nor Brandenberger could agree on which to use, hence their widely divergent views as to whether string gas cosmology can reproduce the scale-invariant spectrum that we see in the CMB. In my opinion, this disagreement reflects the underlying truth that string theory is still a work in progress, which leaves room for these sorts of altercations.

Linde and his colleagues also attacked the model for not solving the flatness problem; recall that a flat universe is thought to be unstable and so we shouldn't expect the universe to be flat, yet it is. String gas cosmology has to assume a flat universe to begin with, whereas inflation drives it relentlessly toward flatness: score for inflation. Inflation, though, does not explain why we live in a four-dimensional universe. String theory assumes that all the extra dimensions of space are curled up and that the strings can wrap around the compact dimensions. If the universe is to expand into its current enormous size, these winding strings have to annihilate each other, allowing only three of the nine compact dimensions to expand to what we see today. In my view, the crowning achievement of string gas cosmology, which (to my knowledge) has not been replicated by any competitor, is that it offers a reason for why we live in a world with three spatial dimensions—namely, that three is the maximum number of dimensions in which such annihilation can occur.[13]

Most other string theorists, however, appear to be unimpressed with string gas cosmology and prefer to stay in the inflationary camp, attempting to show that superstrings could reveal inflation's underlying physics. One proposal, by Gia Dvali and our old friend Henry Tye, is called brane inflation, in which the motion of branes drives the universe's expansion into higher-dimensional space. As they approach each other, the branes act like pumps, injecting more energy into our space, causing it to inflate. As string theory leads to many additional fields, its proponents hoped that one of these fields may be the inflaton. But stabilizing these fields such that the three-dimensional universe grows but the extra dimensions don't is a thorny problem, and as far as I can tell, one that has no satisfactory answer. Perhaps one will be found, but until then I believe it's worth looking at alternatives, such as string gas cosmology.

With string gas cosmology gaining little traction amongst cosmologists, it needed something that could propel it into the limelight. If the model could predict a genuinely novel feature about nature that distinguished it from inflation, the world of cosmology might actually take notice. When Nayeri, Vafa, and Brandenberger computed the spectrum of perturbations from a gas of compressed strings uncurling into our expanding universe, they immediately realized that this process would generate gravitational waves in addition to sound waves. Now, as discussed in chapter 2, inflation does the same thing, so this point might not appear to give one theory an advantage. But these tremors in the fabric of the cosmos would have a peculiar property that could make them clearly distinguishable from the inflationary paradigm. Just as light has a spectrum associated with it, generating different signal strengths at different wavelengths, so do gravitational waves. We need to pay particular attention to whether the waves have more power at longer or shorter wavelengths. Recall that this property is known as tilt. The tilt of the sound waves in the primordial plasma is called scalar tilt, while that of the Big Bang's gravitational waves is known as tensor tilt. Red tilt means more power for longer wavelengths, while blue tilt implies the opposite. Inflation and string gas cosmology typically predict a red scalar tilt. But when it comes to gravitational waves, string gas cosmology predicts a noticeable blue tensor tilt, while inflation generally produces a red tensor tilt. This distinction is crucial, as it gives us a way to experimentally probe string gas cosmology and distinguish it not just from inflation but also from other string-inspired models we shall encounter later in this chapter.

Many physicists say that discovering primordial gravitational waves will validate inflation. But I think the story is more nuanced than what we've been sold. At the minimum, we also need to know if they are tilted to red or blue. The (anticipated) discovery of primordial gravitational waves by BICEP was declared (prematurely) a major victory for inflation, but what few realized was that the result hinted at the crucial color of the tilt. And it was not red but blue, just as string gas cosmology had predicted almost a decade earlier. Brandenberger pointed this discrepancy out to the BICEP scientists before they went to the press, but they ignored him and declared victory for inflation. Brandenberger recalled my friend Brian Keating, a member of the BICEP team, replying politely to his emails with, "I'm too stressed right now but I will have to take a careful look at this." To be fair to the BICEP team, the blue-tilted signal had a statistical significance high enough to make it interesting but not high enough to make it compelling, providing some justification for ignoring it; and the team received over a hundred emails in the days leading up to the press release.[14]

Of course, the BICEP team's claims to have made any detection at all turned to dust, but I believe the story remains a cautionary tale that simply finding the primordial gravitational waves may not be enough to allow us to really tell which model of the early universe is right. Keating would later write a book entitled *Losing the Nobel Prize*, detailing the affair, but making no mention of string gas cosmology.[15] Later on, however, he hosted Brandenberger on his YouTube channel and admitted that he "wasn't being honest" about the claim that B-modes would test inflation.[16]

In the years since the BICEP episode, Vafa has thrown another challenge to the consensus view. Vafa has a reputation for doing mathematically rigorous calculations and is unconvinced by the haphazard guesses of colleagues who simply wanted to get an accelerating universe into string theory. Accordingly, he provided mathematical conjectures for what type of cosmology can come out of string theory; almost all its universes, with a positive cosmological constant, particularly those with inflation, reveal logically inconsistent worlds and so should be referred to as *swampland* rather than *landscape*. Thus, according to Vafa, string theory may not predict a multiverse after all. Moreover, not only does the theory make inflation considerably less likely, it also makes less likely the widely held belief that dark energy is simply Einstein's cosmological constant. I am sure this news is music to cyclic-universe proponents' ears, as they could use this changing dark energy to imply that the cosmological constant may eventually become negative and thus cause the universe to re-collapse. The battle lines of a new controversy have been drawn. But Vafa and colleagues are not the only string theorists to challenge the standard cosmology. Another well-known figure who has done the same is the father of string theory, Gabriele Veneziano.

THE PRE—BIG BANG

Lake Geneva is an enormous body of water as deep as the sea, shaped like a croissant and bordered by snowcapped mountains.

The city surrounding it is home to the European Association for Nuclear Research (CERN), location of the particle accelerator known as the Large Hadron Collider. Veneziano, one of CERN's leading theorists, likes to ski in Geneva's mountains during his time off. One winter day in 1990, there was a shortage of snow, so he concentrated on the early universe. It was then that he realized that string theory might have something to say about this most mysterious era. What drew Veneziano's attention was a type of duality in string theory. Dualities are ubiquitous in physics; they are situations in which two different descriptions of reality are actually equivalent to one another. In string theory, strings can wrap around hidden dimensions, just like a piece of wire can wrap around a straw. There are also free strings that don't wrap around the hidden dimension. The number of times strings wrap around the hidden dimension is called the wrapping number. The energy of a wound string is proportional to the size of a hidden dimension, while the energy of a free string is inversely proportional to it. This fact leads to a very strange result. If we know only how much energy there is, we cannot tell if we live in a world with a large hidden dimension and lots of free strings or one with a small hidden dimension and lots of wrapped strings. The two are mathematically equivalent, as they give precisely the same value for the strings' energy. So, strings, according to Veneziano, "cannot distinguish between a big or a small circle." This equivalence is known as T-duality, and it was already a key ingredient in Brandenberger and Vafa's proposal for string gas cosmology. The same duality would become a founding principle for Veneziano's new cosmology.

We often describe the expansion of the universe by a number called the scale factor. For example, if we say it is one now, it was one-half when the universe was half its current size. As the cosmos expands and galaxies drift apart, the scale factor grows. On the other hand, at the classical Big Bang, the scale factor is zero. This is why physics breaks down at the singularity: dividing by zero is always messy. (Try dividing by zero on your phone's calculator app and you'll see what I mean.) But in string theory, things are different: the scale factor cannot be zero because strings, unlike particles, have a minimum size. Instead, there would be a cosmological version of T-duality, which Veneziano calls scale factor duality; this version forms his basis for rewriting the history of the universe. As we wind the clock backward, we find there is no Big Bang singularity but another dual universe; just as our universe is expanding, so the dual universe is contracting. The hot Big Bang then represents the bridge between the two. It's a moment of transition, not one of beginnings.

For years following the Penrose-Hawking singularity theorem, cosmologists had assumed that the beginning of time at the Big Bang had been proven. But that proof was based on a classical theory of gravity; now, two different string theory models were claiming the universe had no beginning. One confusing aspect of this pre–Big Bang scenario is that it describes a prior contracting universe in one frame of reference, but in another frame, it describes a continually expanding universe. These two different frames are the same frames that caused so much confusion and tension between string gas proponents and critics. But here, the physical predictions from the model do not depend on the choice of frame. In either case, the cosmic energy density goes to zero as we travel back into the infinite past. In this era the universe is in a near vacuum. As we move forward from this ancient state toward our own era, the energy density grows until it reaches some maximum value, often called a Big Bounce, but unlike what is found in the singularity theorems, it never becomes infinite, instead declining afterward. A new field associated with string theory, called the dilaton, plays a crucial role in the pre–Big Bang universe—as it controls the extent to which strings couple to one another, which in turn determines the strength of all fundamental physical forces. Everything is connected. One way I like to picture the pre–Big Bang model is to think of a city in COVID lockdown; interactions are minimal. The dilaton starts out weak and then gradually evolves, so physical interactions increase, just as they do in the city whose restrictions are lifted. As the dilaton dials the strength of interactions upward, the universe becomes increasingly dense until eventually it is transformed into its fiery Big Bang state.

One way to visualize the contrast between inflation and the rival string cosmology models is to think of cosmology as a mountain and the quantum gravity regime as its peak, which is sharp and pointy in classical cosmology but smooth in string theory. Inflation imagines a journey mapped far away from the peak and is ambivalent to what came before it (be it from the peak or elsewhere). String gas cosmology, however, starts right at the peak with a thermal soup of strings but slowly drifts off downhill into an expanding universe, which makes our hot Big Bang. Finally, the pre–Big Bang model is a journey right across the peak, with our ascent and descent describing the same story but told in reverse order. The creators of the pre–Big Bang model (Veneziano had worked closely with cosmologist Maurizio Gasperini) thought that the scale-invariant fluctuations needed to seed galaxy formation would arise from variations in the dilaton field, but on close inspection, they realized they were wrong in this assumption, and Veneziano recalled that he was about to give up on the whole idea. Inconsistency with observations was not the only problem the model faced. Like string gas cosmology before it, it drew the ire of Linde, who claimed that in the pre–Big Bang, "you must start with even larger universe. So instead of solving

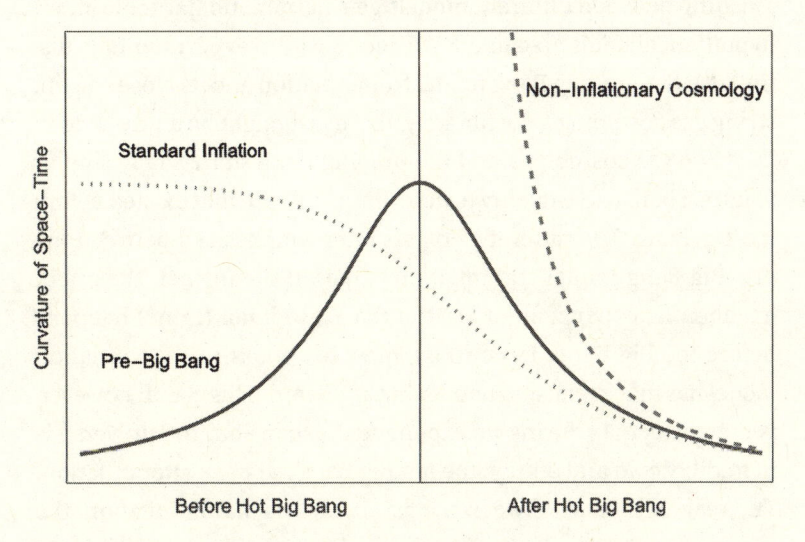

The evolution of the curvature of space-time in inflation and the pre–Big Bang models, contrasted to non-inflationary cosmology. Credit: Niayesh Afshordi.

this problem, you have even greater problem of why the universe is large . . . You don't solve these problems for which inflation will play such an important role and after that you still have your singularity problem to solve." Veneziano concedes that while there are good arguments to support the removal of the singularity in string theory, none of them is rigorous. "It's an essential element of the construction of the pre–Big Bang scenario to be able to remove the singularity. Now, I must say that in spite of many ideas, there is no proof." With no proof of singularity resolution and a model that was inconsistent with observations, all might have seemed lost for the pre–Big Bang. But early in the first decade of the 2000s, inflationary cosmologists were being creative with introducing various new fields into their models. The pre–Big Bang model advocates followed their cue and helped themselves to one of these fields to generate the required scale-invariant spectrum of sound waves in their own model. Having come up with some myself, I've always been a bit uneasy about all these extra fields that get invented in theoretical physics. It is true that they allow a lot of flexibility in cosmological model building and can rescue scenarios that were thought to have been ruled out by experiment. However, distinguishing between different models gets harder and harder as more hypothetical fields are added. We need a way to experimentally distinguish the pre–Big Bang model from inflation and its close cousin, string gas cosmology. Fortunately, there just might be a way to do it.

String gas cosmology and the pre–Big Bang share many similar features: both use T-duality to describe a universe that existed before the Big Bang. The difference that is often emphasized between the pre–Big Bang model and inflation is that in the former, the era of exponential expansion (at least in the Einsteinian frame) happens before the Big Bang. In contrast, most textbooks say, the inflation model has that expansion happening afterward. This claim, however, is not correct, as during its exponential expansion, the universe is incredibly cold and empty; the hot Big Bang happens after inflation. The real difference between the two models is that in inflation, the expansion rate of the universe, while being incredibly large, is slowly decreasing, whereas in the pre–Big Bang, it grows continually until the hot Big Bang starts. This difference leads to a potentially verifiable signal, yet again in the form of primordial gravitational waves.

The year 2016 brought the news of the first-ever direct detection of gravitational waves using giant lasers at LIGO, the Laser Interferometer Gravitational-Wave Observatory. The discovery gave its team leaders one of the fastest ever Nobel Prizes. LIGO is not designed to see primordial gravitational waves but is instead optimized to detect them from colliding black holes and neutron stars (stars so dense they have the mass of the sun squeezed into an orb the size of a small city). The next generation of detectors will not just have more sensitivity but also detect gravitational waves at much lower frequencies. At extremely low frequencies, gravitational waves can also imprint the polarization of the cosmic microwave background. Recall that polarization is the direction of the electric fields in electromagnetic waves and that gravitational waves can change the pattern of polarized light we see in the CMB, creating the infamous Big Bang B-modes that BICEP thought they had detected. The pre–Big Bang model generates gravitational waves of much higher frequency, which should leave no trace in CMB polarization. So, if we found a signal of primordial gravitational waves in LIGO-type instruments that were not accompanied by B-mode polarization, that finding would fit in very well with the pre–Big Bang scenario. The pre–Big Bang model typically generates even bluer tensor tilts than string gas cosmology does. So, a detection of gravitational waves that are blue-tilted but produce little imprint in CMB polarization could be a smoking gun for the pre–Big Bang model. And Veneziano is optimistic that such a signal can really be found.

But both string gas cosmology and the pre–Big Bang models were formulated before Witten's M-theory revolution and the introduction of branes. For a small band of renegade scientists, it was the behavior of these new higher-dimensional structures that would pave the way for yet another new model of cosmology.

THE EKPYROTIC UNIVERSE

Neil Turok was born in South Africa at the height of the apartheid era. His father, a communist and opponent of the regime, was arrested. After three years in jail, he was forced into exile. Turok clearly seems to have inherited his father's tendency to speak his beliefs.[17] As long as I've known him, he has used his formidable creativity to challenge consensus cosmology. Turok was one of the first scientists to investigate the relationship between string theory and the Big Bang. This investigation culminated in a paper in 1988, claiming that string theory naturally led to inflation.[18] But over the subsequent decade, he changed his position, rebelling against inflation and setting a trail that Paul Steinhardt would only later follow. By the end of the nineties, not only was Steinhardt coming around to Turok's antiestablishment views, but they also both shared a fascination with the newly emerging picture of M-theory. Just before the millennium, they decided to host a conference at the University of Cambridge. It would encourage M-theorists to interact with some of the biggest names in cosmology.

Unfortunately, the conference was sponsored by NATO, and its timing couldn't have been worse; their forces had controversially bombed the Radio Television Series of Serbia. NATO had accused the Serbian government of orchestrating ethnic cleansing of Kosovo's Muslim population. I recall my high school years in Iran, where the massacre of Muslims in Europe (and the Islamic republic's [in]actions) was a hot topic. We even had a poetry night at our high school, where I read a poem I had written about an orphaned Bosnian girl. In 1999, it seemed that NATO was determined not to turn a blind eye to such atrocities anymore. However, many Eastern-bloc scientists objected to their use of airpower and threatened to withdraw from the conference. At the time, Phil's good friend was an air force pilot doing the bombing, while Phil's girlfriend, who was from Belgrade, was protesting the bombings.

Despite these tensions, the M-theory/cosmology conference went ahead. By their own account, one talk stood out as being a revelation for both Steinhardt and Turok.[19] Burt Ovrut lectured on a new framework called Hořava-Witten theory, for how higher-dimensional branes interact with each other. Like a person staring into a mirror, the branes in this theory have to be partnered. They cannot exist alone. Our entire universe could exist on one such brane, and it would be joined by another separated by a hidden dimension. What excited the authors (Steinhardt and Turok were joined on the paper by Ovrut and Steinhardt's PhD student Justin Khoury, who later hired me at the Perimeter Institute) was the possibility that these branes might collide. If that were to happen, the brane-worlds would be pumped with matter and radiation, generating an immense inferno. I like to think of two giant hands clapping, making an almighty applause. For the would-be rebel cosmologists, that scenario sounded too much like the Big Bang to be a coincidence and should form the basis for a new cosmological vision. Ovrut agreed. What's more, the branes do not shrink to zero size at the collision point, so there might be no singularity and no beginning of time. The cosmologists decided to name their model the Ekpyrotic universe, after ancient Greece's Stoic conception of the cosmos as having been born in a great purification of fire: the Ekpyrosis.

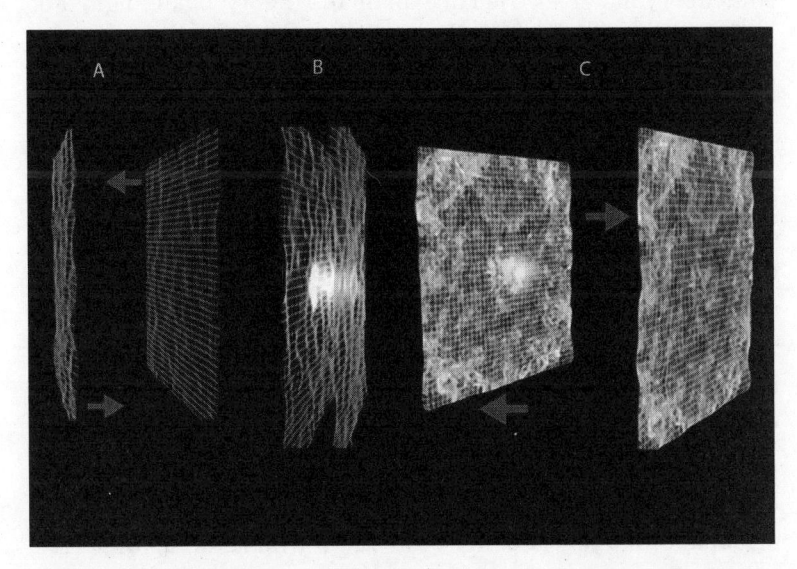

The Ekpyrotic model shows two higher-dimensional branes colliding. The energy of the collision becomes matter and radiation, triggering our Big Bang. In phase A, the branes of a cold, diluted universe are approaching. In phase B, they collide and heat up. In phase C, they expand away from each other. Credit: Phil Halper and Nick Franco, 1185 Films.

If Ekpyrosis was to challenge the mainstream, it had to address the problems inflation solves in an entirely novel way. Like other string models, the existence of a long pre–Big Bang era could alleviate the horizon problem. The monopole problem was solved as the temperature of the collision was much lower than the temperatures needed in Grand Unified Theories to make monopoles. The physicists found that as the branes approached each other, they would begin to ripple like sails blowing in the wind; therefore, not every part would collide at the same time. Some sections would hit earlier than others. These early-hitting sections would generate higher temperature and density regions that would seed the density fluctuations (or sound waves) so impressively achieved by inflation. Unlike string gas cosmology and the pre–Big Bang, which have remained obscure, the Ekpyrotic universe received a huge amount of press attention. If you ask me, Turok and Steinhardt were much better at marketing than Brandenberger and Veneziano were. During one TV broadcast, comedian Jim Carrey tried to explain the theory to late-night talk show host Conan O'Brien. That was until Carrey was interrupted by a call from Stephen Hawking explaining how excited Carrey and Hawking were over the new Ekpyrotic model. Although appearing enthusiastic about the model for TV, Hawking remained steadfast in his support for inflation. In a conversation with Phil, Linde recalled first reading the preprint with his wife, string theory and quantum gravity expert Renata Kallosh. "We were so impressed." He'd looked at the paper and said, "Wow"—so struck was he by encountering a consistent theory in string theory that did not include inflation. However, the wife-and-husband team were ultimately disappointed. "In half an hour, the whole magnificent building just crumbles." According to Linde, the colliding branes would cause the universe to collapse in a crunch instead of expanding from a bang. "You cannot make an error worse than that," he said. In their critique, Kallosh and Linde (along with their colleague Lev Kofman) called the scenario the "Pyrotechnic Universe."[20] And the Ekpyrotic model was about to come under fire.

In 2013, Phil and his wife, Monica, drove more than four hours from the Icehotel in Sweden to chase an especially powerful display of the aurora borealis, dancing overhead in spectacular displays of lights and motion. They finally managed to get some clear sky to marvel at the heavens in Rovaniemi, a remote city in Finnish Lapland. But twelve years earlier, it was the location of a different type of firework, which happened during the fifth international conference of Particle Physics and the Early Universe. Turok and Steinhardt presented the model, hoping it would set the world of cosmology ablaze. But the reaction was as cold as the location. Linde was ready to fight back, and the duo who tried to counter his critique admitted great disappointment at the reception the model received. A particular problem highlighted by the critics was that the Ekpyrotic model had unreasonably assumed that branes started in a smooth and flat state. But inflation takes any initial state and makes it smooth and flat. I think we can say then that Ekpyrosis suffered from the same problem as other string models: the inability to compete with inflation in solving the flatness problem. But as Turok and Steinhardt explored the Arctic countryside, they began to realize that the newly discovered dark energy might provide a remedy for their woes and turn their liabilities into assets.

In the Ekpyrotic universe, the movement of the branes is controlled by a springlike force, operating in a higher dimension. What was being suggested was that this force is felt in our reality as dark energy. Recall that dark energy causes the universe to accelerate in its expansion. It's a super-diluted form of inflation and could therefore achieve similar miracles given enough time. If accelerated expansion happened before the branes collided, they would be smooth and flat at the collision, setting up our universe to be the same. After the collision, the branes would move apart. The springlike force would relax, but eventually, as the universe's expansion diluted the collision's energy and the branes moved far enough away from each other, the springlike force would start to dominate again and the universe would contract, setting up yet another collision, bounce, and expansion. The model then had truly inherited the spirit of the original Ekpyrotic universe of the ancient Stoic philosophers. It would be cyclic, repeating endlessly as the branes smashed into each other, creating an inferno to cleanse the universe. While Ekpyrotic proponents tend to be critics of the inflationary multiverse, the two frameworks have clear similarities. In brane inflation models, inflation is driven by the movement of two opposing branes and the collision heats the universe into a Big Bang. In the cyclic Ekpyrotic model, today's exponential expansion can just be thought of as a version of inflation. And while Steinhardt and Turok seem to loathe the multiverse, their model also has an infinite number of Big Bangs, except the bangs are separated in time rather than space. What's more, there is another world that we collide with. So, I can't help but wonder that the differences between Ekpyrosis and eternal inflation are somewhat exaggerated.

Another problem for the Ekpyrotic model was describing the transition from contraction to expansion; a precise description of the bounce seemed elusive. Alan Guth claims that the model really solves the horizon and flatness problems in the same way as inflation, but that it does so at a different period. As Guth put it, "When I feel like being snarky, I say that the Ekpyrotic model is just inflation plus a bounce, and everything works about it except the bounce." Guth agrees with me that a cyclic universe is not so different from a multiverse and complains that while Steinhardt focuses on the measure problem for inflation (recall that this is the problem of defining probabilities for an infinite set), he ignores the fact that his cyclic model has the very same problem. In fact, any infinite universe will have a measure problem, and while the multiverse of eternal inflation is infinite in size, a cyclic universe is infinite in time—and one cannot just declare the universe finite by fiat so that we can have an easy time calculating probabilities. I also hosted Steinhardt for a colloquium at Harvard shortly after their work, and the physicist Nima Arkani-Hamed (a staunch string landscape supporter at the time) made the exact same point in the hours-long debate that ensued.

However, although the two frameworks have their similarities, one difference that might alleviate the measure problem for

Ekpyrosis is that each cycle only exists for a finite amount of time. In eternal inflation, by contrast, there are an infinite number of bubble universes, an infinite number of which will exist for eternity. But an endless series of cycles raises other problems. Most acute is that of the second law of thermodynamics, which you'll remember requires that entropy or disorder increase over time. But if the universe has always existed, why aren't we in a maximal-entropy state? One possible solution is to appeal to a change in entropy density, that is, how much entropy there is per unit volume of space. As long as the increasing entropy is spread out faster than it can build up, there will always be regions that experience low entropy, even in an eternal universe. Sean Carroll and Jennifer Chen offered a similar solution in a model where baby universes bud off from a parent universe that otherwise exists in equilibrium.[21] In the Carroll-Chen model, the babies can have low-entropy beginnings even if the parent has forever-increasing entropy. The Ekpyrotic universe takes similar advantage of declining entropy density. As each cycle progresses, only the space between the branes contracts. The volume in each brane is expanding. This expansion creates new space for entropy to dilute and thus, according to its authors, generates a truly eternal cosmos. But recall from chapter 2 that a theorem proved by Borde, Guth, and Vilenkin rules out a universe expanding eternally into the past. Since the pre–Big Bang and string gas models have a prior region of contraction balancing out our era of expansion, this restriction isn't a problem for them, but the Ekpyrotic model has a net average expansion over all its cycles and so should fall victim to Borde, Guth, and Vilenkin. Their theorem works by following a hypothetical particle back in time in an (on average) expanding universe. The further back in time one traces its history, the more its measured velocity increases. Since no particle can exceed the speed of light, the authors concluded that the past of any expanding cosmology must be finite. When we put this proposition to Steinhardt, he pointed out the flaw in their reasoning, which is that they are assuming the particle has a fixed mass, but this does not have to be true. He gave the example of the Higgs particle, whose mass changes drastically at high energies, as a result of interactions with quantum fields. I agree with Steinhardt here, who points out a symptom of one of the failing assumptions of the BGV theorem: a classical description based on particles. In contrast, the modern understanding of the microscopic world is based on quantum fields and not classical particles, so it is not hard to imagine that such quantum properties might be needed to make sense of an Ekpyrotic universe—thus undermining the claim that the BGV theorem implies the Ekpyrotic universe must have a beginning.

Perhaps the most distinctive feature of Ekpyrosis is that it's a gentler process. The branes slowly approach each other, and the collision does not move masses in the same violent manner that other models invoke, generating few gravitational waves in the primordial universe. This gentleness does give the model a feature that conforms to our criteria: it is easily testable, even falsifiable. If we see any sign of primordial gravitational waves (as was thought to have happened during the BICEP incident) the theory is wrong. On the other hand, inflation is compatible with both a strong and weak gravitational wave signal, which makes it less predictive, and so here I think Ekpyrosis scores an advantage over inflation.

In summary, string theory is in some ways like inflation. Both started out as radical ideas that are now mainstream. Both have some impressive achievements. Inflation can solve our cosmological problems; string theory can unite gravity and quantum mechanics. But when we look at the details, we see challenging issues. Just as inflation has a measure problem (arising from its unlimited reproduction feature), so does string theory have the seemingly impossible challenge of finding our world in its bewildering landscape of over 10^{500} possible solutions. Ultimately, it's unsatisfactory for physical theories to describe a myriad of hypothetical worlds, each with different values of the constants of nature; somewhere in this landscape, one of those universes must be ours. Inflation at least makes claims of empirical confirmation, which we have seen are open to challenge. String theory, some argue, offers nonempirical confirmation through its formidable, mathematically consistent structure.[22] The Big Bang may provide an arena to test string theory, but alas, it does not currently give us a clear picture of the early universe. Many proponents think strings can be used to solidify inflation, and its landscape reinforces the use of the anthropic principle. String theory creates a vast sea of possible universes, and eternal inflation makes each one real. Others have used the theory to construct alternatives to inflation. These models have common features but also notable differences. All predict that the universe existed before the Big Bang, yet again casting doubt on the notion that the bang was the beginning of all things. But each makes different predictions for primordial gravitational waves. In the case of string gas cosmology, they should be visible to both indirect detections via CMB polarization and direct detection in future satellite missions. In the pre–Big Bang case, a signal should not appear in CMB polarization but should be revealed in direct-detection satellites. In other words, the B-modes that BICEP was thought to have discovered would have been trouble for the pre–Big Bang but not for primordial gravitational waves themselves. For the Ekpyrotic model, nothing should be seen in either type of experiment, as its gentler transition cannot make the sort of primordial gravitational waves observers are hoping to see.

The fact that string theory gives rise to string gas, pre–Big Bang, Ekpyrotic, and brane inflation models (amongst other models we don't have space to mention) leaves me perplexed. It's not that string theory has no answers to the question of what really happened at the Big Bang. It may even have too many answers. Another concern I have is the lack of a clear resolution to the singularity. Yet despite decades of work there is no consensus on how such resolution can be achieved. Sure, there are arguments that strings can't be compressed to zero volume; that's fine. But we'd like to see detailed calculations of what happens at the scorching temperature of the Big Bang, and so far, such calculations are lacking. This weakness has led the Goliath that is string theory to be challenged by a proverbial David armed with a simple weapon: a theory that has no need of unseen dimensions or elusive supersymmetric particles and gives a clear and easy to understand narrative about what really happened some fourteen billion years ago.

✷ 5 ✷

THE BIG BOUNCE

According to a small band of insurgent theorists, attempts to quantize gravity so often fail because they overlook an essential lesson from Einstein. That lesson can be described by an incredible show still running in Las Vegas called Kà, by the performance group Cirque du Soleil. As the show begins, a remarkable sight unfolds. The stage appears out of an empty void in front of the audience, it tilts and rotates, and this movement becomes part of the story. At one moment, it is horizontal; at another, entirely vertical, and the acrobats must climb it. The stage is as much a player in the show as the performers are. Similarly, general relativity tells us that the stage of space and time is not independent of the particles and forces moving on it. It is not a passive background but a dynamic player. A quantum theory of gravity then should not be tied to a fixed background but be background-independent, with the stage itself being a star of the show. In other words, space-time must be described by a theory where it is not just the backdrop where particles exist but becomes a living, breathing, dynamic entity; not just a classical field as in relativity but a fluctuating quantum field subject to the uncertainty principle. The new string theory that was becoming all the rage in the 1980s didn't have this background independence. It described strings moving in a space-time governed by classical physics. For these rebel theorists, this deficiency called for a different approach, one that described one-dimensional loops that are the very fabric of space-time itself. They called this theory loop quantum gravity, and with it they hoped to reveal the secrets of what lies behind the mirage of the singularity, both those at the center of black holes and that of the Big Bang itself. Few people outside gravitational physics have heard of the theory's founder, Abhay Ashtekar. Inside the field, things are different. At one conference, Phil recalls, fellow loop theorist Jerzy Lewandowski gave a speech on how to be more like Abhay: "First, be a genius; second, know physics better than Newton or Einstein; and third, know mathematics better than Hilbert or Cartan."

In Salman Rushdie's novel *Midnight's Children*, all infants born within one hour of Indian independence in 1947 have supernatural powers of thought. Born two years after independence, Ashtekar might not have magical powers, but at the University of Bombay, he was one of the few bright students selected to spend some time at the Tata Institute for Fundamental Research, home of India's first digital computer. In order to breathe some life into a seemingly dry topic, the teachers encouraged the reading of Richard Feynman's *Lectures on Physics* (a book I, too, read when I was a teenager). Feynman, whom we met in chapter 3 when discussing the double-slit experiment, won a Nobel Prize for his contribution to quantum theory and is adored by many for his down-to-earth style and witty repartee. He played the bongos and did physics in a strip club, and when *Omni* magazine named him the world's smartest man, his mother apparently countered, "If that's the world's smartest man, God help us."[1] Despite his genius, he was a giant with a mixed legacy. In 2012, his FBI file was published and showed allegations of abuse by his ex-wife, not to mention the naked misogyny evident in his public writings.[2] Still, he has for decades been considered an icon of modern science and so his text was a favorite choice of instructors looking to spruce up the teaching of physics in India.

Ashtekar, still a teenager, found that one of the problems in Feynman's textbook had an error in it. He wrote to Feynman and surprisingly received a reply: "Congratulations. The book was wrong. You are right. You know the subject well enough and work carefully enough to be able to rely on yourselves [sic]." Ashtekar recalled to Phil, "I think this was a subtle note saying I shouldn't keep writing such letters to him in the future."

After doing a postdoc with Roger Penrose, Ashtekar returned to the University of Chicago, where he met Amitabha Sen. The two quickly bonded, going on double dates and writing several papers together. Sen would eventually be Ashtekar's best man but not before he had left physics to work for Motorola. It was in the mid-1980s that Ashtekar, building on Sen's work, found a way to rewrite Einstein's equations of general relativity in a language familiar to particle physics. This new technique became known as the Ashtekar variables, and theorists soon realized that they could forge a path toward a theory of quantum gravity.

In 1986, a six-month workshop on quantum gravity was organized in Santa Barbara and several talks were given on the new variables. Lee Smolin, who would become another of the founders of the new theory, was in the audience.

As a teenager in the early 1970s, Smolin had dropped out of high school to pursue his dream of creating geodesic domes inspired by the designs of his hero Buckminster Fuller. As he researched the mathematics of curved surfaces, Smolin came across Einstein's work on this subject. He tells Phil that one night, after being stood up on a date, he began to read an essay by the great master on the beauty of physics. That warm spring evening when his life changed, "I was sitting on my parents' front porch, the crickets were singing, and I had this essay of Einstein . . . and I was inspired by his vision of why he wanted to be a scientist. It just swept me up in all my adolescent rebellion and my idealism. The idea that there is a beautiful world of nature out there which completely transcends the inequities and struggles of human existence, and you can transcend to this sort of pure place by trying to learn what the laws of nature are expressed in this beautiful mathematics. I don't believe any of that anymore, but I just fell hook, line, and sinker for that vision." Smolin was on a mission to finish what Einstein had started, to reconcile the two most fundamental theories, quantum mechanics and general relativity. At the Santa Barbara conference, Smolin and Ashtekar had already begun to collaborate to see if a truly background-independent theory of quantum gravity could be developed from the new variables. As the workshop was drawing to a close, the third founder of loop quantum gravity, Carlo Rovelli, turned up unannounced and uninvited.

Rovelli and Smolin were born within a year of each other and both considered themselves radicals. Rovelli had set up a political radio station, refused military service, and been jailed for "crimes of opinion." He traveled the world and took hallucinogenic drugs but eventually returned to Italy to study for a philosophy degree in Bologna.[3] He recalled, "I don't think there's a direct connection between taking LSD in London in the 1970s and doing science twenty or thirty years later . . . But drugs give you experience of an idea which is hard to get at intellectually: that things might be completely different from what we usually assume."[4] As for Smolin, it was reading Einstein that changed the course of Rovelli's life. "I fell in love with physics when I started studying general relativity and quantum mechanics," he told us. "I thought, wow, this is fantastic, it just absorbed me. . . . I thought, *This is so beautiful. Maybe I want to do this for my life.*"

Rovelli journeyed to the Santa Barbara conference. "It was unstructured, unpaid, I had gone by myself, I had no paper published, I had no chance of getting a job. I was just infinitely curious about science and what was known about quantum gravity." There, the trio of Ashtekar the mathematician, Smolin the visionary, and Rovelli the philosopher was formed. They would soon create a new theory of space and time, loop quantum gravity. To explain loop quantum gravity, I like to think of an analogy with my T-shirt, which says "Theoretical Physicist" on it. As I look into the mirror, it seems smooth, almost as if it's made of one piece of fabric. But the experimental physicist in me (as clumsy as he may be) wants to examine it under a microscope. If I were to do so, I would see a complex netted structure of cotton and other fabrics. According to loop quantum gravity, space is the same: it looks continuous from a distance but at short scales, it is woven by discrete, one-dimensional quantum threads. But unlike my shirt, these loops are not made from anything; they are the stuff of space itself. And this point is a key difference between loops and strings: strings move in space, but loops *are* space.

A decade after Ashtekar's breakthrough, loop theorists found they could reproduce a classic result of Stephen Hawking and

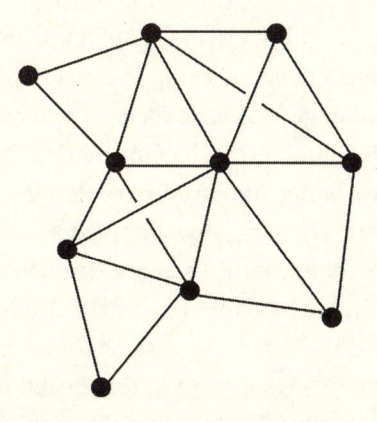

A cartoon of a small sample of space in loop quantum gravity, where discrete areas and volumes are the fundamental units of geometry. This gravitational model contrasts with that of general relativity, where areas and volume are continuous. Credit: Niayesh Afshordi.

Jacob Bekenstein in calculating the entropy of black holes. But string theorists were no less tenacious and were claiming the same milestone. To me, the truth seems slightly less impressive than the marketing. String theorists could only reproduce the famous Bekenstein-Hawking formula for a particular type of hypothetical black hole (one that does not emit Hawking radiation), which as far as we know doesn't exist in nature. Similarly, loop theorists' calculation depended on a free parameter, like a dial that you can tune to get anything you want. Ideally, a free parameter in a theory (like electron mass or charge) has to be fixed either by observation or theoretical consistency. The fewer the free parameters, the more powerful a theory is thought to be. That's not to say there isn't progress, but it's a common theme we will continue to encounter: that we are far from the summit. As I watch these two mountain parties ascend, it's hard to tell who is further ahead. But what's easy to spot is that neither seems keen on helping each other through the challenging climb ahead. A popular scene from the sitcom *The Big Bang Theory* portrays the two rival theories as religious dogmas, as one of its characters (Leonard) is dumped by his girlfriend (Lesley) for preferring strings to loops. "How will we raise the children?" she asks.

One tradition I would love to see more of is the mock debates that happen at the Foundational Questions Institute Conference, founded in 2005 by cosmologists Max Tegmark and Anthony Aguirre. One year, Rovelli debated string theorist Raphael Bousso, except they were challenged to argue for the opposing sides. Bousso, arguing against his actual research in the mock debate, condemned string theorists for being "ignorant particle physicists who don't understand general relativity." Rovelli hit back, saying that "string theory was true by definition." In 2022, Phil hosted a real debate between Rovelli and string theorist David Gross, who had won the Nobel Prize in physics for his work on the nature of the strong nuclear force.[5] The two were keen to tease each other with backhanded compliments. Rovelli compared Gross to Einstein, who had created wondrous physics in the first part of his life only to chase dead ends in his autumn years. Gross praised Rovelli's eloquent popular-science writing and philosophical tracts, suggesting he should probably stick to those.

The rivalry has been dubbed String Wars, and Smolin has been at its forefront, publishing a book called *The Trouble with Physics* that outlined his objections to the dominance of string theory.[6] Ashtekar's views are similar, complaining that the theory tries to do too much without focusing on the vital issue of how to resolve singularities: "Some people made serious statements that string theory would explain everything, including poetry! . . . I think that it is really true to say that a majority of string theorists had and perhaps still have what Alan Greenspan [former US Federal Reserve Bank Chair] called this 'irrational exuberance' in financial markets." Rovelli, now with his loop hat back on, claims, "String theory in my opinion is a wonderful idea but it has not delivered. . . . It doesn't mean it's wrong, but, as a research program, it has lost I think quite a lot of its initial appeal. . . . [L]oop quantum gravity is far less ambitious. I personally think that we are not at the point of writing a theory of everything . . . so we better do one problem at a time." Loop quantum gravity is not trying to unify all four forces, and so it does not aspire to cloak itself in the title of a theory of everything. Rather, it's an attempt to take one force, gravity, and reformulate it in a quantum language. Whether this endeavor is celebrated as a virtue or scorned as a vice rests in the eye of the beholder.

What loop theorists see as an advantage—a focus restricted to only the quantum aspects of gravity—string theorists view as a deficiency. If we have a theory that unifies all forces, quantum gravity will emerge naturally. When string theorists do engage with loop theory, they sometimes focus on the fact that physical theories can be separated according to whether they specify the kinematics (that is, the main characters of the theory—imagine the football and the players) or the dynamics (the rules of the sport). It's claimed that loop quantum gravity struggles to specify the dynamics.[7] Rovelli is dismissive of this issue, saying that it was only a problem twenty-five years ago and that critics simply haven't kept up with the literature. But when I talked to Lee Smolin, he confessed that the theory has been much more successful in kinematics (that is, in making geometry discrete) than it has in dynamics (that is, in making that discrete geometry evolve in a sensible way). When Einstein presented general relativity as a new theory of gravity, he was able to show that it accords with Newton's theory in most cases, differing only in situations where the space-time curvature is significant. A quantum theory of gravity must similarly be consistent with Einstein's theory in all well-tested circumstances (differing only in the most extreme conditions). I would like to see a formal proof of this compatibility with general relativity, but loop theorists can't offer such a proof. When we raised this issue with Rovelli, he conceded that no such verification exists but said there were good arguments to suggest it. To be fair, string theory also has a similar challenge when it comes to accommodating the right number of dimensions or the small cosmological constant that we actually see in the world. In Rovelli's view, spending more years perfecting the theory's mathematics isn't worthwhile. Researchers, he said, have been doing that for decades; the theory is now mature enough that they have to shift their attention to the physical world. Doing so means, amongst other things, tackling the ultimate question of what happened during the Big Bang.

In my view, both theories are speculative. String theory has the potential for a more significant payoff in explaining all the forces of nature, but it costs more: we have to assume unseen entities like supersymmetry and extra dimensions. Loop quantum gravity aspires to explain less but is more minimalistic and thus satisfies our criteria of simplicity better. I guess you get what you pay for. I fear that these two warring camps may never resolve their differences and might instead become like religions into which physicists are born depending on their parentage—that is, on their PhD supervisors. If scientific dogmatism is our nemesis, then the arena of Big Bang cosmology may be our liberator. The problem for physicists is that the energy scales of giant particle colliders are nowhere near the energies needed to probe the quantum gravity regime. But the extreme energy densities at the birth of our expanding universe provide a probe of what would otherwise be untestable. The recipe for peering into the extreme quantum gravity world is straightforward: construct models of the early universe that use some unique features of these new theories, and then compare them with the data from cosmology. We saw string theorists attempt this feat only to come up with a myriad of different models. Some, like brane inflation, try to justify inflation; others, like string gas cosmology and Ekpyrosis, seek to replace it. Loop quantum gravity, by contrast, tells a far simpler and easier to follow story of what really happened at the Big Bang—but, as we shall see, is not immune to controversy or bitter infighting.

In 2004, I received my PhD from Princeton. It was the eve of the String Wars, but I had done my studies in cosmology so had no need to pick a side in this particular battle. Many of my peers leave the field at this point, looking for jobs on Wall Street or in Silicon Valley. I'm told they even have some semblance of job security. But I knew I wanted to stay in science—and so I had to contend with living inside the pinball machine that is the world of the postdoc. Here, freshly minted PhDs bounce from one temporary position to another, hoping to eventually find that elusive permanent position. Postdocs typically last two to three years (with rare, prestigious five-year appointments) or until the grant money runs out. It's a high-stress role compensated by meager pay (albeit still much better than that of graduate students). In 2022, I saw the situation summed up nicely on Twitter by research chemist Tom Wilks, who adapted the Queen song "Bohemian Rhapsody" to address the plight of the postdoc:

Because I'm easy come, easy go
Stress is high
Pay is low.

My first postdoc was a fellowship at the brand-new Institute for Theory and Computation at Harvard. As a fellow, I had the freedom to work on any topic I wished, but I soon began searching for a permanent position. One of my first interviews was at Penn State University. I didn't get the job, but I wonder how my research direction might have changed if I had. Penn State is a major center for loop quantum gravity. Shortly before I interviewed there, a young postdoc who had arrived from Germany started to take the first steps toward using loops in imagining a quantum Big Bang.

FAREWELL TO SINGULARITIES

Martin Bojowald remembers becoming interested in science when he got a microscope as a child. This gift kindled a passion for biology, but his mathematical skills pushed him toward physics. At the turn of the century, he began to examine the early universe with the new tools developed by loop theorists. The view of the Big Bang as the beginning was so entrenched that when Bojowald turned up at Penn State to do a postdoc with Ashtekar, he was offered the option to do another project as he was told there was nothing much to be gained from examining the Big Bang. Or at least that's how Bojowald remembers it. But Ashtekar tells a different story: he recalls hiring Bojowald for his impressive knowledge of cosmology—a direction Ashtekar wanted his research group to go.

Memories are funny things. Far from perfect past records, they are more often than not narratives we tell ourselves. Researchers have even shown that it's all too easy to form false memories of events that never happened.[8] Bojowald and Ashtekar's recollections are at odds, perhaps signaling that the two would eventually clash severely. What we know is that Bojowald made the first steps in building a model of the Big Bang that uses the principles of loop quantum gravity; this model goes by the name of loop quantum cosmology. However, in his influential paper, Bojowald thanked Ashtekar "for suggesting a study of the implications of discrete volume and time close to the classical singularity."[9] As he recalled to Phil in an interview, when Bojowald began modeling the Big Bang using loop quantum cosmology, he noticed something curious: his equations included two independent expressions for the volume of the universe. This point puzzled him for some time, until he realized its physical meaning: the doubled volumes represented two universes with a quantum region in the middle, replacing the infinitely dense singularity predicted by the classical theory.

Bojowald soon found that the quantum geometry of loop quantum cosmology would ignite a period of ramped-up superinflation, implying that loop quantum cosmology might yield inflation purely from the effects of the new quantum geometry. No mysterious inflaton field was needed. But soon, doubts began to creep in. At one conference, Canadian physicist Bill Unruh pointed out that Bojowald's work had hidden unrealistic assumptions about the way the universe evolved. Given a positive cosmological constant in Einstein's equations, the cosmos should expand forever, but according to the early version of loop quantum cosmology, it would re-collapse. Something was wrong. Quantum gravitational effects had to be restricted to the extreme conditions close to the Big Bang—if it were otherwise, we should have already seen them— but in Bojowald's model, such effects could appear literally at any scale. The excitement over superinflation would also have the rug pulled from under it. In Bojowald's cosmology, superinflation was probably too short-lived to remove the need for the inflaton field. Ashtekar recalls doing the calculations and feeling great disappointment. It seemed their breakthrough had been a mirage. Many of these issues were debated at a summer workshop in Vienna. The participants argued about what needed to be done over wine-filled dinners, evening movies, and operas in the park. Eventually, they started to develop ideas about how to put loop quantum cosmology on a sounder footing, resulting in a paper by Ashtekar, Bojowald, and Jerzy Lewandowski in 2003.[10] But soon, Ashtekar realized that results depended on something called a cutoff. A cutoff is a procedure in physics where an arbitrary maximal value for an energy scale is imposed to make a calculation possible. Such a move is acceptable as long as the final answer doesn't depend on the cutoff. But it did.

A deeper overhaul was needed, and Ashtekar sought new talent for his team, a search that resulted in the hiring of two new postdocs. The first was Param Singh, who had come from Pune, India. Singh had had two main jobs there. The first was to work on the physics of the early universe; his PhD on the subject would later win him one of India's highest academic awards. The second was to catch poisonous snakes, especially cobras. "It is quite hard . . . We were not supposed to kill them, we have to catch them, and we have to release them in the forest, so we could not hurt them." He sees parallels in the two jobs, given the controversy surrounding quantum cosmology: "My passion was essentially to play with dangers." Ashtekar frequently traveled to Poland to collaborate with theorist Jerzy Lewandowski; the two had been working on the entropy of black holes but had been stuck on a particularly tricky calculation. Tomasz Pawlowski, still then a graduate student, stepped in and showed them how to do it. Amazed, Ashtekar immediately offered him the second postdoc position to help reformulate loop quantum cosmology.

The new team reviewed the three main problems: that the cutoffs could not be removed without changing the results; that quantum effects could arise at virtually any density when such effects should, instead, be restricted to extreme conditions near the Big Bang (and inside a black hole); and that a positive cosmological constant yielded, under the new model, a collapsing universe rather than an expanding one. But they made no real progress. Singh recalls the somber atmosphere on one particular evening after getting nowhere. "The whole approach seemed to be wrong," he said. "And we just called it a day at like 7:30 in the evening. The next morning, we came back at nine o'clock and the solution was on the board.... Abhay wrote the solution, and we tested it immediately and it solved all the problems." Ashtekar had found and removed the block that was confounding attempts to build a new cosmology. The problem lay with a simple assumption that is commonly used in the field and so seemed innocuous and involved a mathematical object called a metric. Metrics are used in relativity to describe distances in space-time. But cosmologists often use a simplification wherein a reference metric is used to ease calculations. By removing the reference metric and using the real, physical metric to implement a key step instead, all the problems at once went away. The cosmological constant did not cause the universe to collapse, the answer didn't depend on the cutoff, and quantum corrections only arose at the extreme energy densities close to the Big Bang. Ashtekar's face lit up when he recalled those events to us: "To me it was not just joy; I was awestruck at how beautiful nature is. You fix one conceptual thing you were screwing up by making a subtle mistake inadvertently, and then it takes care of everything."

Another drawback of the earlier approaches was that no numerical simulations existed. This time the researchers were determined to do things right and plugged the new framework into a computer program called Mathematica. The run times were extremely long, some taking almost a month, but when the results came back, they could watch a contracting universe evolving toward a singularity. At first, the simulation matched general relativity, indicating that loop quantum cosmology reproduces Einstein's theory at low curvature; so far, so good. But as the universe's curvature increased and the singularity approached, the loop quantum cosmology simulation showed something different, something many at first refused to believe. The collapsing universe stopped and emerged, expanding in a new era. The Big Bang was now a Big Bounce. A contracting universe could have given birth to our expanding cosmos. If we draw a picture of this universe in space-time, it resembles an hourglass rather than the standard ice cream cone of conventional cosmology. The bounce had been hinted at in Bojowald's early work, but he hadn't taken it seriously, and Singh had seen it before in work done with Kevin Vandersloot, but they were not convinced it was real either. As Singh put it, "We thought maybe our code was giving us errors. . . . We just thought maybe this was just the limit of our analysis . . . we had no clue what we were looking at."

In 1983, Alan Guth and Marc Sher had shown that a bouncing universe was impossible, so there was a lot of skepticism.[11] But those results used only classical physics. As we have seen, many string-inspired models also imply a bounce and now, Singh and Pawlowski were sure they had seen the same thing. Guth and Sher's conclusion didn't apply in the quantum world. Singh recalled, "We were convinced more or less because we knew that the code is correct and numerical errors were small, and we might be seeing the real thing. But we were very young too. . . . We were doing hundreds of simulations and then we wrote the code in different ways, and we were getting the same thing."

Ashtekar remembers events similarly. "I thought it was too good to be true, so to say, and I wanted to be very very careful," he says. "The question was whether this was an artifact of our specific choices. . . . We changed various things in the computational models. I kept asking a thousand questions to the postdocs who were doing the simulations to make sure nothing was slipping us up, but the predictions remained robust in all these things." The team was sitting on an incredible result but waded through nine months of further checks before publishing what has become known as the improved-dynamics version of loop quantum cosmology.[12]

Physicists often use two different methods to solve problems. One, known as a numeric approach, uses computer simulations (finding approximate solutions in numerical terms). The other is known as an analytic approach and involves solving equations exactly, traditionally using nothing more than a pencil and paper (though nowadays, computers can carry out analytics too). Singh, Poplawski, and Ashtekar had found the bounce using numerics. To make loop quantum cosmology more robust, Alejandro Corichi, Singh, and Ashtekar created an analytic version, showing that the bounce is not only seen but occurs at exactly the same density as was found in the numerical models.[13] Other groups around the world started to investigate and found the same result. However, Bojowald claims that the model relies on hidden assumptions that cast doubt on its trustworthiness, although two years after the improved-dynamics paper, Bojowald authored a cover article for *Scientific American* with the tagline "Forget the Big Bang, now it's the Big Bounce."[14] The article used a wonderful analogy to explain why quantum effects lead to a bounce. Think of a sponge; it absorbs water but only up to a point. Go past this point and the sponge will switch from water absorbent to water repellent. Space is similar; you can squeeze more and more mass/energy into it, but when space is full, it will, just like the sponge, switch its properties. Gravity will become repulsive rather than attractive, and this quantum repulsion triggers the bounce. This analogy reminds me of another astrophysical process; when stars run out of their nuclear fuel, their cores have nothing left to counteract the compression of gravity and so collapse. However, in all stars except the most massive ones, a new quantum mechanical effect known as degeneracy pressure creates a repulsive force, stabilizing them against further collapse. The great astrophysicist Subrahmanyan Chandrasekhar used this effect to study stellar evolution; now his Chicago protégé Ashtekar and colleagues had found something similar in the case of cosmic evolution. A new quantum repulsive force would overcome the otherwise inevitable gravitational collapse, remove the singularity, and trigger a Big Bounce.

In the new loop quantum cosmology, superinflation was present but still too short-lived to replace the need for standard inflation. However, loop researchers found that the probability of there being enough regular inflation was now close to one rather than being exponentially suppressed as had been claimed before.[15] Moreover, they could use the cosmology results to independently constrain the free parameter that was thought to be crucial to calculating black-hole entropy from loop quantum gravity.[16]

INTERNAL STRIFE

Despite the exciting results, a rivalry began to emerge between Singh and Bojowald, with Bojowald claiming that virtually no information could survive the Big Bounce. He called this phenomenon "cosmic forgetfulness."[17] If true, the memory of the pre–Big Bang universe would be forever lost, like the details of a midnight dream evaporating from our minds in the morning. In quantum mechanics, particles like electrons are described as spread-out waves across space. Quantum cosmologists think of a wave function of the entire universe, spreading out over all possible sizes and shapes. If the wave function is too smeared out at the Big Bounce, we will never be able to figure out what the universe was like before it. Singh collaborated with Alejandro Corichi to prove a theorem that implies that the wave function would stay sharply peaked; independent analysis agreed.[18] Moreover, numerical simulations showed the same thing. Our expanding universe would not be disconnected from the pre–Big Bang contracting era.

The disagreement over cosmic forgetfulness ran deep. While the loop quantum cosmology community has continued to grow—it represents one of the largest for any model we shall examine in this book—one of its founders was becoming increasingly skeptical of the field.[19] As these disagreements became more acute, I believe Bojowald became increasingly detached from the loop community. While reading Bojowald's book about the bounce, *Once before Time*, I perceived what I took to be veiled insults toward Ashtekar, as when Bojowald described him as "an analyst equipped with technical brilliance and unparalleled mastery of the dark art of scientific power play" (which is a skill sometimes I wish I could master).[20]

I discussed this critique with my friend and colleague Lee Smolin, who is a founding father of loop quantum gravity but has also been critical of the simplifications used in quantum cosmology, be it from strings or loops. According to Smolin, Ashtekar's great skill is to take an idea and make it mathematically rigorous. But this process can often lead the idea's original creator to feel slighted, even if that rigor is needed.

However, Bojowald claims he has serious scientific criticisms that are simply unwelcome, a hangover from the days when loop quantum gravity was a small group dominated by the much larger string community. "But now loop quantum gravity is pretty big too," he tells Phil, "[but] somehow the culture hasn't changed." At loop conferences, he says, "there is still a lot of suppression of criticism. . . . It's very political."

Would Bojowald even be welcome at a loop quantum gravity conference? "Of course," says Ashtekar. It's not clear whether such an invitation would be accepted.

When we asked Singh how he felt about the situation, he said, "[Bojowald] was my very close friend in the very early days. . . . I had great respect for him. I joked with him." He added, "When the results of the bounce came out, [Bojowald] was very skeptical—and he has a right to be skeptical, no doubt about it. But even when different groups were giving different results, he was still attacking and attacking. . . . So eventually I think the community started paying less attention to what he was saying. I find it unfortunate. I think it's a misfortune for the community as a whole. . . . The progress was slow because of these internal strifes."

Bojowald hasn't given up on loop quantum cosmology, and he has even used his own version to do battle with Turok and collaborators, who we saw argued that the Hartle-Hawking model was fatally flawed.[21] But Bojowald argues that loops can come to the rescue and create a stable transition from the no-boundary state into a smooth, homogeneous universe. The difference, then, between Bojowald and Ashtekar is not whether the singularity is resolved (both agree that it is) but rather *how*. It's comforting to know that both sides of this debate back our thesis that the singularity can't be trusted, but obviously we want to know precisely what replaces it—and herein lies the new conflict. My sense is that most loop quantum cosmology researchers side with Ashtekar and reject Bojowald's criticisms. But I also believe it's wrong to settle disputes in physics by head count. Simulations and theorems may make an idea plausible and attractive, but only contact with experimental data can make something fact. And hiding in what looks like the random noise of the primordial light, loop researchers believe they have found a signal that may herald the Big Bounce.

QUANTUM GRAVITY IN THE SKY

The main focus for confronting the theory with observations has been the oldest light in the universe, the afterglow of the primordial fireball, the cosmic microwave background (CMB). After the Big Bang the universe was a hot plasma, which—just like the surface of the sun—was opaque. It would take 380,000 years before the universe would cool enough to become transparent; at this point, the photons were set free from the fog of the plasma, and now—billions of years later—we can see them in our telescopes. The primordial hot plasma that emitted the CMB is like a curtain preventing us from seeing the hot Big Bang itself, but by looking at barely detectable temperature fluctuations (one part in 100,000), it is possible to make some inferences about what is hidden from direct view. These fluctuations grew from minuscule ripples in the density of the early universe. An analysis of the details of this process that turns them into predictions for what we see today is known as cosmological perturbation theory. Speaking from experience, this theory often conjures up nightmares, as the calculations are intricate and prone to error; even the biggest names in twentieth-century cosmology, like Hawking, Guth, and Steinhardt, have struggled with the foreboding mathematical machinery needed. Loop quantum cosmology researchers would face the same problems but with the added difficulty of coping with a fluctuating quantum space-time.

After years of struggle, loop theorists had a method they were satisfied with and began to compute what the CMB should look like following a Big Bounce rather than a Big Bang. They claimed their results fit the data better than standard inflationary models do.[22] This finding is related to intriguing anomalies seen in the CMB—places where statistical patterns of hot and cold regions deviate subtly from the predictions of standard inflationary cosmology. Such deviations were seen in NASA's COBE and WMAP probes (more on this point in a moment). In fact, the first project for my PhD that David Spergel suggested was to develop a model to explain these anomalies. But frustratingly, nothing I tried worked. Later, many (including Spergel) would write these deviations off as statistical flukes, even though Planck's higher-resolution maps showed the anomalies were still present. The Planck press release used the headline: "An Almost Perfect Universe."[23] The perfection was due to the excellent fit between the Planck team's findings and the standard cosmological picture. But the "almost" was down to the anomalies. At the forefront of science and discovery, scientists often dismiss or consider new, unexpected results to be flukes, until they can find a good explanation and verify it with independent data. Often no such explanation is found, but when one is, it can lead to seismic paradigm shifts. In this case—if the results of loop researchers can be verified by the most detailed available examinations of the CMB—decades of Big Bang dogma stand to be overturned.

Astronomers give objects in the sky an apparent size in terms of degrees of angle. The sun, for example, has an angular size of about half a degree—just about how big your thumb looks if you stretch your arm. A bright star like Betelgeuse is actually far bigger than the sun, but it is so distant it appears to be much smaller; its angular size is only one hundred-thousandth of a degree. CMB anisotropies (or temperature fluctuations) are also examined in terms of varying angular scales; they are strongest on a degree scale, which is the apparent size of how far sound waves traveled in the Big Bang plasma, as seen from Earth. The first anomaly, which we mentioned in chapter 2, is a surprisingly large cold spot. Some have speculated that this cold spot might have a primordial origin, possibly even the remnants of a bubble collision with another universe. Another is that if we look at larger angles (bigger than sixty degrees), the observed anisotropies will mysteriously disappear. Yet another issue is that one side of the sky seems to show bigger fluctuations than the other. These anomalies are puzzling features that are hard to explain in the standard picture. During the Planck press conference, George Efstathiou, the Cambridge cosmologist presenting the results, went as far as to say, "We have to be open minded that there might be new physics involved in these anomalies . . . that we can track our universe back through the Big Bang to a previous Big Bang phase."[24] According to loop theorists, their model is that "new physics."[25]

The data also shows other correlations that loop quantum cosmology researchers say verify their model. It's easy to get carried away by the excitement of these claims. But now is the time to bring some caution to the story. Science is successful because it is skeptical. No radical results should be accepted without the highest level of scrutiny. Scientists must try and rule out less prosaic explanations before endorsing an exciting one, and I think we are very far from having done so. The anomalies could still be statistical flukes. When we challenged Ashtekar about this possibility, he said the anomalies have a 1 in 1,000 odds of being down to chance. In our line of work, however, that probability is nowhere near unlikely enough to allow us to declare a discovery. As an example, we can think of a coin toss. The odds of getting heads for a fair coin are 1 in 2. Suppose we get heads twice—this outcome could easily be a fluke, as there is a 1 in 4 chance of it happening. What about getting ten heads in a row? Here the odds are 1 in 1,024. But if we throw the coin thousands of times, getting ten heads in a row will be much less surprising. This is why scientists tend to urge caution, as we make so many different analyses of huge datasets that if we don't set the bar high, meaningless accidental patterns will masquerade as scientific discoveries. For example, physicists did not celebrate the discovery of the Higgs boson until 2012, after they broke through the landmark "5 sigma" level of statistical significance. In plain language, reaching 5 sigma means that the odds of the results happening by chance are no more than 1 in 3.5 million (equating to the odds of getting twenty-two heads in a row). Even if the conventional 5-sigma requirement for discovery appears to be overly strict and arbitrary, it has been borne out by historical precedence, as physicists often tend to underestimate their sources of error.

That scientific benchmark is why the CMB anomalies (with odds of 1 in 1,000) are treated as curiosities rather than discoveries. However, if one explanation can resolve several anomalies, its statistical significance increases; loop researchers say their theory meets this standard. But loop quantum cosmology is a simplification with assumptions. If the initial conditions are changed, the bounce will still happen, but the numerical predictions will differ. So right now, we don't know if the model's output is truly showing us a feature of the bounce or if it's merely a consequence of the choice of the initial conditions plugged into the computer models.

Loop quantum cosmology assumes inflation, so any predictions are really the predictions of the two theories combined. In fact, different inflationary models exist, so researchers generally pick an inflationary model favored by the data and then add loop quantum cosmology corrections to it. But while inflation is a credible idea, it is not a verified fact. Singh thinks the community should be less risk averse and look more seriously at how loop quantum cosmology might be embedded in non-inflationary models. As someone who has worked on alternatives to inflation, I'm sympathetic to this point of view. Another scientist who thinks there may be a way to create a model of loop quantum cosmology without inflation is my friend Francesca Vidotto. I vividly recall a drive from a philosophy of cosmology meeting back to Waterloo, Ontario, where the Perimeter Institute is located, when I learned from Vidotto about the dramatic Ashtekar-Bojowald loop quantum cosmology face-off and its detrimental effect on the careers of young loop researchers. As the famous African proverb goes, "When elephants fight, it is the grass that suffers."

Vidotto is now exploring a model of loop quantum cosmology in which the strange interconnections between distant particles, known as quantum entanglement, create correlations in the CMB. She believes this model could reduce the need for inflation or maybe even eliminate it altogether. She has also worked on addressing a problem that Smolin and I have with loop quantum cosmology, namely that it is an oversimplified approach. For her PhD, she pioneered a program creating a cosmology from the full mother theory of loop quantum gravity.[26] This approach goes by the name spin-foam cosmology, and she believes it has the potential to make loop quantum cosmology a more predictive model. But while the theory is still a work in progress, she told Phil that the basic conclusion is unchanged: "No singularities can form in the early universe, so there will be no Big Bang—independently of how you construct your quantum cosmological theory from loop quantum gravity."

Looking for footprints of the bounce is no easy task, but one thing that can help is, simply, raw computer power. Singh and colleagues now have access to enormous supercomputers that can run simulations of the quantum universe. They also bid for time on national supercomputing facilities, which are not cheap; a single figure on a paper can cost $10,000 of computing time. When investigating the evolution of three-dimensional structures, the computational effort rises exponentially, but as computer power increases, loop cosmologists will be able to run more realistic simulations, enabling detailed comparison with observations and, hopefully, finding signatures of the bounce. The sort of computations researchers now do in fifteen minutes would have taken Singh and Pawlowski millions of years to do early in the first decade of the 2000s. Singh described this new supercomputing capacity to us with a wild enthusiasm, calling it "game-changing"; but though he has one eye on the future, he has another on the past, claiming the new simulations counter Bojowald's criticisms of their approach.

Perhaps the most challenging problem of all has to be in calculating asymmetries in the energy distribution of the universe that may arise from the bounce. Recall that standard inflation models predict Gaussian fluctuations—in other words, that there should be approximately as many hot spots in the CMB as cold ones, and that the temperature distribution should follow a bell

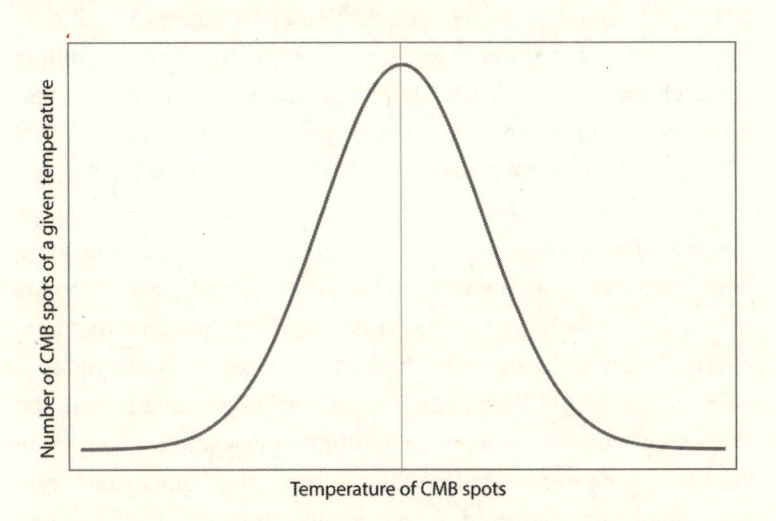

A Gaussian (or bell) curve shows the approximate distribution of temperature for cosmic microwave background spots. Note that this distribution has the same number of hot and cold spots. Credit: Niayesh Afshordi.

curve. Yet, that may not be the case for other Big Bang models. For two years, loop researchers plugged away trying to determine if loop quantum cosmology could give rise to any noticeable non-Gaussianities. What calculations show is that as we go to larger scales, deviations from Gaussianity become stronger. We are limited by the size of the observable universe, which hinders our ability to see the strongest effects. It's like trying to see directly the curvature of the Earth by standing on its surface rather than far above it. It can't be done. We can, however, infer the Earth's curvature, for example by observing that the moon is upside down in the southern hemisphere, or by watching sailing ships disappearing beyond the horizon. And bold explorers of quantum space-time have a way to do something similar.

Again, quantum entanglement comes to the rescue, as it creates connections between the longer wavelengths invisible to us and the shorter ones that are within the grasp of our detectors. These correlations imprint into the CMB, creating both the anomalies that loop researchers say back their model and still more as-yet-unseen correlations in the data. Recently, though, a spat broke out when *Scientific American* published an article claiming that loop quantum cosmology predicted a particular signature of such non-Gaussianities that should have been seen by the Planck satellite. A search for this signature came back empty, leading the popular magazine to run the headline "The Universe Began with a Bang, Not a Bounce."[27] Loop researcher Ivan Agullo complained to Phil that whilst they quoted him praising the new work, they chose to "ignore the crucial part where I discuss its non-applicability to loop quantum cosmology—and then, feeling misrepresented, I contacted the editors." In his letter to the magazine, Agullo pointed out that loop researchers had already published articles suggesting that the supposed signal would have been washed away by oscillations in the primordial universe—meaning that the Planck satellite's observations were indeed compatible with the predictions of loop quantum cosmology. But, according to Agullo, the editors of *Scientific American* refused to publish it.[28] Nevertheless, the model implies that certain other signatures should be seen. When we asked Ashtekar what we should do if they weren't, he did not say he would abandon loop quantum cosmology. Instead, he said he would question whether the initial conditions assumed might be changed. The possibility of revising the model in light of non-supporting observations gives me pause. In my opinion, doing so feels a bit like changing an exam script after the teacher has given the answers out. Ideally, new data should be clearly predictable such that it could throw the entire model out. But making such model-invalidating predictions simply isn't possible for loop quantum cosmology as it now stands, because any predictions depend on the matter fields assumed to have been present at the Big Bounce. To be fair, this problem is present in most other cosmological models we cover in this book, providing raw material for continual debate and interpretation. It remains to be seen if more robust predictions can be found (for this model or for others) or if new data can constrain these assumptions.

The Planck satellite has essentially exhausted the large-scale information available in the CMB. We need an independent check on the claims of loop quantum cosmology researchers. And we do have the possibility of such a check. I'll never forget the excitement of looking through a telescope and seeing a fuzzy patch of light that I knew was the Andromeda galaxy, bigger than our Milky Way, filled with countless stars. Galaxies are netted together to form huge weblike structures that arose from fluctuations in the very earliest moments of the expansion. The structure of galaxies is three-dimensional and so can in fact provide way more information on primordial non-Gaussianities than the CMB—which is more like a two-dimensional flat surface—can. But what the universe gives us with one hand it taketh away with the other. The processes that formed galaxies were much messier than those that formed the CMB, so what you gain from the extra information of three dimensions you lose in its bewildering complexity. Looking for a signal of quantum gravity in galaxy structures rather than in the CMB is like mining for diamonds rather than coal. The latter is far easier to extract but of course the former is worth more.

One problem we shall keep coming back to for any model that features a previously contracting universe is that the entropy (again, roughly a measure of disorder) of the Big Bang was extraordinarily low. But such a low entropy does not seem possible if the pre–Big Bang era was long—in which case things would have been far more disordered than we now observe them to have been. One solution is that the previous universe's entropy is trapped in the contracting branch, unable to influence our expanding universe, so we must reset the entropy clock at the Big Bounce. This paradigm allows the universe to be truly eternal into the past, the prior contracting era being a mirror image, in time, of our expanding universe. Just as our cosmos will never end, the contracting universe that is our twin may never have begun.

From the outset, I think the loop community has certainly done many things right. They've developed the theory from first principles, built it up carefully, and then begun to apply it to cosmology. Then they did the work, computed CMB properties from the theory, and made predictions that the data confirmed. But as you may have guessed by now, I'm not ready to close the case. The fact is, the loop model's picture of the Big Bounce and the predictions that come from it rely on many simplifying assumptions that can be challenged. And then there is internal and external strife over the basic soundness of the theory. However, other predictions from loop quantum gravity—like the suggestion that black holes also bounce—might help us get a clearer picture. Here again my take is that the theory is not currently developed enough to make unambiguous predictions. But I know that my friends like Rovelli, Vidotto, and others are working hard on this research, and we shall see where their work leads.

One thing I find intriguing is that the idea of a bounce does keep popping up. We saw first that the Hartle-Hawking no-boundary proposal implies a bounce, at least in the real-time axis. Then we found that multiple models inspired by string theory also imply a past contracting universe. There's some irony in the possibility that, in the end, these two bitter rivals—string theory and loop quantum gravity—might be telling us a similar story. Maybe this alignment is a hint that loops and strings can be united? Loop theorists frequently come from a general-relativity background, trying to create a quantum version of Einstein's gravitational theory. String theorists, on the other hand, come from the tradition of particle physics and try to incorporate gravity into quantum mechanics. Some theorists, like Jorge Pullin and Rudolfo Gambini, have found hidden connections between the two approaches.[29] One of my favorite YouTube cartoons shows caricatures of Carlo Rovelli and string theorist Nima Arkani-Hamed arguing only to find their research programs are "the same bloody thing."[30] Maybe the two will meet in the middle at the Big Bounce. It's a story that is not new; Big Bang giants like Dicke and Gamow also endorsed a bouncing universe. But after many years in the wilderness, the idea is clearly having a revival. The Big Bounce is more ubiquitous as an idea, though, than the single-shot contraction-followed-by-expansion envisaged by the likes of Brandenberger, Veneziano, and Ashtekar. If such a process could happen once, maybe it could happen again and again and again, fulfilling a truly ancient idea of a cyclic universe.

✨ 6 ✨

DÉJÀ VU AND THE CYCLIC COSMOS

When I was a child, I learned of the myth that deep in the heart of paradise lived a brilliantly colored, solitary bird. So beautiful was its song that the sun itself would stop to listen. Every thousand years, the bird gathers a pile of wood and sings so passionately that it ignites the pile, killing the bird in a burst of flames. But an immortal can never truly die, and soon the bird is resurrected from the ashes. No one knows precisely how old the legends of the phoenix (ققنوس in Persian) are, but they date back at least to ancient Greece. The idea of death and resurrection infuses many ancient beliefs. The Stoics thought the universe itself underwent cycles, as do many Eastern traditions, such as Jainism and Buddhism. In Hinduism, even the creator God, Brahma, undergoes cycles of death and rebirth. His lifespan is a mahā-kalpa, lasting 311 trillion years. These enormous epochs seem right at home in the world of contemporary cosmology. Trillions of years from now, the last stars will be gone, and the glow they produce that illuminates the cosmos will fade into oblivion. Think of everything you know and love: the Earth, the sun, even giant galaxies evaporating into nothingness. It's a future I find to be almost unbearably bleak. But could the darkness end with new light? Might the universe be reborn like the phoenix?

Ironically, cyclic cosmologies are having their own resurrection moment. In fact, there are so many versions of the cyclic universe in today's scientific literature that we cannot possibly cover all their myriad versions. We have already reviewed the Ekpyrotic model, but we shall see how its authors seek to detach it from its M-theory roots and throw away the original narrative of colliding branes, keeping the bare bones of a slowly contracting and expanding universe. We'll also meet the curious Baum-Frampton model, which describes the universe ripping itself apart only to pick up the pieces and begin again, attempting to address a recurring puzzle: Why was entropy so low at the Big Bang? High entropy is always more probable than low entropy. Explaining the low entropy of the Big Bang has thus motivated many of the scientists we shall meet to create cyclic cosmologists. The question is also an obsession of Roger Penrose, one of the founders of the singularity theorem that supposedly proved there couldn't be anything before the Big Bang. But like Hawking before him, Penrose has undergone his own U-turn, having become persuaded that the Big Bang was not the beginning after all and that the universe can undergo infinite cycles. Here, we shall try to cut through the hype and expose the strengths and weaknesses of these new cyclic modes. We shall see to what extent they can either rival inflation, or in some cases incorporate it, constructing a credible scientific version of an ancient myth.

Modern cyclic theories of the cosmos date back to Alexander Friedmann, one of the fathers of Big Bang cosmology. In perhaps the world's first popular modern cosmology book, published in 1923,[1] he described the possibility that the universe might contract down to zero size before rebounding into expansion. But many of my colleagues believe that a cyclic cosmology has a fatal flaw: the slow buildup of entropy. As we've discussed, you can think of entropy as a measure of disorder. More precisely, entropy accounts for the number of ways there are to make something. For example, there are many ways to reshuffle a deck of cards, but there is only one which gives the original order. Think of just the picture cards: we could shuffle them and retain their original order (say, *jack, queen, king*) but we could also get *queen, jack, king* or *king, queen, jack*, and so on. In short, the probability of getting the original order back with this subset of cards is 1 in 6, so there are more ways to create a random sequence than there are to restore the original order. You could thus say that the original order has low entropy, while all the other orders have high entropy.[2] Just as the order of the cards becomes more random as we reshuffle them, the second law of thermodynamics tells us that entropy almost always increases with time. So, naively, one might expect the law to rule out an eternal cyclic universe, as eventually we would reach a state of maximum disorder in which no structures could possibly exist in the universe. But what if disorder in fact knows no bounds? In a scenario featuring a sequence of expansion and contraction, bangs and crunches, entropy could still grow but be chasing a moving target—as each cycle could see more expansion than contraction, creating new space for more entropy and thus never reaching a maximal-entropy state. However, if each universe were bigger than the last, then going back we might eventually reach a cycle of zero size within a finite time.[3] So even a cyclic universe would need a beginning; immortals, it seems, still have to be born in the first place.

Another issue concerning the entropy of the universe has always puzzled me. We can understand this problem by picturing films running forward or backward in time. For a while, my sons were obsessed with watching Black Eyed Peas music videos on YouTube in reverse, so much so that they could recite the songs backward. As bizarre as these reversed videos sound, hundreds of them apparently exist. It turns out that it's very easy to tell the difference between the reversed version of a music video and the normal one. But now let's imagine a different video. Think of a pendulum. A video of a pendulum gently swaying back and forth would look identical whether it was played forward or backward. Our universe seems to be more like the flashy Black Eyed Peas video and less like the boring pendulum. We can tell the difference between running a video of events running backward or forward in time because entropy increases in the forward direction. We see eggs break but never unbreak. Time has a definite arrow, and it points to the future. But the fundamental laws of physics make no distinction between the past and future and so are more like the pendulum. This conundrum is a deep one and lies at the heart of physics. Fundamentally, the past and the future should look the same, but they don't. Many cosmologists have in the past simply ignored this mystery.

The arrow of time exists because entropy was extraordinarily low at the Big Bang and has been growing ever since. But why was it so low in the first place? We infer the low-entropy state of the Big Bang, so one might expect that relic of the Big Bang, the CMB, to show this low entropy in all its glory. But it doesn't. In fact, the perfect black-body spectrum that COBE first saw in the early 90s is high entropy, as high as it can possibly be. Roger Penrose calls this problem the "mammoth in the room." It can be resolved by assuming that the low entropy of the gravitational field would overwhelm the high entropy in matter fields. But even that assumption still leaves us with the mystery of why the gravitational-field entropy was so low. High entropy is always more probable, so the low entropy of the gravitational field seemed extraordinarily unlikely. One might argue that this low entropy is delicately fine-tuned and appeal to the anthropic principle to solve it. While that approach might work for certain physical parameters (for example, if electron mass were much smaller or larger than we observe, life would be impossible), the same line of reasoning won't work here. The entropy of the universe could have been enormously higher, and life would still have been possible. So, it seems something else forced the universe into a low-entropy state, and that conclusion is what motivated Roger Penrose to begin his journey to undo the Big Bang singularity and replace it with a cyclic cosmos.

Penrose's work on singularities won him a Nobel Prize in 2020. But years before, he started to doubt whether these mathematical monstrosities truly existed. As he told us, referring to his collaborator Hawking, "Stephen was absolutely brash about using the word *singularity*," and claiming that singularity research proved the universe had a beginning. In the late twentieth century, as Hawking ascended to rock-star status, such pronouncements were treated as indisputable facts. The discovery of dark energy in 1998 implied that the expansion of the universe was not slowing down but speeding up. If this acceleration continued, a universe that re-collapsed would be impossible. Thus, the presence of a singularity at the Big Bang and a universe filled with dark energy seemed to sound the death knell for cyclic cosmologies. But like the phoenix, what is immortal can never truly die. Soon, new versions of a cyclic universe would be reborn from the ashes. Steinhardt and Lehners even named one theory the Phoenix universe.

CONFORMAL CYCLIC COSMOLOGY

In the 1970s, Penrose moved to an appointment as professor of mathematics at Oxford and took over the supervision of the students of his old mentor Dennis Sciama. One student who stood out was Paul Todd, who pointed out a rather curious fact that might have radical implications for the universe's origins and fate. His reasoning focused on what is known as conformal geometry, which is concerned with the properties of shapes that stay constant as their size changes. A triangle's angles are independent of its size, and so the angles are said to be "conformally invariant." Every measuring object in the universe, be it a ruler or clock, has to be built of particles with mass. Todd proposed that, in the extreme heat of the early universe, mass might not exist, so the entire cosmos at that time could be characterized as conformally invariant. In such a world our conventional notion of time disappears; the relations between events will stay the same but the distances between them become meaningless. As an analogy, think of a chess game. We could write down the order of each move and what time it occurs. Now delete the times of the moves but preserve their order. You'll get the same winner. This is the kind of information that physicists call "conformally invariant." And when we switch to this new language, something truly amazing happens: the singularity that formed the basis for believing the Big Bang was the beginning disappears. At first, this conformal rescaling was considered just a tantalizing mathematical trick—but new physics would start to convince some physicists that it was real.

Following the pivotal discovery of the accelerating universe in 1998, Penrose started to ponder the universe's far future given a cosmological constant. With all matter decaying away and the last black hole evaporating into the void, he reasoned, there would be nothing left but radiation, with no way to build a clock or a ruler.

The universe would become conformally invariant once more, losing its sense of scale. The cosmos would "forget" how big it was. Small and large would become equivalent. The universe is born in a Big Bang, expands until all matter disappears, and then undergoes a conformal rescaling to become another Big Bang. This process would, according to Penrose, repeat endlessly into the past and future. As Penrose says, "That's the thing hard to get your mind around: how is it that this very big, cold, rarefied universe can be equivalent to a very hot, dense smaller universe? But they're physically equivalent if you don't have a scale . . . I used to say, 'This is a crazy idea,' more or less to prevent other people from saying it first. The idea was that it's not so crazy because the squashed-down infinity is very like the stretched out Big Bang, extremely like it." He dubbed this "outrageous proposal" CCC: conformal cyclic

In conformal cyclic cosmology, the universe is cyclic but never re-collapses. Each eon expands until all mass disappears. At this point, the laws become scale-free, and the cosmos can rescale (shrink) to make a new Big Bang. The right-hand side represents a coordinate system that expands with the universe. Credit: David Yates.

cosmology. When I remember my first reaction to hearing of CCC, I think of quantum pioneer Niels Bohr's statement upon learning about the neutrino: "We are all agreed that your theory is crazy. The question that divides us is whether it is crazy enough."[4]

CCC belongs to the early twenty-first century, but part of its inspiration can be found in a wood carving made back in the 1950s by the Dutch artist M. C. Escher. His *Circle Limit* series features a disc with interlocking angels and demons (see color plate section), appearing large in the center and becoming gradually smaller as the boundary

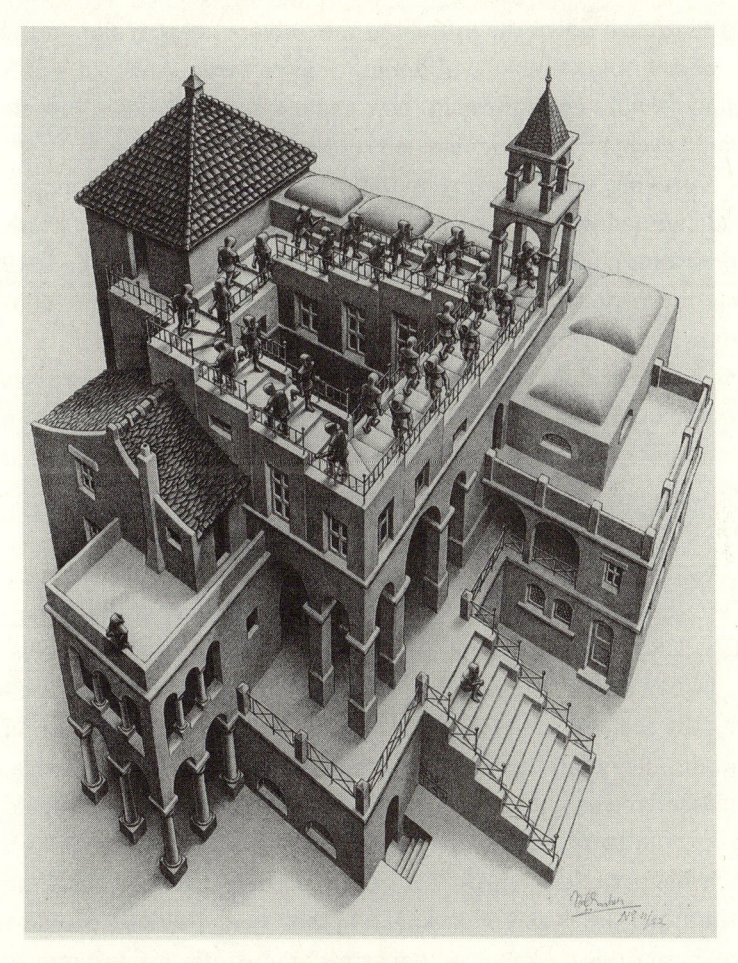

The Penrose staircase as depicted in *Ascending and Descending* by the artist M. C. Escher. Copyright: M. C. Escher Company—The Netherlands, 2024.

approaches. Technically it's a depiction of the hyperbolic plane, a strange mathematical world where infinity can be squashed down to a finite surface, and it's this very geometry that lies at the heart of relativity. Penrose has an intimate knowledge of Escher's art, as he and his father created "impossible objects," which Escher used in subsequent images. Perhaps the most famous example is the Penrose Staircase (featured in films such as *Labyrinth* and *Inception*), which depicts monks forever climbing a staircase that loops back on itself without increasing in height.

Penrose was suggesting that a conformal rescaling could similarly squash down the infinity of the universe's expansion in the same way that Escher had done in his pictures. And just as his monks would cycle through their endless ascent, so the universe would cycle through endless eons of expansion.

A pre–Big Bang state is something that many, including Penrose, had claimed was impossible. But he had always admired Dennis Sciama for changing his stance on the steady state when the CMB was discovered. "He said, 'OK, I was wrong.' . . . I have tremendous admiration for him. It's very unusual for people to say, 'The talks I've been giving to you are wrong; steady state is wrong, there was a Big Bang.'" Penrose would follow his mentor's footsteps but in the opposite direction. His work was thought to have proven that the universe had had a beginning; now he would give talks arguing it hadn't.

What excited Penrose so much about CCC was its ability to explain the puzzles concerning the low-entropy start of the expansion. The universe's entropy exists almost entirely in black holes, and when they evaporate, their entropy vanishes with them, resetting the scale in time for the next cycle. The far-future expansion mimics the inflationary epoch without the need for a hypothetical inflaton. And as a bonus, the transition would also create a new field that would acquire mass after the Big Bang and fill the universe with invisible particles. Penrose calls these particles erebons, after the primordial Greek god of darkness, Erebos. Such particles could be the mysterious dark matter that astronomers infer exists but have never seen.

But for CCC to work, Penrose needs to make at least two radical assumptions that my colleagues and I struggle with. Firstly, CCC assumes that in the far future, all matter decays into radiation—and while certain particles do decay, I see no evidence this is true of all matter. Of course, there is nothing wrong with making conjectures as long as we recognize them as such. Still, that assumption seems conservative compared to Penrose's second suggestion: Embedded in quantum mechanics is the unitarity principle, which implies that information is always conserved. This principle is often described using the example of a burning book; you might think that once the book is reduced to ashes, all the information is gone, but according to quantum mechanics it still exists even if we can't conceive of accessing it. Information conservation is a tenet of quantum theory, but Penrose was ready to sacrifice it in order to save Einstein's beautiful theory of relativity. CCC requires that information be destroyed. If information survives, universal entropy can't be reset for the next cycle. That black holes destroy information seems to be the inevitable consequence of relativity, as nothing can escape a black hole except Hawking radiation—which, at first glance, lacks information (we will come back to this point in chapter 9). But unitarity says that such information destruction is impossible; in quantum mechanics, information is always conserved. The conflict between quantum mechanics and relativity regarding the fate of black holes is known as the information paradox, or the firewall paradox—the latter name alluding to the possibility of observers burning up at event horizons. This paradox forces us to conclude that at least one of our deeply cherished theories has to be modified. The view among most of my colleagues is that relativity will be amended, not quantum mechanics, and information will be preserved in black holes. But Penrose is not alone in thinking it's the other way around. Many of my relativist friends also are unhappy with treating unitarity as dogma, and I try to keep an open mind. Penrose doesn't mince his words, saying that his critics simply have "faith" in the supremacy of quantum theory. But he would not focus his attention on the information paradox to persuade the wider community of CCC. Instead, he turned directly to the CMB for signs of the past eons.

In November 2010, a curious headline appeared from the BBC: "Cosmos May Show Echoes of Events before Big Bang."[5] The story quickly spread around the globe. Penrose and his Armenian collaborator Vahe Gurzadyan claimed that black-hole collisions in the previous eon had sent out gravitational waves that imprinted as circles in the CMB. But Gurzadyan was not the first scientist to look for such circles. Two years earlier, David Spergel had agreed to search for them on Penrose's behalf. Or, more specifically, Spergel had asked my friend and college classmate at Sharif University, Amir Hajian, to search for him. To Penrose's disappointment, Spergel and Hajian came back empty-handed. But in the fall of 2010, Penrose attended a celebration of the great Indian astrophysicist Subrahmanyan Chandrasekhar. There he met Gurzadyan, who deployed an alternative methodology and found the elusive circles. When Penrose and Gurzadyan got the referee reports for their monumental claims, he found that one was quite positive, one was fairly neutral, and one was dead against the idea. Norwegian astrophysicists Hans Eriksen and Ingunn Kathrine Wehus were two of the most outspoken critics of Penrose's CCC claims. In an interview with *Wired* magazine, they were asked to explain why they didn't see a signal when Penrose and Gurzadyan did. Their answer was blunt. Wehus said, "We do it correctly, and they do not."[6] Let me explain:

Enormous confusion persists about the claims and counterclaims of CCC advocates versus CMB cosmologists. So, let's clear one thing up: there do exist circles in the sky that have lower temperature fluctuations than others, as Penrose predicted. The issue is how to interpret them. What most of us think is that this finding is consistent with random fluctuations in the CMB, and so doesn't represent any significant evidence of a previous eon. (Remember the old adage, made famous by Carl Sagan: "Extraordinary claims require extraordinary evidence.") Look at any random series of data, and you will find patterns; it doesn't follow that they mean anything. An April Fools' online preprint in 2011 searched for evidence in the CMB for various facial expressions, concluding "that the CMB is sad, good and disapproving."[7] Whether this article was a satire on Penrose and Gurzadyan or on the remarkable appearance

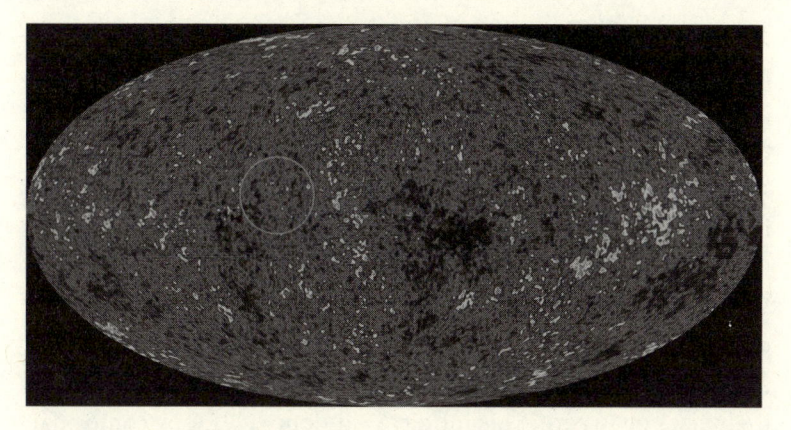

The cosmic microwave background, as detected by NASA's WMAP spacecraft (five-year data release), with Stephen Hawking's initials, "SH," circled. Credit: NASA / WMAP Science Team.

of Stephen Hawking's initials in some CMB maps (another freak random coincidence of noise) is left to the reader's imagination.

Despite the CMB community's doubts, Penrose would not be dissuaded. At the fiftieth anniversary celebrations of the CMB at Princeton in 2015, he was invited to present his results. He recalled, he told Phil, people in the audience saying, "You know, Penrose used to do good work. . . . now he's doing all this stuff about things before the Big Bang. Why should we be paying attention to it?" Penrose continued, "I felt really upset about this; they should give me a chance to talk." After the meeting, Penrose says, Canadian cosmologist Douglas Scott approached him, offering to redo the analysis. But he too found no evidence of a previous eon. The search for external validation seemed hopeless; until that is, Penrose was invited to Poland.

After an uprising and then complete destruction by the Nazis in 1944, Warsaw would exhibit its own cyclic behavior as the charming old section of town rose from the ashes and was rebuilt brick by brick. When Penrose arrived there to receive an honorary degree, his hosts invited him to tour the majestic Tatra Mountains region. His guide was Paweł Nurowski, an expert in conformal geometry and a former collaborator of Ezra "Ted" Newman, one of the giants of relativity. Newman expressed his confidence in Penrose's model at a conference in Oxford, by giving it impossibly higher odds of being true than any of its competitors, joking, "If I take the ratio of the likelihood of Roger's CCC model over any of the other [models of the early universe] I would say it's a number that is approaching infinity."[8] Nurowski, initially intent on shooting down Penrose's model, turned—with more research—from critic to evangelist. What persuaded Nurowski (which I sympathize with) is that the only part of geometry that goes singular (or blows up to infinity) at Big Bang is the conformal part; that shapes look the same but their sizes go to zero. This infinity is undone in CCC by conformal rescaling. Convinced of the model, Nurowski decided to search for its associated circles. When looking for structures in the CMB, one must distinguish between what should be there by chance and what should not. He needed to create a multitude of simulated maps and then compare them with the real one. But he had no idea how to do it. Fortunately, just down the corridor was Krzysztof Meissner, who used his considerable particle physics experience at the LHC to create a thousand simulated maps of the CMB. Initially the team thought they had found strong evidence for "non-statistical" (their lingo, for "Eureka!") rings in the CMB maps.[9] But upon peer review, they discovered the same effect happening in random simulations, thus undermining the claim that these rings are unique to the sort of sky we should see from CCC. As a result, their conclusions were softened in the published article.[10]

Later, Phil caught up with the outspoken Penrose critic, Scott, who claims that Penrose "talked to several different people, some of whom looked into [these circles in the CMB] but didn't find anything and told him that. And he wasn't prepared to listen to the answer, so it's very Donald Trumpy . . . I think without it being Roger Penrose it wouldn't have had any attention." But when analyzing the Polish group's claims, Scott admits, "They made a much more careful job of this than Gurzadyan did. And now the difference of opinion that we have with them is more subtle." The subtle difference is known as the look-elsewhere effect. We can explain this effect by thinking about Scott's license plate number, which happens to include his initials. But this inclusion is not necessarily some startling coincidence. His number plate could also have had his initials backward or his wife's initials or one of his parents' initials, and when you include all these possibilities you find that it's not so amazing after all. In other words, if you have a loose enough definition of *something unusual* and then you look in a large enough dataset, you are bound to find something unusual by pure chance. The critics alleged that Penrose and colleagues appeared not to have taken this point into account correctly. A more recent study using different techniques also failed to find evidence of CCC-inspired circles.[11] However, even Scott is happy to concede, "I don't think it disproves CCC because I don't think it was a prediction . . . [I]t was a sort of a concoction of, well maybe this would happen and maybe we should look into that." I agree with Scott here that, as far as I can tell, Penrose circles are not real predictions of the CCC framework, not until that framework actually gives an explanation of where the rest of the CMB fluctuations come from.

Scott has been at the forefront of CMB data analysis for over two decades, so it wasn't surprising that when Phil hosted a debate between Penrose and Guth about CCC versus inflation, it was Scott's papers that Guth kept referring to to knock CCC down.[12] It's clear that the CMB community remains largely dismissive of claims for empirical evidence of CCC, while most are persuaded that inflation did occur. As Scott puts it, "something like inflation is something like proven." However, as even Penrose agreed, CCC is something like inflation—as the exponential expansion in the future of our eon, driven by the cosmological constant, can be thought of as just inflation happening before the next Big Bang. Even Guth himself now says that inflation happened before the hot Big Bang, and so perhaps there is less of a gap between CCC and inflation than one might imagine. Although I'm skeptical of some of the details of CCC, I do think Penrose is right in his assertion that a conformal rescaling (or conformal *something*) might be key to understanding the Big Bang. It is perhaps his own "spectacular realization" that is yet to be broadly appreciated. In chapter 10, we'll see how I have recycled some of Penrose's CCC to build a new model of the Big Bang in which the universe becomes its own mother. But I'm not alone in following Penrose's quest to solve the mystery of our low-entropy beginning and finding a cyclic universe at the end of our journey. Two other physicists who did so, in particular, happened to live in the small college town of Chapel Hill in North Carolina.

THE PHANTOM MENACE

Just as CCC takes advantage of the observed accelerated expansion of the universe, so does the Baum-Frampton model, named after its creators Paul Frampton, a particle physicist, and his former graduate student Lauris Baum. Normally Frampton assigned topics to his students, but Baum kept quizzing Frampton about the mystery that had fascinated him as a student. It was the very same puzzle that had propelled Penrose to create CCC: the enigma of why the Big Bang had such low entropy when high entropy is always more probable.

During his discussion with Baum, Frampton realized that dark energy, the very thing that had supposedly killed off cyclic cosmology, might provide an answer to the entropy puzzle. Whether this energy of empty space is really a cosmological constant or not is still unknown, but if it is, the universe cannot re-collapse into what's called a Big Crunch (that is, a Big Bang in reverse). We cosmologists characterize the nature of dark energy using a parameter called w. If w is exactly -1, it suggests that the energy of empty space remains constant through time and space. Its repulsive gravity, acting as an eternal stretching force, will ensure the universe will continue to expand forever. If w is greater than -1, dark energy will decay and could lead gravity to act attractively again, as it does in the world of our familiar experience. One might expect that such a w-value is needed for a cyclic cosmology, as it could lead the universe to re-collapse. But in 2003, Robert Caldwell and collaborators wondered what might happen if w were less than -1.[13] This scenario, they claimed, would lead to "phantom dark energy," implying an unstoppable repulsive force. The energy density of the universe would relentlessly increase, becoming infinite in a finite time. It would act like a merciless destroyer of matter; the universe would be split into countless causal patches disconnected from each other. This phantom menace, as they called it, would tear galaxies, stars, planets, and even the very atoms that form them apart in what is known as the Big Rip. It was while pondering this violent end to the cosmos with Baum that Frampton says "the really brilliant idea" emerged. "I fell out of my chair," he said. "That's the universe; our universe is going to contract out of it. Now this solves an enormous number of problems because the typical patch, almost all the patches, will have no matter and no black holes, just dark energy, so it's empty." Baum and Frampton called this idea the come-back-empty principle. Using modifications to Einstein's equations, the pair showed that these empty patches of space, created by the Big Rip, could bounce back to form new universes empty of matter and radiation. Crucially then, these new universes would have no entropy. Thus, the low-entropy start of our Big Bang universe would not be some bizarre coincidence. Instead it would be the inevitable result of the slate having been wiped clean, the new universe birthed from an empty patch, torn from the previous universe during the Big Rip. Structures would be generated only after the bounce by inflation. Frampton claims the model is "inspired speculation," as the come-back-empty principle is, in his view, the only way to really solve the low-entropy mystery of the universe and allow for an eternal cyclic universe. Of course, Penrose says the same about CCC.

One thing that puzzled me about this model is that in it, just as the universe is undergoing increasingly violent expansion, the empty patches suddenly bounce. Such a turnaround has to be physically well motivated, and my first suspicion was that this bounce was put into the model by hand and therefore jury-rigged to create the outcome the authors wanted. However, as I read their paper, I could see some motivation for it in corrections to Einstein's equations inspired by brane-world scenarios—and, in fact, the same modification lies at the heart of the loop quantum cosmology models.[14]

Of course, no one can create a radical new theory without receiving a backlash from rival cosmologists, and soon after the model was published, Xin Zhang claimed that the casual patches would actually reconnect before the bounce, ruining the come-back-empty principle.[15] When we put Zhang's criticism to Frampton, it seemed to us that he was not sympathetic. "It's mathematically wrong," he said. "He made mistakes in his mathematics. My advice to the editors was to say it's a waste of paper to publish either his comment or my response." While I personally don't agree that cordial scientific debate is a waste of paper, I do agree that—as far as I can tell—Zhang's criticism is wrong: it confuses the physical definition of a causal patch.

After first proposing the model, Frampton worked with another graduate student, Kevin Ludwick, to see if it could also apply to a model known as the Little Rip, in which the energy density goes to infinity only after an infinite amount of time. Ludwick showed that the Little Rip wouldn't be a problem for the model but eventually concluded that the basic premise was flawed, claiming the patches would not necessarily be devoid of all matter. So, entropy would rise between each cycle. While tension certainly arose between the two physicists, Ludwick described Frampton as a good advisor who cared about his progress. Nevertheless, as he was entering his final year, Ludwick found Frampton's usual attention waning. One meeting seemed forever postponed, and eventually Chapel Hill staff started to ask about Frampton's whereabouts, but nobody knew where he was. The reason for Frampton's absence turned out to be something Ludwick could never have imagined in his wildest dreams.

In 2011, Frampton began an online relationship with someone using the name Denise Milani. Following months of messaging, the two agreed to meet in Bolivia, where Milani, a bikini model, was supposedly on a photo shoot. But Frampton was delayed, and Milani did not show; instead, she said she had to fly on to Belgium. She would send a ticket for him to join her via Argentina. Frampton showed no animosity and even agreed to her request—to carry with him a suitcase that Milani had apparently left behind. It had none of the designer trappings one might imagine of an international model, but instead was plain and dull. He took the bag to Argentina, where he was told Milani would arrange a flight so he could join her in Belgium. But no ticket came. After more than two weeks of this wild-goose chase, Frampton headed home to the United States with Milani's suitcase in hand, hoping she might visit him to retrieve it. You can probably see where this is going; the suitcase did not belong to Milani, nor had she ever corresponded with Frampton. The luggage actually belonged to South American drug smugglers posing as Milani, and inside it were two kilos of cocaine, a fact Frampton claims he only discovered once he was caught with the drugs in Argentina. He was arrested and eventually sentenced to almost five years in jail, where he witnessed the harsh conditions of South American prisons and even a brutal murder.[16]

Back at Chapel Hill, Ludwick said he was "kind of blown away." "I felt bad for him," Ludwick says, "I prayed for him. . . . I did what I could to support him." Ludwick was soon in a meeting where staff members were advising students on their future. He recalls, "Someone asked a question and said, 'What if my advisor is taking a sabbatical and I just can't reach them?' And I thought about raising my hand and saying, 'What if my advisor is being detained in a South American prison?'" Even from jail, Frampton continued to supervise students by phone, and Ludwick would send the latest physics papers on CD to Frampton, who had access to a computer but no internet. His university put Frampton on unpaid leave and later fired him. However, in 2015 the North Carolina court of appeals ruled that the university had violated its own policies and thus was obligated to compensate Frampton for his back pay.[17]

Later, Frampton went into house arrest, and I remember him joining a physics conference in Miami remotely via Skype in December 2013. I was then surprised to see him in person at the same conference the following year (after an early release), where he was advertising his new book, *Tricked!*, an account of his ordeal. Since being released from prison, Frampton has continued to work on his cyclic model and has made significant changes. Crucially, he has argued that the come-back-empty principle can be realized without any phantom dark energy and requires w to be almost exactly equal to minus one. He's also come to believe that the model might not require inflation. "I'm allowed to change my mind, aren't I?" he protested in an interview when we challenged him on the subject.

From one perspective, dropping phantom dark energy from the model might be seen as an advantage, showing that the come-back-empty principle is more generic than first thought. But it also removes a key observational constraint on the model. Discovering that dark energy really is a phantom would not have proven the Baum-Frampton model, but it would at least have raised my confidence in it, as most of my colleagues assume that dark energy is simply Einstein's cosmological constant. Without such a development, I see little prospect for discriminating between the Baum-Frampton model and other cosmologies. As Frampton himself admits, any cyclic cosmology, be it his own or CCC, will have an expansion history that is so similar to the standard model it will be very difficult to tell them apart. However, I'm not sure all is lost; perhaps we could test the modified equation the model uses to predict the evolution of the universe. Others have built their own cyclic cosmologies that require this violent end to the matter in the universe. We might even think now of a wide class of models under the bracket "phantom bounces" (originally proposed by my friends Will Kinney, Katie Freese, and her student Matthew Brown), although according to Frampton's latest thoughts, his model no longer belongs in this category.[18] So even as the idea of phantom cycles might have died in Frampton's mind, it has been resurrected in the model building of others. In more recent years, Frampton has remained focused on particle physics, an area that recently made a discovery that not only might imply another version of a cyclic universe but also threatens our very existence.

TUROK'S FAREWELL

In 2012, CERN announced that their 24-kilometer particle accelerator had detected the long-sought Higgs boson. This particle is an excitation of the Higgs field, which bestows fundamental particles with mass. After the elation of discovery, physicists noticed something strange about their new particle. The mass of the Higgs boson seemed to imply that the amount of energy in the vacuum of empty space was only metastable, like a vehicle balancing on a cliff edge.

Just as a gust of wind could knock such a vehicle to its doom, so a large enough fluctuation would lead our universe to transition to a lower energy state. Such a transition would release an enormous amount of energy (imagine a nuclear blast that only gets stronger as it expands), obliterating everything in its aftermath. As beautifully depicted in the movie *Oppenheimer*, the first atomic-bomb scientists worried that they might create a chain reaction that would ultimately ignite the atmosphere and destroy the world. A vacuum transition would act similarly but, in this case, annihilate the entire universe. However, after this cosmic cleansing, it's conceivable that the cosmos could begin anew.

In 2008, Neil Turok became the director of the Perimeter Institute. He was already good friends with Paul Steinhardt, from their time at Princeton and their work on the Ekpyrotic model, so Steinhardt became a regular visitor. Around the same time, Itzhak Bars, a string theorist known for his peculiar passion for two dimensions of time, was on sabbatical at Perimeter. I remember the three of these physicists discussing their different visions of the early universe over lunch at Perimeter's famous Black Hole Bistro. Following the 2012 discoveries at CERN, they converged on an idea: the "Higgs Bang," which suggests that this metastable Higgs could create a contracting era, thus leading to a cyclic cosmology.

But just like the vacuum energy of the Higgs field, Turok's enthusiasm for the Higgs Bang was only metastable. Maybe, as his collaborator Jean-Luc Lehners put it best, this tenuous excitement was simply due to the fact that Turok has "a hundred ideas a day" and had become more fascinated by an even newer idea, that the universe was not cyclic but underwent a one-shot bounce leading to the hourglass structure we have seen before. The twist in the new idea (known as the mirror universe) is that the hourglass now has a vanishing neck, reaching zero size in the middle, and is thus singular (akin to Hawking and Penrose's theorems), though time runs forward on either half of the hourglass, reversing direction at the singularity. The proposal is a more precise realization of Julian Barbour's Janus point, which we discussed in chapter 3, and it embraces a singular origin of time. Steinhardt and Turok's partnership, the focus of the anti-inflationary rebellion, was no more. Turok would leave Perimeter amid the pandemic to lead the newly established Higgs Centre for Theoretical Physics at the University of Edinburgh. (His books are still left behind next door, while I am writing these words.) As we say in Persian, two beggars can sleep on a rug, but two kings can't fit in a land. Steinhardt would find another collaborator from Europe, someone who lacked Turok's experience but more than made up for it in enthusiasm and fresh insights. Together they would rebuild the Ekpyrotic model that had tried to overturn inflation, but this time they'd do it without appealing to exotic speculations like branes colliding in higher dimensions.

EKPYROSIS UNPLUGGED

When I was a PhD student at Princeton early in the first decade of the 2000s, the inflationary pioneer Slava Mukhanov decided to visit for a one-year sabbatical. I recall he told me that he was going to work with Steinhardt on a cosmology textbook. However, Mukhanov ended up writing the book on his own, I presume because the two men couldn't see eye to eye, a dynamic that would escalate exponentially in the following years. A decade later, Mukhanov took on a new physics PhD student, Anna Ijjas, who had already received a doctorate in philosophy. Philosophers have a great tradition of finding counterarguments to any established position. Perhaps that's why it's often said that if you had three philosophers in a room, they wouldn't agree that there were three philosophers in a room. They might debate what is meant by a *room* or a *philosopher*. This difference in approach may also be why scientists are less familiar with the criticisms of inflation than they should be. With her philosophical training in hand, Ijjas sought counterarguments to inflation and found a long-forgotten paper that even its author, Roger Penrose, said "nobody [had] paid any attention to." The paper claimed that inflation could not arise from generic initial conditions. Eventually, Ijjas would present the issue at Harvard, in front of another of the theory's founders, Paul Steinhardt. As he recalled in an interview with Phil, "She gave a seminar on inflation and I came to it, really ready to give a lot of trouble." But to his surprise she quickly turned to the problems of the theory and when he posed advanced questions, "she answered them all. And then I thought, *who is this person? She's an advanced postdoc of some sort or she's an assistant professor?* No, I learned she was a graduate student," Steinhardt said. The confusion had arisen because the speaker had been introduced as Dr. Ijjas, but Steinhardt did not know that her doctorate at that point was in philosophy, not physics.[19] That a physics graduate student in her first week at Harvard was able to offer such penetrating insights amazed Steinhardt, and their meeting marked the beginning of a collaboration that would rewrite his earlier attempts at a cyclic universe.

With the LHC's failure to find supersymmetric particles, fundamental physics went into a crisis mode. While superstring proponents still hold hope that these predictions of their favorite theory may yet be discovered at higher energies, others became more dismissive of the framework. An association with string theory came to be seen, in some circles, as more of a burden than a blessing. Physicists had always thought string theory could resolve the singularity, and Steinhardt had hoped his Ekpyrotic model, which—as we have seen—envisioned the Big Bang as the collision of higher-dimensional branes, might succeed in doing so. To add insult to injury, it never did. And so eventually Steinhardt and Ijjas sought to detach the Ekpyrotic model from its string theory roots. If the pair were to throw away the inessential baggage of colliding branes and keep the core principles of a slowly contracting cyclic model, they would have to rethink what it was that made its rival, inflation, so successful in the first place.

To understand how inflation seems to solve so many cosmological problems, let's return to the standard analogy of the universe inflating like a balloon. During inflation, the balloon expands rapidly but the horizon (basically how far we can see) barely changes, so everything looks like it's becoming flatter. For Steinhardt and Ijjas's cyclic model, the balloon slowly contracts but the horizon shrinks quickly. So, although the universe is getting more curved, the amount of curvature you can see shrinks. What is important is the relationship between how fast the balloon is changing size versus how much of the balloon an observer can see. And the argument is that a shrinking horizon would be just as efficient at throwing out irregularities like cosmic curvature and monopoles as a rapidly expanding universe would be. This equivalence (or duality) was first noted by my colleague Latham Boyle while he was studying for his doctorate at Princeton under Steinhardt. The two, collaborating with Turok, argued that a bouncing universe would therefore have no need for inflation. Later, Turok hired Latham at Perimeter, back before the two Ekpyrotic heavyweights parted ways. Not everyone, of course, is convinced that a cyclic universe can replicate the success of inflation. We put these points to the father of the theory, Alan Guth, and he replied, "If they can make a model like that work then that's a viable alternative to inflation.... This period of slow contraction has to be followed by a bounce, and the bounce is the hard part." He's also skeptical about claims regarding the cyclic model's ability to reproduce the right patterns seen in the CMB, recalling that "they insisted that their original [Ekpyrotic] model produced the right spectrum of density perturbations, just like inflation, and immediately several papers were written arguing that that just wasn't true—and about ten years later or so, Steinhardt admitted it wasn't true." It was indeed the case that the original model had identifiable problems, and just like inflation, the Ekpyrotic scenario (through the blood and sweat of many of my friends and peers) has gone through several iterations to reach its current stage.

Roger Penrose, the figure who inspired Ijjas on her journey to find an alternative to inflation, is no fan of bouncing cosmologies. A fast-contracting universe, he argues, would end in a tumultuous coagulation of black holes and irregular regions. In short, it would become a mess that looks nothing like our smooth, homogeneous universe. Steinhardt admits that he had never heard of the problem before early in the first decade of the 2000s, when he started thinking about bouncing cosmologies; he admitted to Phil that his ignorance on the matter was related to the fact that he had "never taken a cosmology course." But to his surprise, the cyclic theory itself already had an answer to the problem. In order to discover if the fast-contracting-universe result was the same for the slow contraction assumed in their model, Ijjas worked for almost four years writing the code that would simulate a slowly contracting universe. With Steinhardt and leading numerical relativists (scientists who use sophisticated algorithms to solve the challenging equations of Einstein's theory on supercomputers) as collaborators, they were finally able to see whether a bounce could reproduce the success of inflation. Their results showed that a contraction that bounced into an expansion would not be plagued by the pandemonium that Penrose and others envisaged but instead could result in our smooth, homogeneous universe. Just as inflation is modeled as being driven by a hypothetical field that permeates all of space during the exponential expansion, a bit like the Higgs field does today, so cyclic theorists also imagine a new field that drives the evolution of the cosmos. But unlike inflation, this field is assumed to be still present and to drive the accelerated expansion that was discovered at the end of the twentieth century. This field is what astronomers describe as dark energy; however, in this framework, it is not a cosmological constant but slowly changes its value over time. And, crucially, it can alleviate the messy contraction that would otherwise destabilize a cyclic universe. In the slow-contraction model, dark energy is an evolving field with negative pressure in our era but positive pressure in the contracting era. This new pressure term ensures the contraction is slow enough to make the model robust to initial conditions. But adding hypothetical fields was a development that made some cosmologists uncomfortable with inflation. And the new cyclic scenario could not escape the same issue.

A novel feature of Ijjas and Steinhardt's model is that they claim to avoid using quantum gravity to describe the beginning of our expanding universe. Steinhardt had previously argued that quantum gravity was essential, but he is now abandoning the idea because the bounce is assumed to happen at densities well below the Planck density. To me, that sounds like wishful thinking; to realize this classical bounce, Steinhardt and Ijjas had to break a taboo, namely violating the null energy condition (that observers cannot see negative energy densities). Violating this condition is often considered a problem, as doing so could lead to violent instabilities that would prevent the universe taking on the smooth, homogeneous form we observe in the CMB. Most of my colleagues would agree that quantum mechanics and relativity have to be unified somewhere, and the singularity theorems of Penrose and Hawking tell us where: in the heart of a black hole and at the Big Bang. Trying to resolve these mathematical oddities without quantum gravity may end up being the Achilles' heel of some of these new cyclic models.

The gentle process in Ijjas and Steinhardt's slowly collapsing universe would generate negligible primordial gravitational waves. If we were to detect such waves, the model would be ruled out. That's a falsifiable prediction, which is great. But a prediction for something we won't observe is always less impressive than a prediction of something we will. Cosmologists are impressed with inflation because it seemed to predict features such as a flat universe, which were confirmed only by later observations. Cyclic proponents will need something similar if they are going to convince us to abandon not only inflation but also a quantum Big Bang. While the computer simulations that Steinhardt and Ijjas have been running show no primary gravitational waves (which come from quantum fluctuations of geometry), there is a hint of something subtler: what Steinhardt calls secondary gravitational waves, emitted by collisions of lumps of the hypothetical field they use to get the bounce in their model (much as colliding black holes emit gravitational waves seen by LIGO). These waves possess a very different spectrum to those of the sort of primordial gravitational waves generated by simple inflationary models. They should not show up in any near-term experimental searches for gravitational waves, but the ambitious projects on the drawing board that we shall discuss in our final chapter might just have a chance to detect them. Another possible experimental probe is to search for changes in dark energy; even the smallest decay over time would be a tremendous discovery that could (yet again) shake the world of cosmology to its core and might provide evidence for cyclic behavior. Lastly, new measurements of the mass of fundamental particles might be able to confirm whether our vacuum really is metastable; if it is, it will be able to tunnel to a new vacuum compatible with a collapsing universe, a plus for cyclic models. Metastability might also, however, mean that our entire universe could be wiped out at any moment, so there are swings and roundabouts to the possibility (ups and downs for our American readers).

Without some new experimental data, it will certainly be a struggle for cyclic theorists to challenge the hegemony of inflation. And the reasons for this tribulation may not be entirely scientific. Ijjas has had to face the challenge of arguing for a minority opinion in the male-dominated world of physics. Steinhardt has also stressed the need for the community to promote more diversity, not just in its gender and racial makeup, but of its scientific thought. Cosmology, he claims, is dominated by three groups. One is his own small renegade band of inflationary skeptics, who consider the multiverse to be that model's fatal flaw; another consists of inflationary multiverse proponents; and the third, perhaps largest, group is experimentalists and observers, who focus on testing inflation, which Steinhardt says they do "without critically examining whether those tests make sense now that it has become clear that inflation produces an unpredictive multiverse."

THE ENTROPY PARADOX ONCE MORE

A key focus of cyclic proponents is the low-entropy origin of the Big Bang and the belief that inflation cannot solve the problem. According to Steinhardt and Ijjas, when the universe contracts, the entropy is thrown out of the horizon, allowing the next cycle to begin with entropy reset. The ejected entropy has to go somewhere, though, and a forever-expanding universe can provide a dumping ground for it. For this proposition to work, the universe must expand more than it contracts in each cycle, so that the dilution in entropy density due to additional expansion can cancel out any new generation of entropy. One complaint I often encounter is that this state of affairs leads to a universe that shrinks into the past and eventually reaches a starting point of zero size, negating the motivation for the cycles to begin with. But if the universe is infinitely large, then it could shrink into the past without ever reaching a point of zero size. This idea may seem counterintuitive. But infinity is not simply a really big number, it's an unbounded set. Taking all the odd numbers away from the unlimited set of all numbers still leaves all the even numbers left, and there are infinitely many of them as well. So even infinity minus infinity can still yield infinity. By analogy, an infinite universe that shrinks into the past can stay infinite. Like quantum mechanics, this notion messes with our intuitions. For the domains of infinite sets are not the worlds of everyday experience that shaped the evolution of our minds. So we have no right to trust our intuitions in either the quantum or infinite realms.

It should not come as a surprise to learn that cyclic advocates believe the universe has always existed. But what a lot of people don't know is that the father of inflation himself agrees. Guth argues that if we trace the eternally inflating multiverse back in time, we will inevitably come to a region of low entropy, and on the other side of that region will be another domain of inflating space but with the arrow of time reversed, much like Janus universe models. However, some philosophers have argued that this hourglass universe still has a beginning, as the middle point where the entropy is lowest marks the origin of time. In fact, they frequently cite Guth to prove the universe had a beginning. He, however, says, "I often call the middle the logical beginning in the sense that that's where we begin our description. If somebody wants to leave off the word *logical*, I would add it back in." Think of particles of gas in a box. These particles are subject to random motions, and so it's possible, although very unlikely, that they could all suddenly find themselves in the corner of the box. This arrangement would be a low-entropy state, but it would not be the beginning of the particles. Similarly, the low-entropy state in the middle of the universe's evolution does not represent its origin. Ironically, this new thinking was motivated by Penrose's critique of inflation.

As one of Guth's former collaborators, Sean Carroll, explained, "Penrose's arguments are fascinating because he was basically critiquing inflation before inflation was invented." In particular,

Penrose and Carroll claim that inflation makes the entropy problem even worse, as in that model, entropy is even lower before inflation than it is in the standard cosmology. In order to really solve the problem of why entropy was so low at the Big Bang, Carroll sought a way to link the far future to the remote past. But such a scheme need not follow Penrose down the cyclic path; instead, Carroll (with his co-author, Jennifer Chen) suggested that there could be quantum fluctuations that cause baby universes to pinch off from a mother universe and then undergo inflation to make them expand and cool—giving us the universe we see. As Carroll explained, "There in my mind, that's where inflation is actually doing something really, really important," as it can rapidly grow small low-entropy seeds into new universes that look just like ours. These baby universes would be born with low entropy, but the mother universe's entropy always increases because it *can* always increase—and so an eternal universe need not defy the second law of thermodynamics.

That the universe had no beginning is a view shared by yet another inflationary founding father, Andy Albrecht. Having helped pioneer the alternative paradigm of varying speed of light, or VSL

An artistic impression of the Carroll-Chen model, explaining how a mother universe—the middle region—can give birth to new universes with low entropy that evolve like ours. As in the Janus universe, observers on either side of the middle believe the opposite region is in their past. Credit: Jason Torchinsky.

(which we will explore later), Albrecht now seems to have pitched his tent firmly in the middle between cyclic theories and inflation. He was, like Ijjas and Carroll, inspired by Penrose's essay highlighting the mystery of the universe's low-entropy state. But it was another, even older source that really put the issue into focus for him. Ludwig Boltzmann was one of the founders of thermodynamics, mocked by fellow scientists for his belief in "unobservable" atoms. Boltzmann realized that while entropy almost always increases, driving the universe to a lifeless equilibrium state, a rare fluctuation toward low entropy was inevitable given enough time. Our universe may have been born from one such fluctuation. But a single galaxy would seem more likely to fluctuate than a whole universe, and even more probable would be just a solar system. As a reductio ad absurdum, some suggested that just a brain would be the most likely object to fluctuate into existence. Since the cosmos is not dominated by these "Boltzmann brains," clearly our universe could not have been born by such a random accident. But Albrecht argues that in inflation, things are different. Instead of a whole ready-made universe, filled with stars and galaxies fluctuating into existence, the vacuum of empty space could transition into a subatomic inflationary seed that would grow into the grandeur of the universe. And surely such a seed would be more likely to form than a complicated brain. After this tiny seed brought our universe into existence, there would be time for evolution to slowly form complex structures without the need for improbable fluctuations. Every person, planet, star, and galaxy would enjoy a brief life but eventually fade from existence, and the universe would return to the vacuum equilibrium state from which it came. A baby, fusing back with its mother, to begin the process anew. Albrecht called this model de Sitter equilibrium cosmology, after the Dutch physicist Willem de Sitter, who first described a universe filled with nothing but vacuum energy. Unlike the Carroll-Chen model, where the baby universes never rejoin with the mother, de Sitter equilibrium is a cyclic cosmology (though its cycles don't necessarily follow regular intervals like the seasons but are more like economies that have cycles of boom and bust). The way I like to think about de Sitter equilibrium cosmology is as eternal inflation, turned upside down. In eternal inflation, most of the time the universe is inflating at a very high expansion rate, with rare excursions to a Big Bang universe, where we can come to exist. In contrast, in de Sitter equilibrium, most of the universe is empty with dark energy, with rare excursions that kick-start a high-scale inflation that we need to explain the CMB.

In this chapter, we saw how cyclic cosmologies of ancient cultures have been reincarnated in the modern language of twenty-first-century physics. We have Penrose's forever-expanding CCC, Baum and Frampton's come-back-empty principle, phantom bounces, Steinhardt and Ijjas's bouncing universe, the Higgs Bang, and Albrecht's de Sitter equilibrium model. What I like about CCC and bouncing cycles is that, according to their proponents, they can be falsified by the discovery of primordial gravitational waves. Baum-Frampton and phantom bounces should also be refuted if dark energy is found to decay; since our future is our past, testing cosmology today can teach us about our origins, which is truly amazing. But I would like to see a deeper connection of the models

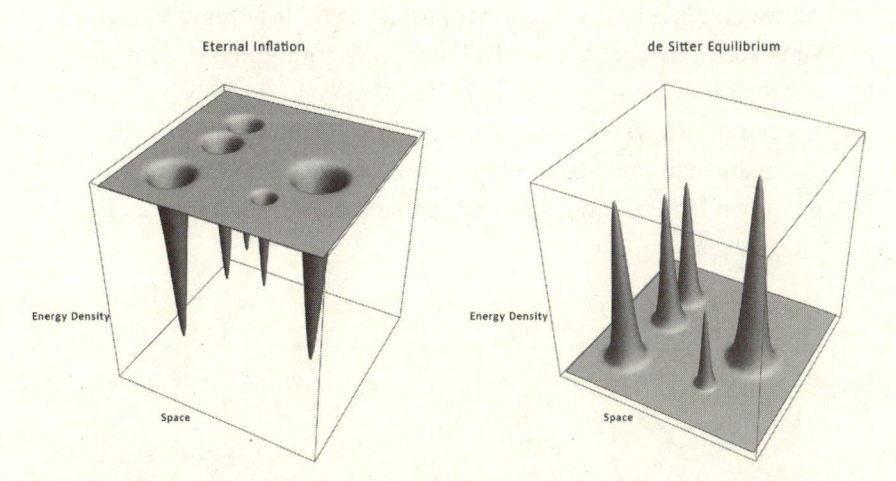

In de Sitter equilibrium cosmology (*right*), the universe is mostly in a low-density "de Sitter space" (dominated by dark energy), with the exception of rare upward excursions of energy density to a hot Big Bang. In eternal inflation (*left*), the universe is mostly in a high-density inflating phase, with rare excursions downward in energy density when inflation ends and a hot Big Bang ensues. Every cone is a snapshot in Big Bang cosmology. Credit: Niayesh Afshordi.

to more fundamental physics, and other ways of testing the ingredients of these proposals. What would be truly wonderful is to understand the nature of dark energy at a more fundamental level. If it's really a cosmological constant, CCC and de Sitter equilibrium would be in, and phantom bounces and slow contraction would be out. But then we would need some other independent evidence for those models still remaining on the table. The circles in the sky of CCC remain unconvincing to me and my colleagues. The surviving suspects should ideally compute the temperature variations in the CMB sky, but to my knowledge, they haven't done so. I shall describe my own attempt at this computation in chapter 10, where I push the whole cyclic theme to a different level. But cyclic cosmologies do have the potential to explain the deep mystery of our apparently improbable beginning. Ironically, the puzzle of the Big Bang's low entropy was thought to rule out an eternal universe, but recently, the pendulum has begun to swing in the opposite direction. I find it striking to think of how many of the leading architects of today's cosmology now believe the universe to be eternal into the past. It's a concept that was thought to be dead but is now arising from the ashes; for ideas have a habit of reproducing themselves just as universes do in the cyclic scenario. But reproduction and repetition are not the same thing. For another class of models, the universe does not retrace its past but explores a continually branching structure, giving birth to progeny universes through a process that is forged deep inside that other place where singularities supposedly lie: black holes.

7

BORN FROM A BLACK HOLE

November 1, 1755, was a busy morning for the people of Lisbon. The Catholic society was gathering in the city's cathedrals to celebrate All Saints' Day. Candles were lit and flowers hung. But as services began, a monster was stirring in the Atlantic. A massive slip of the Azores-Gibraltar fault line would trigger one of the most violent earthquakes in modern history. As the story goes, the flaming candles were tossed into the decorations of dried flowers, creating a firestorm that scorched the city. A 9-meter-high tsunami only compounded the devastation. While the Lisbon earthquake was an unprecedented tragedy, one positive result was that it marked the beginnings of modern seismology. At the heart of this movement was perhaps one of the greatest scientists you have never heard of. John Michell was born in England on Christmas day, 1724. He was educated at Cambridge University, where he taught subjects as diverse as Greek, Hebrew, theology, arithmetic, geometry, and philosophy. We have no images of what he looked like, but the diarist William Cole described Michell as "a little short Man, of a black Complexion, and fat."[1] After the great Lisbon earthquake, Michell theorized that these mighty convulsions come in waves of energy with elastic compression; he was able to identify where the epicenter was and correctly suggested that the tsunami was caused by movement of the undersea floor. But Michell's foresight was not restricted to geology. He made so many advances, from creating the technology for measuring Newton's gravitational constant to discovering double stars and pioneering astrostatistics, that his works have been described as sounding "like they are ripped from the pages of a twentieth century astronomy textbook."[2] But if Michell is to be remembered for just one discovery, it should surely be his realization that stars with a large enough mass would create so much gravity that not even light could escape. They would stop shining into the heavens and instead become objects of utter blackness; he called them "dark stars."

Although Einstein had unveiled the fundamental equations of relativity in 1915, he struggled to find exact solutions to them. The first to do so was Karl Schwarzschild, whose extraordinary talents enabled him to have two scientific papers published before he turned sixteen. Although he was forty years old when the First World War broke out, Schwarzschild volunteered to serve in the German army, keen to show that Jews could be German patriots. Amongst the death and destruction, Schwarzschild began to play with Einstein's theory. His work implied that any object compressed tightly enough into a sufficiently small space could become something akin to Michell's dark star. For the sun, we would have to squeeze its radius down to about 3 kilometers. Since the sun has a radius of 700,000 kilometers, it shines a beautiful yellow light into space, a radiance that life on Earth depends upon. This special distance is known as the Schwarzschild radius, an ironic moniker as his surname can be translated into English as "dark shield." However, Schwarzschild refused to believe that any object could become so compressed, describing his calculations as "clearly not physically meaningful."[3] Tragically, he wouldn't live to see how visionary his ideas were, dying of a rare skin disorder, aged just forty-two.

Almost two decades would pass before the issue of dark stars would resurface. On a boat from India to Britain, a young student named Subrahmanyan "Chandra" Chandrasekhar, whom we met earlier, began to apply the principles of relativity to the fate of stars. He then discovered something very surprising: any cold object bound by gravity and made of matter must be less massive than a maximum limit, or otherwise its particles would have to fly faster than the speed of light. This upper bound is known as the Chandrasekhar limit. If the core of a star runs out of fuel to burn, it starts to cool down, but if its mass exceeds Chandra's limit, then it has no choice but to succumb to gravitational pull and collapse into who-knows-what.

Chandra came from a family of intellectuals. His mother had translated famous European novels into Tamil. The year he arrived in Cambridge for graduate studies, his uncle, Sir Chandrasekhara Venkata Raman, became the first Indian person to be awarded the Nobel Prize in physics. It was not surprising then that Arthur Eddington, who had gained fame as the man who had proved Einstein right, took an interest in Chandra's work. He sat on Chandra's PhD oral defense, but it soon degenerated into a farce as Eddington and Fowler (Chandra's supervisor) spent most of the defense arguing with each other.[4] These fights continue today when scientists meet who have strong (but conflicting) convictions and who enjoy an audience. By 1935, Chandra presented his findings to the Royal Astronomical Society, but Eddington launched an ambush, rubbishing the young Indian's conclusions in a surprise follow-up talk, saying, "I think there should be a law of Nature to prevent a star from behaving in this absurd way!"[5] At a subsequent meeting, Chandra recalled that Eddington made his work "into a joke. I sent a note to Russell [Henry Norris Russell, who was presiding], telling him I would wish to reply. Russell sent back a note saying, 'I prefer that you didn't.' And so, I had no chance even to reply; and accept the pitiful glances of the audience."[6] Four years later, Robert Oppenheimer (the father of the atom bomb) and his student Hartland Snyder further developed the theory of gravitational collapse and showed that a star with a sufficiently large mass that had run out of nuclear fuel would implode in on itself, not just becoming dark, but compressing to infinite density, pressure, curvature, and temperature: a singularity (in the Oscar-winning movie *Oppenheimer*, he proudly hands out copies of this article to his students).

As we saw in chapter 1, these early singularity theorems assumed the star was perfectly symmetrical, so the results were not taken seriously. It would not be until the 1960s that Roger Penrose would prove that the unrealistic assumption of perfect symmetry didn't matter; the center of dark stars represented a singularity where physics as we know it would eventually come to an end. This discovery was both profoundly puzzling and incredibly exciting. As John Wheeler said, "No field is more pregnant with the future than gravitational collapse, no more revolutionary views of man in the universe has one ever been driven to consider seriously than those that come out of pondering the paradox of collapse, the greatest crisis of physics of all time."[7] At a conference in 1967, Wheeler asked the audience for a catchier term than *gravitationally completely collapsed object* and someone shouted, "How about *black hole*?" The name stuck.[8] Michell's work had long been forgotten, but ironically, it was the subject of his original research on binary stars that finally confirmed that black holes were real. Astronomers observed stars orbiting around invisible companions with enormous masses, which in retrospect, were sure signs of black holes. Less than half a century later, my colleagues combined radio dishes across the globe to make an Earth-sized telescope that would yield images of the silhouettes of monster black holes at centers of galaxies. But the fact of black holes' existence could not answer the mystery of what lay at their center.

Penrose had developed powerful proofs that a singularity must form within a black hole. But any proof is only as good as its assumptions and I am not alone (arguably, not even in the minority) in wondering if one of those assumptions might be relaxed, revealing something even stranger hiding in the darkness. Both a black hole and the Big Bang have singularities, according to general relativity and to the Penrose-Hawking theorems. For the brave cosmologists who explore the frontiers of knowledge, this fact is a clue, a subtle suggestion from nature that these two mysteries of physics may be different sides of the same coin. If the singularity is to be resolved at the Big Bang, then it should also be resolved in a black hole, perhaps implying that these hitherto-thought-of cosmic destroyers have a motherly side to them, giving birth to new universes.

In 1972, Raj Pathria (an Urdu poet and physics professor who used to sit a couple of doors down from my office) and I. J. Good (a mathematician who had worked with Alan Turing at Bletchley

Park in the cracking of the German Enigma codes) both speculated that black holes might generate new regions of space. Good was trying to find a way to reconcile the steady-state model with the Big Bang and imagined that the continual creation of matter in the former model could be realized if black holes gave birth to new universes. He didn't only work on black-hole singularities; he imagined artificial intelligences so advanced that they make ever-improving versions of themselves, leaving humanity in the dust. Ironically, this idea has become known as the technological singularity (perhaps the rise of large language models, made famous by ChatGPT, is a sign of that vision becoming reality). Pathria also questioned whether the universe was unique, reasoning that if our cosmos were birthed from a black hole, others would surely be created. As we saw in the case of inflation, it seems hard for Western scientists to develop ideas without finding that a Russian beat them to it. Sure enough, as early as the 1960s, Igor Novikov also proposed that black holes birthed new universes, a spark of an idea that was later developed in the 1980s by inflationary pioneer Slava Mukhanov and colleagues. The basic hypothesis was that the singularity would be replaced with a bounce, but as matter cannot escape a black hole, it must create a new space-time region. Quantum gravity pioneer Bryce Dewitt was fascinated by these proposals and encouraged his associate Lee Smolin to pursue the idea, saying, "This is something I think about a lot. See if you can do something with it." What Smolin would do was not simply create a model of a black-hole Genesis but also challenge what would soon become the new orthodoxy of the multiverse and anthropic reasoning.

COSMOLOGICAL NATURAL SELECTION

I first met Smolin many years later, when he gave a colloquium at the Harvard-Smithsonian Center for Astrophysics, where I was doing my first postdoc. He had just published his book *The Trouble with Physics*, attracting a lot of publicity while upsetting many string theorists. At the time, he was strongly advocating for loop quantum gravity as a competitor to string theory. A year later, when I arrived at the Perimeter Institute for my second postdoc, naturally I stopped by his office to introduce myself and explore new areas of collaboration. It turned out that he was more interested in what I did! We have had many conversations since, and I have noticed that Smolin turns his attention to a new problem—from economics to biology to cosmology—every six months. One of those many conversations involved the most massive neutron star known, which had recently been discovered by astronomers. Little did I know that Smolin's search for a link between biology and cosmology was what led him, ultimately, to ask this question.

Recall that some physicists argue that many of the constants of nature found in the standard model of particle physics are delicately fine-tuned for life. Change either the mass of the particles or the strength of the forces and stable atoms won't form; life, and complexity more generally, would be impossible. Not everyone is convinced; Smolin's close colleagues Carlo Rovelli and Abhay Ashtekar remain skeptical of such reasoning. I do, too. But Smolin believes this issue is profound, comparing it to the problem of why gene sequences are so much fitter in organisms than a random selection of genes would be.

Smolin had been impressed with what he had read from evolutionary biologists like Richard Dawkins, Lynn Margulis, and Stephen Jay Gould and had had a radical thought: what if Darwinian principles were more fundamental than anyone in physics could have imagined? What if the universe underwent some form of natural selection? The critical ingredient for evolution to work is differential reproduction. Those organisms better at reproducing will dominate the "fitness landscape." Complexity can arise naturally and gradually without intelligent input, and the cosmos might reproduce by way of black holes giving birth to baby universes. But perfect replicas give evolution nothing to work with; random copying errors are essential to the process. Thus, the baby universe must have different values for the constant than their parents. John Wheeler had already suggested this possibility in a mechanism he called "reprocessing," where the constants of nature get jumbled up during the violent Big Bang epoch. While sailing on Skaneateles

Lake in Syracuse, New York, Smolin started to form an idea. He said, "I had been reading the evolutionary biologists and thinking about the standard model . . . This idea all of a sudden hit me as I made a tack in the sailboat." Smolin realized there was one extra ingredient necessary for his model to work. That is, the changes in the constants of nature must be small from one generation to the next; these slight variations allow nature to find peaks more easily in the landscape representing life, permitting worlds like our own.

Smolin now had a Darwinian model for the universe with the final ingredient in place; he called it cosmological natural selection. In this framework, universes that are better at making more black holes will have more progeny and so be "fitter" in an evolutionary sense. The cosmos then is not fine-tuned for life but for black holes. As stars give birth to both black holes and living organisms, life becomes a happy byproduct. When Carl Sagan said, "We are star stuff," he was referring to the fact that our bodies and environment arose from materials in exploding stars; the stars died so that we could live. But if Smolin is correct, there is a far deeper meaning to this famous phrase. Dying stars don't just give rise to chemical elements. They also form the very fabric of the cosmos that surrounds us.

Smolin argues that if cosmological natural selection is right, our universe should be optimized to produce black holes; change any of the constants of nature, and fewer should result, a model that he argues leads to testable predictions. As we discussed above, Chandra has shown that every stellar remnant that survives the death of its parent star has the largest possible mass. The most massive of these remnants are known as neutron stars. If a neutron star happens to exceed this limit, the gravity of all the matter will outcompete the outward pressure coming from the interior of the star. A sufficiently large mass remnant has no option but to collapse to form a black hole. If cosmological natural selection is true, the cosmos should be selecting for a smaller maximum mass for neutron stars. If there were a larger limit, more neutron stars and thus fewer black holes would be created. Recall that the whole point of the model is that the universe is optimized for black-hole production. However, just as biological natural selection doesn't imply that every human should have countless children, cosmological natural selection doesn't mean that all stars should form black holes (they don't). However, if there is a way to make more black holes by small changes in the constants of nature, then the universe should take that course. When Smolin was looking for a way to test cosmological natural selection, he found an argument from the astrophysicists Hans Bethe and Gerald Brown that relates the maximum mass of a neutron star to the mass of a subatomic particle called a strange quark. If the constants of nature can vary, then lowering the mass of the strange quark will lower the maximum mass of neutron stars, giving us more black holes. From these considerations, Smolin concluded that the lowest he could push this limit was to a point about twice the mass of the sun, although when Phil challenged him on how big the uncertainties were on such a prediction, Smolin confessed he did not know. Recently, astronomers detected neutron stars slightly larger than Smolin's upper limit.[9] Does that mean the end of cosmological natural selection? As we shall see, the situation is not so clear.

Smolin markets cosmological natural selection as an alternative solution to fine-tuning puzzles, in contrast to the standard "anthropic" explanation that asserts that we live in a multiverse generated by eternal inflation and have to find ourselves in one of the rare universes hospitable to life. He tells Phil that the anthropic-multiverse model is "a cop-out; it doesn't make any predictions and it doesn't explain anything. It creates an infinite number of other universes." The number of universes in cosmological natural selection is also vast, possibly infinite, and so at first glance, you might struggle to see the difference. But the critical distinction is that in eternal inflation, it's assumed our reality is a rare patch of life in a sea of dead universes. In contrast, in cosmological natural selection, universes with life are the rule, not the exception. That difference is what Smolin claims makes his model predictive and scientific.

Given Smolin's anti-anthropic stance, it's not surprising that the anthropic advocates have hit back hard. Vilenkin, one of the fathers of eternal inflation, claims that a larger cosmological constant, consistent with inflation, would create more opportunities (as well as infinite time) for quantum fluctuations to make black holes out of empty space, thus undermining the cosmological natural-selection principle. After all, if you can conjure a black hole from nothing, there is no evolutionary advantage to gradually tuning constants to optimize their production. Smolin suggests that Vilenkin's argument could also discredit many well-established theories of science; if we wait long enough, pretty much anything can form as a fluctuation from the vacuum.

Leonard Susskind, another leading anthropic-multiverse advocate, has also raised several objections to cosmological natural selection, most notable of which was that eternal inflation is simply much more efficient at generating universes than cosmological natural selection is. Even if cosmological natural selection is a feature of reality, it cannot compete with eternal inflation in creating new universes. Another objection is that if offspring fail to inherit the genes of their parents, Darwinian evolution would be dead in the water. Similarly, Susskind alleges that even if a black hole gives birth to a baby universe, it would have no genetic memory of its parent (that is, of the parent's physical constants) and so natural selection would fail. "Why would Lenny say that? He doesn't know," replied Smolin when we put this problem to him. But the way I see it, Smolin doesn't know either; it remains, as he admits, only a conjecture. Rüdiger Vaas has also pointed out that the predictions of cosmological natural selection depend on the assumption that our universe is a typical member of the ensemble, but it need not be so; after all, evolution takes time to find peaks in the landscape of possibilities. Lastly, I think we can question whether the science of neutron stars is understood well enough to allow us to be confident that Smolin's predictions are reliable.

Going back to my conversation with Smolin about the most massive neutron star observed, I remember that he was very worried that it would rule out his cosmological natural selection prediction. At the time, I was trying to understand what neutron star structures can teach us about gravity and was dismayed to learn that the predictions of different nuclear physicists for the pressure of nuclear matter can vary by an order of magnitude. I tried my best to put his worries to rest by suggesting that maybe there could be different phases of nuclear matter, some making massive neutron stars, and others making lots of baby black holes. That idea seemed to make him happy, although it possibly also made cosmological natural selection less predictive. Such is the irony of the scientific method.

To be confident that black holes can give birth to new universes, we need a more unified approach to fundamental physics, one that can inform us whether the basic conjectures of cosmological natural selection are correct. To probe deep into a black hole requires a quantum theory of gravity; only such a theory can unlock the secrets of a black hole's hidden interior and reveal whether other universes are born from its core as cosmological natural selection predicts. Since Smolin has been a key figure in the development of loop quantum gravity, one might imagine that that is where support for cosmological natural selection might arise. And to some extent, his intuition that bounces replace singularities has been confirmed by the theory he helped develop. But some of Smolin's key collaborators remain unconvinced of cosmological natural selection. Carlo Rovelli, Francesca Vidotto, and others have used loop theory to model black holes, and what they find is that matter does eventually bounce back into our universe (as a black hole disappears) rather than birthing other universes. Yet others in the loop community side with Smolin. For example, Parampreet Singh's research suggests that infalling material quantum tunnels into a new region of space-time that may have different constants of nature—precisely what cosmological natural selection needs. Which of these narratives is right is unclear, as the calculations are incredibly complex. By contrast, the question of bounce in cosmology must be much simpler than bouncing black holes are, because unlike black holes, the universe (on its largest scales) appears uniform. But, as we saw in chapter 5, there are bitter disagreements within the loop quantum gravity camp about the nature of this cosmological bounce, so one should not be surprised about the situation with loopy black holes. Even if cosmological natural selection is incompatible with loop quantum gravity, a rival theory just might lend weight to Smolin's model.

When I first came to Perimeter as a postdoc in 2009, I had a giddy feeling of being a kid in a candy store. There were seminars on all sorts of fantastic things, from the nature of time to quantum foam and dark matter, happening in rooms with fancy names like the Black Hole Bistro or Alice and Bob (the infamous protagonists of quantum cryptography discussions). One person whom I kept running into in all the seminars was Rafael Sorkin. Sorkin and Smolin were colleagues in Syracuse in the '80s and '90s, but after Smolin helped start Perimeter, he managed to convince Sorkin to retire from Syracuse and come to Canada. Since we kept running into each other and both had little responsibility (me as a postdoc and he as a retired professor), we had many long discussions over the years about topics ranging from politics to black holes, dark energy, and the foundations of quantum mechanics. It turned out that, like a physicist's physicist, Sorkin had played a key role in developing many important topics in classical and quantum gravity. It was natural then, that when I started my own group as faculty, I asked him to come to our group meetings. He happily obliged and ended up converting many of my students to his gospel, otherwise known as causal set theory—a model that might just put Smolin's black-hole speculations on a firmer theoretical footing.

Causal set theory is a proposal for quantum gravity developed primarily by Sorkin and his collaborators. In the same sense that we now know that an apparently continuous fluid, like water, is made out of microscopic atoms, they proposed that what Einstein conceived as continuous space-time is fundamentally granular; atoms of space-time can be modeled as a set of points with fundamental before-and-after relations. These points are often sketched as little balls (depicting space-time atoms) connected by lines (representing their possible before/after, or causal, relationship, should they have one). Sorkin's granular picture gives rise to a discrete notion of time, which is distinct from Einstein's continuous space-time and allows us to ask what happened before the Big Bang, in a less restrictive framework. This model may sound remarkably similar to loop quantum gravity. But the difference can be illuminated by asking the question: what is fundamental? As an analogy, we can consider mixing colored paints. Red, yellow, and blue are primary colors. Green "emerges" when you mix yellow and blue; it is not hiding inside either color. Similarly, we can ask: what are the fundamental building blocks of reality, and what emerges out of those basic constituents? Causal set theory asserts that the order of events (known as the causal order) is fundamental. The causal relations between the points of the causal set are taken as primitive, analogous to primary colors, with geometry being an emergent property. Whereas, for example, in loop quantum gravity (at least as it was originally proposed), geometry is fundamental, and time is emergent. Recently, though, my collaborator and I have argued that casual sets themselves might be emergent, too. I pointed out with Dejan Stojkovic that we can consider the atoms of causal sets as the points of intersection of fundamental strings in a four-dimensional space-time. In this sense, we argued, causal sets may potentially emerge from a more fundamental string theory.

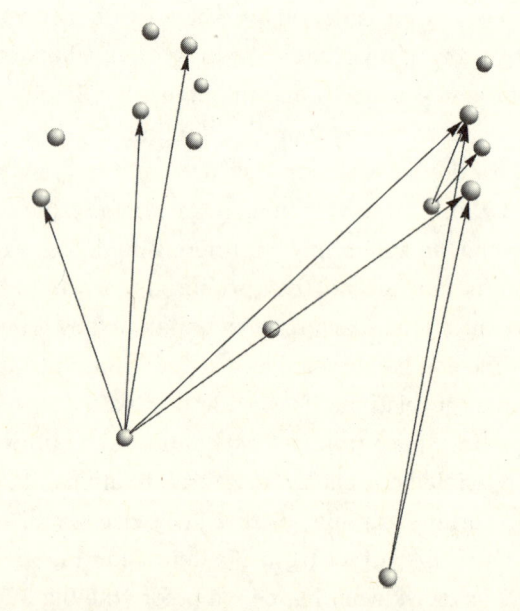

An illustration of the quantum gravity theory known as *causal sets*. The balls depict the atoms of space-time, and arrows represent their causal relationships that go forward in time. While this causal set has fifteen space-time atoms, our observable universe is expected to have 10^{240} atoms. Credit: Niayesh Afshordi.

The Heart Nebula is some 7,500 light years away from Earth. Phil and I share a love of amateur astronomy, though we both grew up in terrible places to view the night sky. I remember using my telescope as a middle-schooler on a rooftop in northern Tehran. Phil captured this image from his home in central London.

A total solar eclipse photographed by Phil in 2017. Some authors have credited unexpected eclipses with changing ancient history, and scientific expeditions have used solar eclipses to confirm Einstein's theory of general relativity. Phil and I have both been lucky enough to witness total solar eclipses, and their majesty blew us away, even though we knew what was coming.

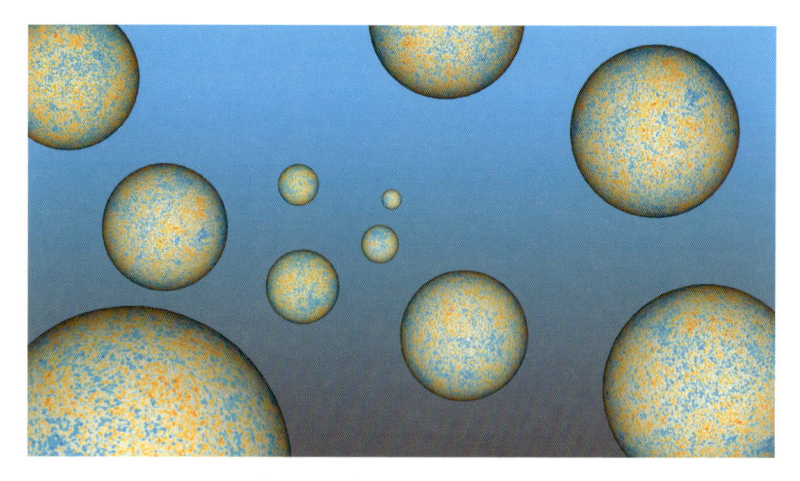

An artistic interpretation of eternal inflation, which suggests an infinite reproduction of bubble universes, or a *multiverse*. To make such a process work, one only must assume that inflating space expands faster than it decays, as is the case in almost all inflationary models. Credit: Niayesh Afshordi.

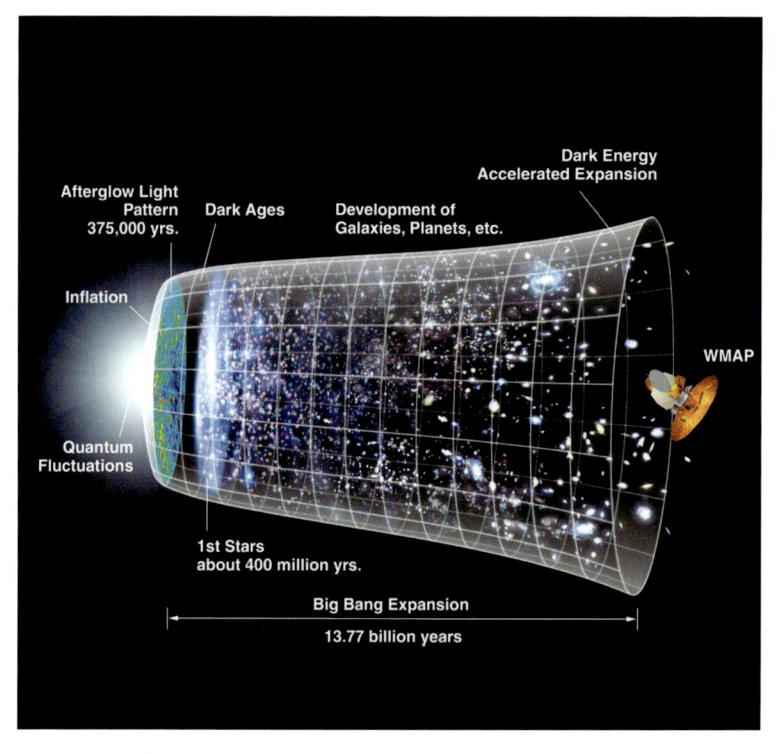

The timeline of the universe, according to NASA's WMAP team. Credit: NASA / WMAP Science Team.

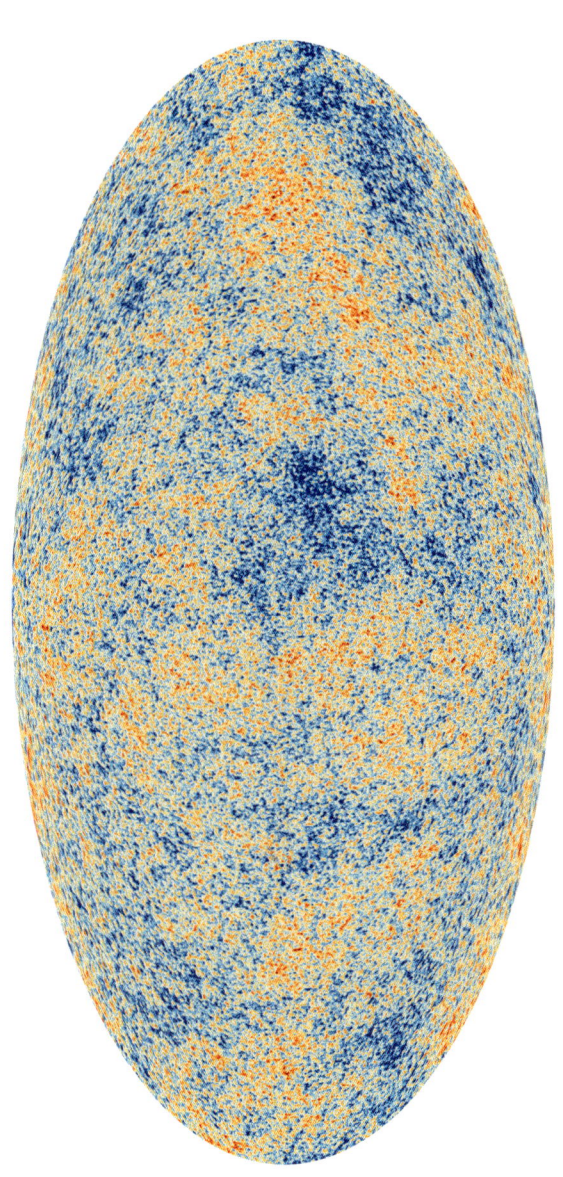

The oldest light we can see is the cosmic microwave background, emitted nearly four hundred thousand years after the hot Big Bang. It is almost perfectly uniform, with tiny variations in temperature at 1 part in 100,000, which are shown as red (hot) and blue (cold) spots. They are a treasure trove of information for early-universe cosmology. Multiple successive satellites have studied these fluctuations; shown here is the most recent map produced by the European Space Agency's Planck probe. Copyright: ESA and the Planck Collaboration.

An infinite hyperbolic plane is depicted in *Circle Limit IV: Heaven and Hell* by the artist M. C. Escher. Roger Penrose suggested that a similar rescaling could squash down the infinity of the universe's expansion in the same way that Escher had done in his artwork. Copyright: M. C. Escher Company—The Netherlands.

Our proposal for an alternate origin of the universe tells the story of the cosmos emerging from a higher-dimensional black hole. The paper, published in 2013 with Razieh Pourhasan and Robert Mann, had the right number of buzzwords—*white hole*, *black hole*, *Big Bang*—to make the front cover of *Scientific American*. Credit: Kenn Brown, Mondolithic Studios. Reproduced with permission. Copyright: Scientific American, a Division of Nature America, Inc., 2014.

The Perimeter Institute for Theoretical Physics in Waterloo, Ontario, Canada, where my colleagues and I research and debate the origins of the universe, among other big questions. Credit: Phil Halper.

A depiction of our view of the universe. In the center is the solar system; as we progress outward, we reach the Milky Way and other galaxies; finally, the fiery outer ring represents the cosmic microwave background, the oldest light in the cosmos. This illustration is a more artistic version of a logarithmic map of the universe created by astrophysicists Richard Gott III, Mario Jurić, David Schlegel, Fiona Hoyle, Michael Vogeley, Max Tegmark, Neta Bahcall, and Jon Brinkmann. Credit: Pablo Carlos Budassi.

Shrinking rose: an image from the last chapter of Beth Gould's PhD work on a cyclic universe, entitled "Periodic Time Cosmology," which posits a model in which the entire universe runs on a time loop. The picture illustrates how a pattern can be invariant under rescaling, shrinking by a factor of 2 toward the top left corner—in other words, this pattern looks the same if we rescale it and then let it evolve from the Big Bang to the infinite future. While this model means that every structure should have infinitely many rescaled copies across the cosmos, there is no guarantee that we can see all of them, or even more than one. For example, if the light disk in the picture were our cosmological horizon, we would not see the structures that are being repeated. Credit: Beth Gould and Niayesh Afshordi.

In this deep-field image from the James Webb Space Telescope, almost every object in the picture is a galaxy. In fact, some of the galaxies appear multiple times, the result of a massive dark-matter-halo "lensing" light emitted by these galaxies like a funhouse mirror. Credit: NASA, ESA, the Canadian Space Agency, and the Space Telescope Science Institute.

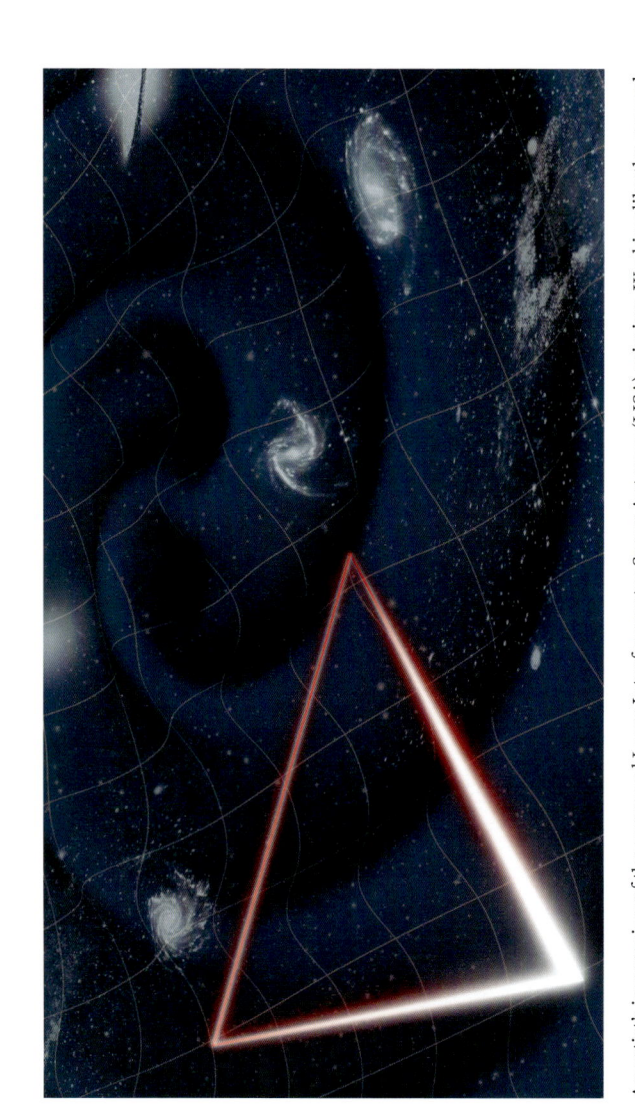

An artist's impression of the proposed Laser Interferometer Space Antenna (LISA) mission. Working like the ground-based Laser Interferometer Gravitational-Wave Observatory (LIGO) but on much larger scales, powerful lasers are sent between satellites millions of kilometers apart as they orbit the sun. LISA can measure tiny ripples in space-time geometry due to gravitational waves that change the distance between satellites by a fraction of the size of an atom. However, only a successor mission like Big Bang Observer is likely to detect primordial gravitational waves with the potential to resolve the mystery of what really happened at the Big Bang. Copyright: ESA.

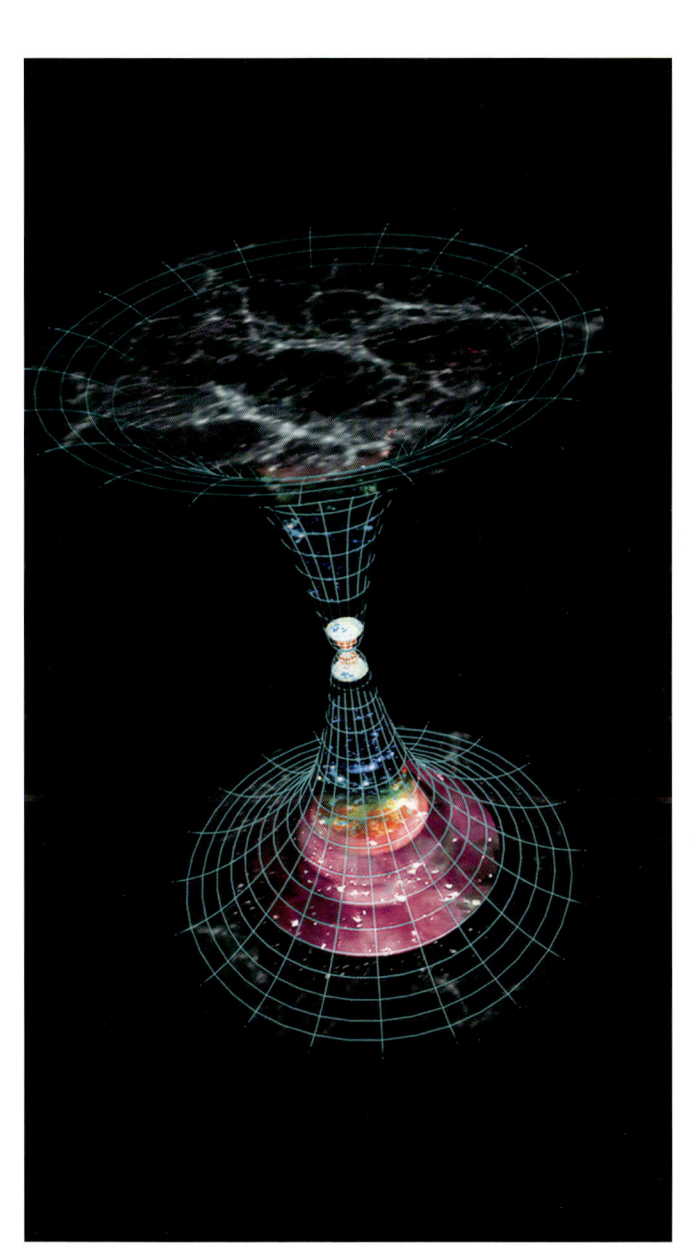

An hourglass shape for the universe is a common theme for many contemporary Big Bang models. Some models have a reversal of the arrow of time in the middle; these are known as Janus Universes. Others have a maximum density creating a repulsive force to replace the Big Bang singularity with a Big Bounce. Credit: Nick Franco, 1185 Films.

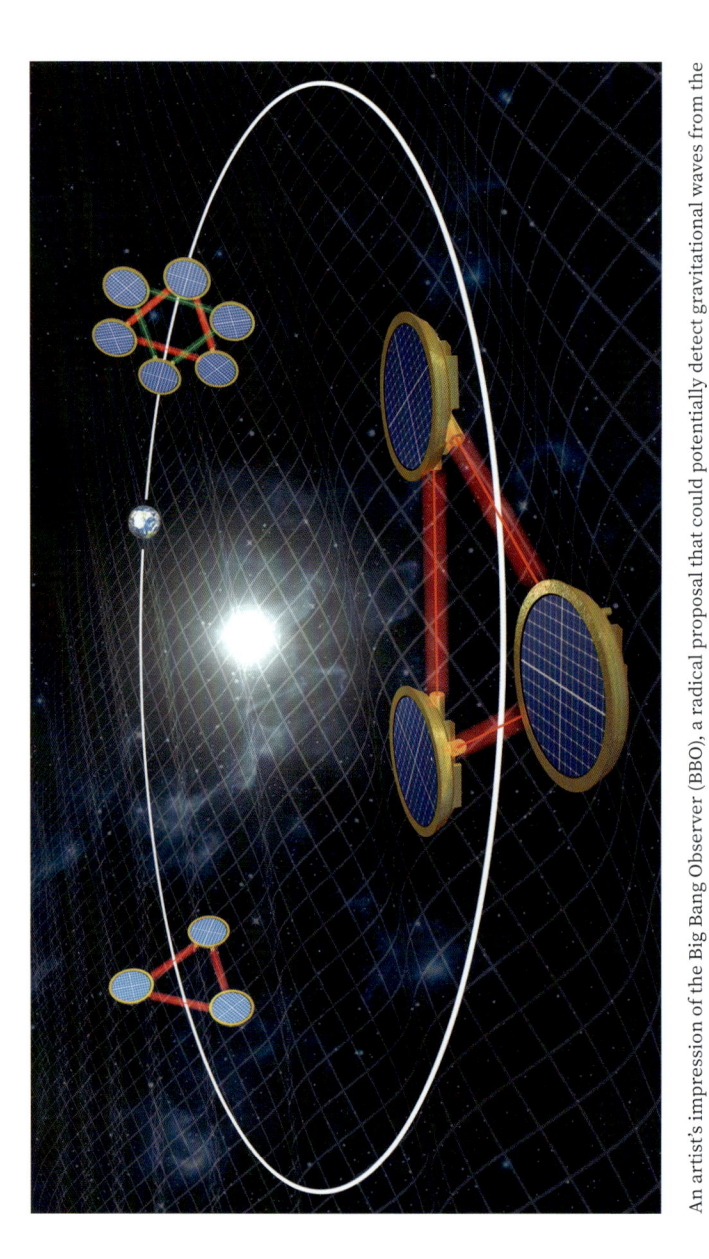

An artist's impression of the Big Bang Observer (BBO), a radical proposal that could potentially detect gravitational waves from the Big Bang. BBO uses more powerful lasers than LISA and requires twelve satellites rather than three, generating ten-thousand-times-better sensitivity. Credit: Christian Unterdechler.

But among so many proposals for quantum gravity, why should you take causal sets seriously? Depending on your mileage, motivation may come from the puzzle of dark energy. A decade before its groundbreaking discovery, Sorkin had noticed something strange. If, according to his causal sets, the continuous space-time was actually a stream of discrete quantum sparkles (or space-time atoms), then Heisenberg's uncertainty principle leads to a prediction for fluctuations in vacuum density, which (we now know) was remarkably close to the density of dark energy, discovered in 1998.[10] It was the breakthrough that shocked the world and resulted in a Nobel Prize for its discoverers. But strangely, few even noticed that Sorkin had already predicted dark energy's existence and density. Sadly, causal set theory remains obscure to this day. Maybe its obscurity has persisted because, as we noted in chapter 4, Steven Weinberg had made a similar prediction using anthropics and the multiverse. One may imagine that it wasn't easy for Sorkin to get his prediction noticed when Weinberg, arguably the most acclaimed physicist on the planet, had stolen his headline.

Weinberg's use of the multiverse is often invoked to explain the apparent fine-tuning of the constants of nature. Just as we shouldn't be surprised that someone wins the lottery if enough tickets are sold, then we should expect that at least one universe has life-permitting constants in a huge sea of possibilities. But if cosmological natural selection can be linked to causal set theory, the very same observation that so astonished the physics community might provide support for an alternative explanation for fine-tuning. This possibility is what Fay Dowker (a former student of Hawking), and her graduate student Stav Zalel set out to examine in 2017. Using causal sets, they found that the parameters change from the mother universe to the baby, exactly as Smolin and Wheeler had conjectured decades before. Their finding doesn't, however, mean that causal set theory predicts cosmological natural selection, only that it seems to be compatible with it.

But what does causal set cosmology look like, and is there a way to test it? Sorkin and his collaborators have suggested that the model that explained the discovery of cosmic acceleration in fact leads to fluctuating dark energy and a cyclic universe over time. He also thinks there may have been a beginning event in which there were zero members of the causal set, realizing Vilenkin's vision of a universe from nothing. Weinberg's solution of an anthropically selected cosmological constant may be the consensus, but it predicts we should measure dark energy as a cosmological constant, while Sorkin's dark energy fluctuates over time, clearly providing a way to distinguish the two hypotheses. Recall the Hubble tension from chapter 1, which arises from disagreement between different ways of measuring cosmic expansion rate; if this disagreement represents new physics rather than a measurement error, then a fluctuating dark energy could explain it. Sorkin and I worked with our PhD student, Nosiphiwo Zwane, originally from Swaziland (now Eswatini), to show that several observational mismatches (like the Hubble tension) may be addressed in Sorkin's model, but much work needs to be done to see which of these cosmological pictures is favored. If it is Sorkin's, then his model could profoundly rewrite the way we see our universe. No longer would we wonder why the value of dark energy is so small compared to what is predicted by quantum mechanics. For in this scenario dark energy is not a cosmological constant as most physicists assume; it is the result of fluctuations of space-time itself. Moreover, if causal sets imply cosmological natural selection, then we can explain why the universe has the life supporting properties that some find so remarkable.

Renowned biologist Richard Dawkins has called for an extension of evolutionary tenets into fields other than biology, claiming that principles such as variation, selection, and inheritance should lead us to a "universal Darwinism." He wrote, "There remain deep questions, in physics and cosmology, that await their Darwin."[11] But if cosmological natural selection is right, physics has already found its Darwin in Smolin, and the principles Darwin discovered might explain not just the origin of species but the very origin of our expanding cosmos.

TORSION BOUNCE

While most of my colleagues and I assume that a quantum theory of gravity is needed to understand the Big Bang, some believe that relativity misses another important effect. Einstein's theory describes space-time as being able to warp, expand, and contract. But one of the world's greatest mathematicians, Élie Cartan, thought it could also twist, a process we call torsion. Despite some interest from Einstein, Cartan's proposal (known as Einstein-Cartan gravity) went ignored for decades. It was mathematically easier to set the torsion of the universe to zero, and doing so seemed to work. But in the 1960s, Dennis Sciama and Tom Kibble realized that while the energy density of matter curves space-time, the intrinsic angular momentum, or spin, of particles can twist it, generating torsion. Sciama, as we saw in chapter 1, was highly influential, mentoring Penrose, Hawking, and other notable cosmologists. Kibble was one of the co-inventors of the Higgs mechanism, and many believe he should have shared the Nobel Prize for its discovery. So having both of these luminaries work on Einstein-Cartan gravity helped breathe life into this once-forgotten theory.

One scientist especially interested in the Einstein-Cartan approach is the Polish American physicist Nikodem Poplawski. In 2019, I invited him to give a talk at Perimeter. The first response I recall getting from my colleagues was, "Have you seen his Twitter feed?" I wasn't sure of their concern, but I suspect they were referring either to his rabid anti–string theory stance or to his unusually-right-wing-for-academia views. More generally, a 2016 study of American professors found the ratio of liberals to conservatives was 6 to 1 (reaching as high as 28 to 1 in New England).[12] Whether his conservatism held back the reception of his talk I'm not sure. Poplawski has built on work that dates back to 1973, when Andrzej Trautman, a renowned Polish relativist, argued that in cases of extreme energy density, torsion causes space-time to stiffen like water turning into ice.[13] This stiffening leads gravity to transform from an attractive force into a repulsive one and causes the universe to bounce. Estimates show this effect becomes noticeable when the density of matter is a gargantuan 10^{93} grams per cubic centimeter (for comparison, a neutron star has at most a density of about 10^{17} grams per cubic centimeter), a level known as Planck density.

But what is the Planck density? Recall from chapter 3 that Hawking and Bekenstein discovered that the entropy of black holes is proportional to the area of their event horizons. In standard physics, the entropy of a box counts the number of particles within that box. So, as a black hole shrinks due to Hawking evaporation, it gets smaller and smaller, lowering its entropy. But entropy cannot get smaller than one, as you can't have less than one particle in a real object. Imagine something with the mass of a speck of dust squeezed into a Planck length. Such an object would be the smallest black hole allowed by the laws of physics, but its density would be a gigantic 10^{93} grams per cubic centimeter; this is the Planck density.

We do know two places that might experience such unimaginably dense states: at the center of a black hole and at the Big Bang. A collapse to maximum density followed by a bounce paints a picture remarkably similar to that of loop quantum cosmology, but interestingly, it is derived in a totally different manner. Poplawski thinks this similarity should raise our credence that black holes bounce to form new universes and has spent much of his career trying to build a model that can confirm this scenario. One of the greatest difficulties, though, is showing that the bounce is robust against many of the simplifying assumptions of the early torsion models. And while Poplawski agrees the new model has much to overcome, he thinks he has discovered features of it that make the story more compelling.

Torsion, Poplawski claims, can solve a mystery that has perplexed physicists for decades. In quantum mechanics, the vacuum can fluctuate particles into existence, but such particles always appear in perfectly balanced pairs of matter and antimatter. Yet, the universe seems to be almost entirely matter. So, where did the antimatter go? Torsion introduces an asymmetry in nature, allowing matter to dominate over antimatter. Furthermore, its repulsive force can allow inflation to occur without any hypothetical inflaton field.[14] Inflation from gravitational repulsion following a big bounce is exactly what loop theorists thought they had found originally, before they abandoned hope that it could explain inflation; the effect was too short-lived to solve the horizon and flatness problems. While presenting for my colleagues and me at the Perimeter Institute, Poplawski stood by the notion of inflation without an inflaton and showed calculations he had performed with Indian physicist Shantanu Desai that gave remarkable agreement with the latest CMB data.[15] But these "predictions" were made after the results were already known, so Poplawski's presentation didn't exactly set the room on fire.

Of course, as we have learnt, there is no theory of the Big Bang that doesn't have its own set of problems. As to Poplawski's torsion bounce scenario, the torsion effects only become important when we reach Planck densities, just as we lose faith in the classical notions of space and time. So, is the bounce a real thing, or just an artifact of pushing a theory too far beyond its limits? Moreover, its similarity to inflation only holds at the level of accelerated expansion; whether (infamously notorious) perturbations behave the same way is yet to be shown. Recall the nightmares that inflation, loop, and string gas people had to endure to calculate the variations in temperature in the CMB from their theories.

If you venture beyond this point in the book, you will notice some of the Big Bang proposals that I have helped develop. Yet I needed all the help I could get to recover fragments of those long-lost cosmic memories. So, I am happy to take the blame for their shortcomings and give credit to all those brave cosmologists who dared to push this frontier.

OUT OF THE WHITE HOLE

I have always thought that there is some hard-to-pinpoint commonality between the simplicity of the Big Bang and the simplicity of black holes. The singularity is one obvious similarity. Right? Yes and no. The singularity of the Big Bang was in our past, an idealization of a bygone era that is forever out of our reach. Of course, things that are hard to see can be arbitrarily (and artificially) simplified. Nonetheless, all our cosmological data is consistent with a universe that was incredibly flat and uniform (space was flat, though space-time was not). On the other hand, the singularity of a black hole, out of reach due to the presence of event horizons (a phenomenon dubbed "cosmic censorship" by Roger Penrose), can be arbitrarily messy. In fact, physicists and mathematicians still argue about the structure of black-hole singularities, using exceedingly complex and incomprehensible calculations (here, I'm speaking from personal experience). Moreover, the singularity of black holes is toward the future (if you fall through the event horizon), while the Big Bang singularity was in our past. So, both phenomena incorporate the S-word, but the similarity might stop there.

To imagine a closer connection between these two cosmic infinities, we can think of a uniform spherical ball of dust expanding into the empty vacuum of space. If the speed of the outermost dust particles is less than the speed it takes to leave the system (the escape velocity), the ball will stop expanding at some point and collapse under its own gravity. To an outside observer, this collapse looks exactly like a black hole with its own event horizon. The dust ball will eventually form a singularity; any cosmologists living within that singularity would consider it a Big Crunch. However, the past of this expanding ball of dust operated very much like a Big Bang, starting from infinite density. In fact, the whole picture is time-reversible; if we record a movie from Big Bang to Big Crunch, and play it in reverse, it looks exactly the same. Indeed, we could be those cosmologists living within the dust ball, not knowing that our entire universe will transform into the mother of all black holes when it re-collapses. Keep in mind, this scenario only holds for our hypothetical example; the real universe will not re-collapse as long as the dark energy causing its accelerated expansion is a cosmological constant.

One point of terminology: in the same way that a black hole has an event horizon that nothing can escape from, there must be a "white-hole" event horizon that everything *must* escape from, as our movie was reversible. However, these white-hole horizons are only an artifact of our imagination, as the real universe is not time-reversible: the second law of thermodynamics tells us that entropy only increases with time. However, increasing entropy doesn't imply that white holes cannot exist, but rather that they should somehow be in place at our moment of creation. A white hole would simply look like an explosion, just like the Big Bang!

So, could the Big Bang be simply a gigantic white hole? In fact, the answer, some may argue, is *yes, by definition*, as we are all emerging out of this singularity. But here's the rub: while the Big Bang is incredibly smooth (to 1 part in 100,000), a star is not. It has a dense core and an atmosphere, and it's surrounded by empty space.

So, our reversible movie of the universe is unlikely on two accounts: violating the second law of thermodynamics and assuming a uniform ball surrounded by an abrupt edge. This incredible uniformity is of course a challenge for all models of the Big Bang (captured via the flatness and horizon problems), which as we have seen are addressed in different scenarios to varying degrees of success.

It was around ten years ago, when I was still a young professor in Waterloo, that I started to think more deeply about black holes. My background was in cosmology and astrophysics, but I was kind of a novice when it came to black holes. However, interesting mathematical developments connected the properties of gravitational waves near the horizon of black holes and those of incompressible fluids (like water in a river).

And do we have some strange fluids in need of some understanding? We do: dark matter, dark energy, and possibly other cosmic fluids. So, I wondered whether some of these fluids that we have a hard time identifying might possibly be projections of higher-dimensional gravitational dynamics. In particular, recall from chapter 4 that string theory has suggested that we (being made out of matter particles) may be confined to a three-dimensional brane, while gravity can venture into higher dimensions. These suggestions are known as braneworld scenarios. At the turn of the century, a trio of physicists out of New York University, Gia Dvali, Gregory Gabadadze, and Massimo Porrati, suggested that there could be different types of gravity: one that, like matter, is confined to the brane, and another (more primal) gravity that is not. This scenario is known as DGP gravity, named after its inventors.[16] The citizens of the DGP braneworld are sufficiently equipped with their own matter and gravity that they may barely notice the stealth higher-dimensional gravity (except at very late times or large scales, according to the calculations). In fact, for a while, it was thought that this stealth gravity (and not dark energy) could be responsible for the recent onset of cosmic acceleration. David Spergel, my PhD advisor who was also the lead theorist on the WMAP experiment, shared some proprietary information with me. He suggested that I should look into whether DGP gravity could explain why CMB temperature fluctuations, on large scales, mysteriously transform from being highly correlated to being random white noise. Alas, doing this calculation was beyond my technical capabilities at the time. A few years later, others showed that the DGP model exacerbates this lack of correlation and is also inconsistent with other cosmological observations.

But while DGP gravity didn't explain the mystery of the CMB sky, I saw the potential for something else. What if DGP gravity and a black hole made the Big Bang? In other words, might DGP gravity be just the framework to describe our universe as emerging, not out of an ordinary three-dimensional black hole as Smolin and Poplawski suggest, but rather from a four-dimensional black hole living at the center of a spherical braneworld? To help develop this idea, I would turn to a brilliant student whose journey to break free from the confines of the past brought back memories of my own attempt to exceed the escape velocity of political upheaval.

In 1999, my wife, Ghazal, and I left Iran to come to the US. Our departure was a particularly traumatic one, not just because we were leaving our families and homes behind or because we had just married two weeks before our flight. It turns out that the chaos of US–Iranian politics makes for a very bumpy ride. First, thanks to the Iran hostage crisis two decades earlier, the United States has no consulate in (or diplomatic relationship with) Iran. Accordingly, Iranian students who want to study in the United States need to apply for visas elsewhere, often in neighboring countries. Moreover, consular officers can turn down visa applications for arbitrary reasons. So, Iranians often have to apply multiple times, in different countries; if we're lucky, we eventually get a US student visa. In our case, I have a distinct memory of waiting in the US consulate in the United Arab Emirates, watching footage of Iranian student protests and unrest on CNN, taken from the very street where our own Sharif University in Tehran was located. This protest was the first of many against the closure of a reformist newspaper by the hardline Iranian judiciary. To add insult to injury, after questioning us about the basics of cosmology, the US consular officer told us that he wouldn't issue a visa as we were good students and unlikely to leave the US. We did finally manage to get visas (this time in Ankara, Turkey), but going to the US now meant that we were also leaving behind our friends, family, and country in turmoil. Little did we know that even this difficulty was nothing compared to what the future had in store for us.

Fast forward a decade, and there was a new set of protests, this time over an (allegedly) fraudulent reelection of a populist president, Mahmoud Ahmadinejad. This episode was a much bloodier and extended one, with a new generation of young student protesters who had not experienced the horrors of the 1979 revolution or the eight years of the Iran–Iraq war. One of them was a young PhD student, Razieh Pourhasan, who later managed to leave Iran to visit my colleague Robb Mann to work on a research project. Despite the generation gap, we quickly bonded over survival stories from our home country, as fish out of water seek any common droplet to breathe. We also happened to talk physics over lunch at the Perimeter Institute's Black Hole Bistro. For obvious reasons, Razieh didn't want to go back to Iran, so she managed to eventually transfer to become a PhD student at the University of Waterloo. Around this time, I started musing about black holes and the Big Bang.

Both Razieh and Mann had a lot of experience with black holes and extra dimensions. I, on the other hand, was a novice in those areas but knew my cosmology. So, we started to work together. Turns out black holes and the Big Bang were a match made in heaven.

When a conventional three-dimensional star forms a black hole, the core collapses inwards, but there is a shell of outer material that gets ejected into space. What is left is a black hole with a singularity at its center and an event horizon at its edge—the point of no return for any observers brave enough to cross. This black hole has one more dimension than its event horizon, so in our universe the event horizon of a normal black hole would be a two-dimensional flat surface. If our universe formed from the collapse of a black hole, we might be living on the shell that was ejected from its parent star. As we observe that our universe is three-dimensional, then any hypothetical black hole that formed our universe at its event horizon would have to have one more dimension: it would, itself, have to have been formed from a four-dimensional star.

Recall that I argued that an inhomogeneously exploding star, which can be idealized as a white hole, looks nothing like our extremely uniform Big Bang. But what if we lived on a three-dimensional DGP braneworld that surrounded a four-dimensional black hole, or white hole? A spherical white hole would yield a perfectly symmetrical cosmos, just like ours. The fact that we see our space to be flat and not spherical is not necessarily a problem, since if the radius of the braneworld sphere is huge, we cannot easily measure it within our cosmological horizon (just as I cannot easily measure Earth's radius from the geometry of streets in Waterloo, Ontario). Another interesting property of the black holes in Einstein's theory of relativity is known as the no-hair theorem. A star might be extremely complex, but once it falls into a black hole, all that is left for outside observers to see is a set of three numbers: energy (or mass), angular momentum (or spin), and electric charge. Just as bald heads cannot be distinguished by their hairstyles, there is little to distinguish one black hole from another, as all their properties can be summed up by just these three numbers. So maybe the no-hair theorem should really be called the three-hair theorem. Furthermore, recall that Hawking and Bekenstein had discovered that the horizons of these black holes have temperature and entropy. Now imagine a four-dimensional star collapsing into a black hole. If a three-dimensional DGP brane, through an as-yet-unknown quantum process, were to emerge from the inferno of the horizon of this four-dimensional black hole, it would have just the right properties to spawn our universe. It would have a smooth horizon, thanks to the no-hair theorem; the brane would be expanding as it came out; and lastly, it would be full of hot radiation, thanks to the temperature and entropy of the black hole. The cosmological Big Bang singularity is simply a mirage here in this scenario. It never happened.

What was even more exciting about this idea was that it provided a process that would yield the same pattern of primordial fluctuations that we see in the CMB. If we imagine that our three-dimensional brane also had a four-dimensional hot atmosphere, the thermal fluctuations in the atmosphere would generate fluctuations in the gravitational field. One might then expect that these gravitational fluctuations would naturally be imprinted onto our braneworld at its inception. This imprinting would generate some regions slightly denser than others, differences that would eventually form the CMB's tiny temperature variations and that would eventually evolve into the majestic galaxies that grace our telescopes. Our model is more in line with string gas cosmology, where structures have a hot origin, and in contrast to inflation, where they arise from a cold quantum vacuum. I admit that the whole idea was very speculative, although not really more so than any other model of the Big Bang. This model was very different from all its predecessors, but it was still composed of ingredients that we are familiar with in other contexts: black holes, branes, extra dimensions, and a hot atmosphere, to name a few. Many gaps (and, frankly, a great deal of wishful thinking) remained in the plot, ones that I hoped would be filled over time. After all, you don't expect (or hope) the first paper on a new idea to be the last.

So, on September 5, 2013, we put out our paper: "Out of the White Hole: A Holographic Origin for the Big Bang."[17] While I didn't think much of it then (I had been in the business of speculation for some time at this point), apparently our paper had just the right number of buzzwords (white hole, black hole, Big Bang), to capture the imagination of science journalists. It started with a friend from my time at Brown, Zeeya Merali, a former student of Brandenberger who left academia after her PhD and is now a very successful science writer. She interviewed me and then published a news piece for *Nature* entitled "Did a Hyper-Black Hole Spawn the Universe?" That article started a deluge of public attention, which culminated in *Scientific American* asking us to write a popular article for them. "The Black Hole at the Beginning of Time" appeared in the August 2014 issue and was featured on their cover.

In the meantime, in spite of the public attention, the reception from the cosmology community, that is, my colleagues, had been muted. You see, the wheels of science turn slowly, and progress often happens in small increments. These increments, you do not read in news headlines, but they are what, unsurprisingly, make up most of our day-to-day lives, only interrupted once every few years by rare genuine leaps, or (more often than not) fleeting hypes. Our model, which we ended up calling "Out of the White Hole," was sufficiently different from everything else cosmologists were doing at the time that no small increment could let them work on it. A more substantial investment in studying the model would require more time, and of course, motivation. With my first son turning four and another just born, I had very little of the former but plenty of the latter.

So, with our new PhD student, Natacha Altamirano, Robb Mann and I decided to do the hard work: land the rocket that we launched with Razieh. Could we turn these speculative ideas into concrete, testable predictions? Natacha was incredible at doing hard calculations and very determined—but even so, designing the atmospheres of the four-dimensional star whose demise spawned our universe, studying its sound waves, and assessing how those waves would impact a three-dimensional braneworld was not an easy task. The trouble was not necessarily that the game was hard to play, but rather that we had to invent the rules at every step, hoping that the final outcome would make sense. Investigating the new model felt like going down a series of rabbit holes and dreading at every turn whether our previous choices were wrong. In spite of this challenge, we powered through and ended up with a prediction for the precise shape of the patterns we expected to see in the CMB sky on different scales. We then teamed up with another PhD student of mine, Beth Gould, who was great at testing different models of the Big Bang against observations, and . . . (drum roll) . . . our model didn't make sense. Not only was the model not a very good fit to the data (although, technically, it was still consistent to the data within a generous margin of error), we needed much more entropy than was allowed within the horizon of a four-dimensional black hole (according to Stephen Hawking). So, had we just summarily falsified our own proposal, or had we picked the wrong rabbit hole somewhere along the way?

It is a bitter feeling to build a theory, bit by bit, with a team of smart, hard-working students and then test it against data, only to see it burn to ashes. The experience is horrid and evokes different reactions, even amongst the best of us.

In the biblical story of the binding of Isaac, God commands Abraham to offer his son Isaac as a sacrifice. In the Muslim version of the story, it is Ishmail, another son, who is the subject of Abraham's dream, and who submits to God's will to be sacrificed. While in both the Bible and the Quran the story has a PG ending, with Abraham's son being spared by an angel, I sometimes wonder whether the real story was R-rated and a parable for the life of a theoretical physicist. For we spend so many sleepless nights developing our theories, perfecting them in every possible way, and then days reveling in their success and beauty, just as we do for our children. Yet, the scientific method compels us to submit our theories to the cruelty of Occam's razor: only the simplest model that fits the data will survive. No wonder so many of us hold on to our beloved theories even in the face of conflicting observations. Others simply do not entertain theories that can be tested with data. Maybe it's primal parental instinct that leads us to want only what's best for our children: that they live long and prosper. Unfortunately, indulging such desires may not be what's best for the scientific method. We should champion our theories, help them grow and flourish. But—as most parents learn sooner or later—a time comes to let our offspring go, releasing them to confront the world on their own. For our scientific theories, things are far harsher, as we are expected to let them die if they fail. Pruning models from the branches of possibilities is progress, but it's often painful to accept that progress when the ideas being pruned are yours.

In real life, Razieh went on to work on an extraordinary range of topics on the cutting edge of string theory. She was a force of nature, starting fruitful collaborations with top researchers she'd met casually in conferences. But she ended up doing postdocs in Iceland and then Italy, while raising a baby with her husband—who worked on a different continent, in Ottawa (a story that, unfortunately, is not uncommon for many junior scientists in the current funding environment). This arrangement took its toll on Razieh, and she ended up leaving academia to take up a data science job for the Canadian government so that she could live with her young family. Having observed Razieh's experience (and, frankly, too many similar examples), Natacha did not even attempt to apply for postdocs beyond her PhD. I tried really hard to convince her to stay, but she was too disillusioned with academia and its skewed incentive structure, in spite of her tremendous success in essentially all the diverse fields she entered. She continues to do amazingly well, but now does so outside the academy, and again to our great loss.

By this point, you may wonder, "Did a hyper-black hole spawn the universe?" I still believe this question is excellent, and maybe if we had gone down a different rabbit hole in our explorations, I would have had a satisfying answer for you. Nonetheless, as we have seen, there are still adventurous physicists out there who entertain this possibility in its various incarnations. But ultimately, we each have only one life to live and need to pick our battles. Sometimes we just have to let go, and for me, there was a new war brewing on the horizon.

Circling back to John Michell, it turns out that his two most famous inventions, modern seismology and black holes, were about to meet again, some 265 years after Lisbon's deadly earthquake. This new adventure, however, would require traveling faster than the speed of light.

8

SPEEDS OF LIGHT

The remains of our past are scattered throughout the world, whether they be buried far underground or hidden in deep space. They are invisible yet call us to recover the memories we have forgotten. Each one of these relics, be it old pottery or a dinosaur fossil, has its own array of experts keen to examine it. A prehistoric excavation might have paleontologists, paleobotanists, and ecologists. Cosmology offers an analogous situation. Some cosmologists specialize in the early universe, focusing on inflation and possibly alternatives. Others work on understanding the accelerated expansion that began a few billion years ago. And quantum gravity is the realm of theoretical physicists working on string, loops, or other paradigms hidden in the shadows, any of which might help us peer beyond the mirage of the Big Bang singularity. But like the story of the rise and fall of a dinosaur kingdom, many of us dream of a hypothesis that can link the findings of these divergent disciplines into a single thread. One candidate for such a narrative has many names, but it first came to my attention in the guise of VSL, or varying-speed-of-light theory. This theory postulates that in the extreme conditions close to the Big Bang, the speed of light is not the constant that Einstein's theory of relativity implies but becomes increasingly rapid, possibly even infinite. The constancy of the speed of light is written into the very fabric of relativity. So, this theory seems like heresy. But, as we shall soon discover, a varying speed of light can not only solve the problems that inflation claims to tackle but can also connect the various pieces of our jigsaw, uniting early and late time cosmology with quantum gravity into a single story. It's a prospect I find intriguing, but I must admit that at first, I was one of the reactionary physicists unwilling to countenance such a strange idea.

JOÃO IN THE PERSIAN GULF

Just as the story of cosmology began in conflict, so did my "Faster than the Speed of Light" adventure. It is hard to imagine a stretch of water more synonymous with conflict than the Persian Gulf. Apart from the endless wars that seem to be fought there, even its name holds conflict: some countries, mainly to its south, prefer to call it the Arabian Gulf, while others take a middle ground and call it simply the Gulf. However, the Iranians to the north consider this terminology an affront to their national pride, hence the phrase *Forever Persian Gulf* on government-sanctioned maps and in songs. Often lost to the astute observer of weekly news of mayhem in the Middle East is that for most of its inhabitants, daily life is relatively unremarkable. So much so that they even have time for relationships—and some, too, even have time to entertain grand questions, such as that of the universe's origin. From 1996 to 1999, I was an undergraduate student studying physics at Sharif University in Tehran, Iran. There, I met another physics undergrad, Ghazal Geshnizjani, who was also interested in cosmology; we fell in love and ended up marrying just as we finished college. Around the same time, one of our professors, a well-known relativist named Reza Mansouri, saw the excitement brewing worldwide in observational cosmology and decided to learn about theories of cosmic structure formation. Paradoxically, some college professors learn topics by teaching them to unsuspecting students, a group that in this case included me. However, solving Einstein's complex equations of general relativity in real-life cosmological settings was brutal. Even though we had a textbook, none of us (including our prof) had the faintest idea what was happening. So, we invited the world experts in cosmology to give lectures at a school in the Persian Gulf in the winter of 1999. After all, who could resist the chance of going to the beach in January?

The site chosen for the workshop was Kish Island, a small coral paradise nestled off the southern coast of Iran. Despite its tropical charms, however, the thought of visiting one of the most conflict-ridden parts of the world appeared to deter some of the more senior invitees. However, it did not faze their younger protégés or some of the more adventurous cosmologists. One of them was Robert Brandenberger, father of string gas cosmology. Brandenberger was influential in launching our careers, helping Ghazal and me move to the US and supervising her PhD. Another interesting character I met was João Magueijo, who was visiting us from Imperial College, London. As Brandenberger had with his string gas cosmology, Magueijo was pushing an alternative to inflation, and his ideas would be my first introduction to VSL and its rewriting of relativity. Breaking up with Einstein, though, is a painful and challenging separation. Most physicists may not believe in one God, but they definitely believe in one speed of light! But if a VSL theory could explain cosmological puzzles, it might stand a fighting chance. One reason Guth got excited about inflation was that it solved the horizon problem. Just as your pizza doesn't cool down right away, as it needs time to equalize with the temperature of the air, opposite sides of the universe need time to come to thermal equilibrium. But recall that our observable universe is too vast to allow light to travel from one side of it to another in the standard Big Bang theory; there simply has not been enough time. So there appears to be no way for the cosmos to become as uniform as we observe it to be today. Guth's solution, inflation, proposed that the universe had been much, much smaller at some very early time, small enough to allow light to easily zip through it. Magueijo's solution, by contrast, proposed that light at that early time had traveled much more quickly, allowing signals to be exchanged at arbitrarily vast scales. This trick was a neat one, but inflation requires no blasphemies against Einstein. Moreover, perhaps inflation's greatest triumph is explaining the origin of the tiny inhomogeneities in the universe. These unevennesses grew to form galaxies from subatomic quantum fluctuations stretched by the exponential expansion. VSL did not have a plausible mechanism by which to achieve the same wonders (at least, not yet). So, I remained unimpressed, and I was not alone.

"I've never felt thrilled about changing the speed of light," said Andy Albrecht in an interview with Phil. Albrecht was not just one of the founders of inflation but also Magueijo's collaborator in his first paper on VSL, published in 1999.[1] In his 2003 book, Magueijo recalls his trials and tribulations in promoting the theory and breaking through the relativistic orthodoxy.[2] This task was not easy (as evidenced by the words of his collaborator, let alone those of his critics). He also entertained concepts that seemed pretty fantastical. In *Star Wars*, ships travel across the galaxy by moving through corridors called hyperspace lanes, which allow faster-than-light travel. Magueijo wondered if there might be remnants of the early universe's higher speed limit still present in space today that might achieve this effect. As we've seen, cosmologists speculate that exotic objects known as cosmic strings could be left over from the extreme conditions of the Big Bang. These strings might be just the ticket to realizing the ultimate sci-fi dream of interstellar travel. The idea was that inside the string, the ancient conditions of the high-speed universe would be preserved, and so a spaceship could enter a cosmic string free from the restrictions imposed by Einstein.

I vaguely followed Magueijo's musings while doing my PhD at Princeton but was mainly preoccupied with the unprecedented CMB maps that were being charted by NASA's WMAP satellite, firmly establishing the premise of an inflationary cosmos and a very dark universe. At that time, I thought the case for VSL was simply not there. Other physicists were more brutal in discounting Magueijo's theory. For example, George Ellis, the preeminent South African relativist and another former student of Dennis Sciama, wrote a scathing critique of Magueijo's ideas in *Nature*.[3] Ellis points out that the constancy of the speed of light is well-built into the foundational structure of physical laws, and that you cannot arbitrarily make it variable in one place or another without a much more fundamental revision of the laws of physics. Ellis claimed that it was not even clear what a variable speed of light means: "It is then not possible for the speed of light to vary, because it is the basis of measuring distance." Inflationary pioneer

Andrei Linde made a similar comment, telling us that "light cannot travel with speed equal to 1.5 speeds of light." I found these criticisms disingenuous. Just because we don't precisely understand how something may work doesn't mean it's impossible. The speed of light might vary relative to other velocities, like the speed of gravitational waves. This idea was yet to be fleshed out, and of course, we needed a complete and self-contained theory, but such a theory was to come. Magueijo's reaction was less diplomatic; he told Phil that Ellis, one of the giants of relativity, will "never learn anything."[4] I wanted no part of this sort of infighting, but life has a way of leading you in the direction you least expect. My thoughts were not of the early universe and solving the horizon problem but on late-universe physics and dealing with the cosmological constant. But I would soon discover that these disparate riddles had a strange connection.

A WALK IN WISCONSIN: THE STORY OF CUSCUTON

In the fall of 2004, I defended my PhD and started a postdoctoral fellowship at Harvard, in Cambridge, Massachusetts. Around the same time, Ghazal also finished her PhD and started a postdoc in Madison, Wisconsin, under the supervision of a young cosmologist, Daniel Chung, who primarily worked on the early universe. However, we made the best of our two-body problem, with each of us spending one-third of our time at the other's home institution while only spending one-third of our time apart. As a byproduct, we flew every other week and had no savings. Whenever I visited Madison, I would hang out with the high-energy physics group to which Ghazal belonged. We usually walked from the physics department to the old Memorial Union building, which overlooked the beautiful Lake Mendota, for lunch. During one of these walks, in early winter of 2006, Ghazal, Daniel Chung, and I started pondering, "What would happen if signals traveled infinitely fast?"

It may sound like we were already pondering a VSL cosmology and joining the small rebellion against inflation, but we were being more conservative than that. As good relativity students, we all knew that "nothing can propagate faster than the speed of light." But the mystery of the universe's accelerated expansion—driven by a cosmological constant or perhaps by changing dark energy—requires some explanation. We found that it's possible to modify Einstein's equations in a surprisingly minimalistic manner to generate something that looks like a cosmological constant in its uniformity yet evolves in time. The most uniform fluid is one with the fastest possible speed of sound, that is, an infinite one. We were proposing then a new field that acts as a kind of universal fluid that permeates space and mimics a cosmological constant by allowing infinitely fast sound waves. This proposition doesn't conflict with relativity because it still does not allow information to be sent faster than light, which is what relativity really forbids. However, the existence of such a fluid revives the memory of the nineteenth-century conception of luminiferous ether that we described in chapter 1. Just as ocean waves require liquid water and sound waves need air molecules, so it was assumed that light waves require their own medium: the ether. But relativity showed that light was special; it requires no medium, and when experiments confirmed Einstein's theory, the ether was abandoned. Our model, though, had resurrected the ether from the graveyard of failed scientific conjectures. As long as ether remains dark (unable to interact with light, like dark matter or dark energy), then it won't violate any experimental constraints.

We named the new field Cuscuton, a term derived from *cuscuta*, the Latin name for a parasitic family of vines (which Morgan Freeman called a "true serial killer," as it tends to kill its hosts). The rationale for the name is that a fluid with an infinite speed of sound does not have any freedom to move except to follow whatever it is coupled with, much as Cuscuta always follows the plant whose life it is sucking out. This field's implications gave corrections to Einstein's equations, the very equations that imply a singularity at the heart of black holes and crucially at the Big Bang. At the time, it was not clear that we could use them to understand either of these physics cases at the extremes; that argument was to come. But we were not the first to play with ideas of modifying the very laws of gravity that modern cosmology is built from and were about to come under fire from another scientist with similar ambitions.

WELCOME TO PERIMETER!

Having submitted our Cuscuton paper, Ghazal and I moved to the Perimeter Institute.[5] My first surprise happened in August, even before we arrived. Since I was about to start this position, I used my affiliation in a paper we submitted about Cuscuton. The next day, I received an email from someone called John Moffat, whom I had not heard of. He expressed serious misgivings about our results, as he thought he had already proposed a similar modification of gravity. He was also surprised that (being affiliated with Perimeter Institute) I hadn't talked to him about them. Copied in on the email was our new postdoc supervisor, who had hired Ghazal and me at the Perimeter Institute. Moffat's email wasn't exactly the "welcome to the Perimeter Institute" message that I was expecting, as it felt like I was already in trouble with the boss, even before my first day of work.

I later learned that Moffat's attempts belong to a class of theories that attempt to tweak Einstein equations to do away with the invisible dark matter that cosmologists believe populates the universe. By contrast, our original Cuscuton model was focused on explaining dark energy. Magueijo described receiving a similar email from Moffat back in 1999, accusing him of "copyright violation" when he first wrote his VSL paper with Albrecht.[6] Unknown to Magueijo and Albrecht, Moffat was the true father of VSL, the first to publish a scientific paper (in 1993) on solving the cosmological riddles using faster-than-light travel, a model he called the superluminary universe.[7] It was a beautiful paper, years ahead of its time and totally ignored by the community, so much so that it was only cited twice in the six years leading up to Magueijo and Albrecht's work. According to Moffat, even the journal *Physical Review D*, which would publish Cuscuton and VSL papers years later, had declined to publish his "Superluminary Universe" paper.[8] It's a shame Moffat is not better known, as his life story is as idiosyncratic as his physics.

Moffat grew up in Britain, where he experienced the horrors of war firsthand when he and his family were shelled by German bombers during a stroll on a boardwalk in coastal England. As a teenager, he trained in fine arts with Russian abstract painter Serge Poliakoff before enrolling in a PhD in physics with Nobel Prize–winner Abdus Salam and steady-state theorist Fred Hoyle. He managed all this while having no prior qualifications in science at all. Remarkably, he taught himself the required material from books in his local library. At the age of twenty, he struck up a lengthy correspondence with Einstein regarding Einstein's attempts to unify physics. Eventually, Moffat became a physics professor at the University of Toronto. Even since retiring from his professorship, Moffat continues publishing papers with his former students and collaborators, having settled at the Perimeter Institute as a full-time researcher. In the years following our initial email encounter, I have spent many hours talking to Moffat at various seminars, in group meetings, or over lunch. He is full of colorful stories about giants of modern physics, but he's also a big fan of highlighting the flaws of the establishment, be it in politics, the Perimeter administration, or standard cosmology. He follows new developments better than anyone I know. To paraphrase the regular exchange in our weekly seminars, Moffat keeps pointing out to everyone that the emperor has no clothes, and others respond: "Yes, we know, this is a nudist resort!" Of course, Moffat knows that too, but he obviously doesn't like it. Moffat continues to be a contrarian and a revolutionary to this day, but having grown up in the aftermath of the 1979 Iranian Islamic revolution, I know from experience that continuous revolution is not necessarily a good thing.

As astrophysicists, we are trained to solve problems and seek to understand the details of the inner workings of natural phenomena. These details are often boring, but ultimately each of them is an integral part of the complete puzzle. When Ghazal and I arrived at Perimeter, we had little idea that it was amid its own revolution. In his book *First Principles*, Howard Burton, Perimeter's founding director and an academic grandchild of Moffat, details his odyssey to start a new theoretical physics institute during the late '90s and early in the first decade of the 2000s.[9] As a recent physics PhD, he was approached by the co-founder of Research in Motion (RIM), Mike Lazaridis, to start an institute to determine how people will communicate a hundred years from now. RIM would go on to build BlackBerry, which was arguably the first smartphone, forever changing how we communicate. The name "Perimeter Institute for Theoretical Physics" would symbolize the goal of pushing the boundaries of physics, but it was also an homage to RIM (the perimeter of a circle being its rim). Then, just before we arrived at Perimeter in 2007, Burton left when his contract was not renewed. According to one account, "in conversation, Burton [came] as close as he [could] to saying he was fired."[10] Our early years at Perimeter were a tortured balance between novelty and mainstream, the old founding vision espoused by Burton and the new vision espoused by the replacement director, Neil Turok, co-founder of the Ekpyrotic/cyclic cosmology model outlined in chapter 6 (yes, indeed it's a very small world). Out of this conflict, I happened to discover my inner John Moffat. I was about to realize that our Cuscuton field, which we had invented to deal with the accelerated expansion of the universe, was sitting at the crossroads of VSL cosmology and a new theory of quantum gravity that might shed light on the riddle of what really happened at the Big Bang.

MUKHANOV VERSUS MAGUEIJO

I would say that both João Magueijo and Slava Mukhanov (the Russian champion of inflation, who first correctly computed the spectrum of its quantum fluctuations) have strong, quirky personalities. For example, Magueijo, who is a professor at Imperial College, London, has a footnote in his 2003 book that reads, "I have at times considered launching a devastating terrorist strike against [Imperial College's (IC's) administration] staff and building. IC's average IQ would increase significantly."[11] This comment came only two years after the 9/11 terrorist attacks in Manhattan. Mukhanov, on the other hand, is an outspoken defender of inflation (though, paradoxically, not a fan of the multiverse). Robert Brandenberger once revealed to me that Mukhanov had, one day, suddenly shown up at Tufts University, in the Boston area, where they were both visiting eternal-inflation guru Alex Vilenkin. When Brandenberger inquired as to what they owed the pleasure of his visit, Mukhanov answered that he had been en route to Princeton, but while reading WMAP's latest paper on the plane he had become too upset to visit the WMAP team who were based there, so he'd taken the next flight to Boston instead. In particular, WMAP reported that they had detected a varying degree of deviation from scale-invariance (known as running), which was not easily explained by inflation. Mukhanov claimed that these results had been obtained using "dirty astro," leading the researchers in question to come to a "non-reliable nonsense" conclusion. As this running required inflationary models to be so finely tuned, that would have been "the end of cosmology as real science." Fortunately for cosmology, Mukhanov was right, and WMAP's running literally turned into dust, just as BICEP's B-modes would a decade later.

Interestingly, both Magueijo and Mukhanov were visiting the Perimeter Institute when I arrived there in 2007. So, if you thought of VSL versus inflation as a war, you might also think that I had just walked onto the battlefield. Turns out this warfare was asymmetric, and little did I know I was about to join the rebellion. If I remember correctly, sometime in late 2007 or early 2008, when Ghazal and I had just arrived at the Perimeter Institute, I overheard Magueijo talking to Moffatt, whose office was next door: "This is called Cuscuton."

"What?" Moffat said.

"Cuscuton . . ."

Of course, I couldn't control myself and had to walk in. It turned out that Magueijo was working on the latest incarnation of his (and, as you may recall, Moffat's) VSL model and had realized that it would approach our Cuscuton theory as it approached the Big Bang singularity. In March 2008, Magueijo published the paper, entitled "Speedy Sound and Cosmic Structure," using Mukhanov's own methods to show that a VSL theory that approaches Cuscuton at early times does the same things that inflation does, that it gives a near scale-invariant spectrum of sound waves, as observed in the CMB. It takes this scale-invariant spectrum to seed the structures of galaxies we see today.[12] This development was a major milestone and so you might think that it served as my call to action. Two pieces of our cosmological puzzles had been linked together, Cuscuton providing a proposal to model dark energy and VSL offering a rival to inflation to solve the puzzles of the early universe. But at that point I still was not taking VSL too seriously. The infinite speed in a Cuscuton theory is fine because you cannot transmit information with it, but I was still uncomfortable with superluminal signals, which were sacrilege according to Einstein's gospel. Magueijo, who according to the *New York Times* was the "bad boy" of physics, was not afraid of blasphemy, telling Phil, "The minimum criterion for solving all these problems was to add thirty-two zeros to the current speed of light and say, 'This is the speed of light in the early universe.'[13] But this is like the minimum. It could be anything above that, and if you don't want to be too fine-tuned, you might as well just say, 'Well, it's infinite.'" I was also obsessing over black holes and dark energy and looking for a faculty job to get closer to Ghazal. So, I had too many distractions. But soon, a heretical proposal by a famous string theorist would pull me back in, putting the final piece of the puzzle in place to reveal an entirely new vision of the Big Bang.

HOŘAVA GRAVITY AT THE HELLI HOTEL

In January 2009, shortly after we all returned to the Perimeter Institute from our holidays, a strange paper by Czech string theorist Petr Hořava appeared on the *arXiv* server.[14] Hořava is a former postdoc with Witten at Princeton (you may recall that the Ekpyrotic model was based on the Hořava-Witten theory of higher-dimensional branes). However, turning his back on his roots, the basic idea of Hořava's new quantum gravity proposal was to sever the ties that bind time to space at high energies by breaking the symmetries (known as Lorentz invariance) of relativity at the Big Bang. Doing so would open the door for our new picture of the birth of the universe. Just as inflation envisioned a phase transition for the energy of the vacuum, like liquid water turning into ice, so Hořava was proposing a change of phase for space-time itself. As we move to higher-energy densities, the relativity of time and space, as described by Einstein, fades away, melted by the extreme energy densities of the Big Bang. This fading means that the speed of light is no longer a constant. In fact, as we approach the Big Bang, the speed of light increases without limit. Such limitless increase was exactly what Moffat, Magueijo, and other VSL proponents had been suggesting all along. One of the only reports on Hořava gravity in the popular press was an article in *Scientific American* written by my friend Zeeya Merali; it ends with theoretical physicist Gia Dvali claiming that this "pathology" of an infinite speed of light caused him to give up on such speculations.[15] As a powerful advocate of inflation and of attempts to link it to string theory, I guess he wasn't keen on alternatives to inflation appearing in a quantum gravity theory. Ironically, what Dvali considered a "pathology" was, in the eyes of the protagonists of this new VSL rebellion, the whole point. Others were rejoicing as now there was a rival to inflation that seemed like a direct prediction of a more fundamental theory of physics.

Hořava was not the first physicist to play with ideas like relativity modifications that broke the connection between time and space. My colleagues and friends John Moffat, Ted Jacobson, Lee Smolin, and João Magueijo (amongst others) had considered similar possibilities, both in theory and in observations. For a while, some people even though breaking Lorentz invariance was a prediction of loop quantum gravity. However, like VSL, these ideas were considered fringe, to be frowned upon at best. Loop theorists eventually argued that their theory respected Lorentz invariance, and the proposal was banished by the community. Similarly, string theorists took principles of relativity and quantum mechanics as divine revelations. Hořava was the first mainstream, influential string theorist (to my knowledge) who deviated from this dogma. Furthermore, he constructed a mathematical theory using the same geometric language as Einstein's relativity, ensuring that he recovers general relativity on large scales. The proposal was revolutionary and cited over 2,500 times, more than any other single-author monograph since its publication in 2009.

The full title of Hořava's paper was "Quantum Gravity at a Lifshitz Point." But where is a "Lifshitz point" and what is "quantum gravity" doing there? Here, to understand the nature of a quantum vacuum and how Hořava proposed to change it, I will use an analogy that I would like to call the "Helli Hotel," which you may also recognize as an homage to the "Hilbert Hotel." The latter was an imaginary hotel that has infinitely many rooms, first described by the mathematician David Hilbert and popularized by Big Bang pioneer George Gamow. The Hilbert Hotel has many strange properties, like having space to admit new guests even when it is full. Such a thing is possible because an infinite hotel cannot run out of space; as Gamow put it, "in exactly the same way that an infinite hotel can accommodate an infinite number of customers without being overcrowded, an infinite space can hold any amount of matter and, whether this matter is packed far tighter than herrings in a barrel or spread as thin as butter on a wartime sandwich, there will always be enough space for it."[16]

Back in Iran, I used to go to a Helli high school (named after the famed fourteenth-century Islamic scholar). The all-boy high school had branches in different cities all over the country that were loosely run under a single umbrella organization. During summer holidays, individual Helli schools would organize trips to other cities, and often we ended up sleeping in the carpeted prayer halls of other branches of Helli schools. Students could use these prayer halls for all sorts of things (other than prayers), even sleeping. So, when students from other cities visited, all they needed were some pillows and blankets, and voila, you had a dormitory. In fact, the smaller the kids, the more of them could sleep within the same area. You can now imagine a mathematical limit, where you can fit infinitely many students in one room if each one is, for example, half as small as the next one. You could then have a few big students, with more smaller students sleeping in between them, and so on. This scenario is what I call the Helli Hotel. The students in my analogy represent the virtual particles in the vacuum of quantum physics; there are infinitely many of them; they are currently sleeping but can be woken up, given the right conditions (turning virtual particles into real particles). Now, imagine that you are at the Helli Hotel, waking up in the middle of the night and trying to make your way to the bathroom, across a dark hall filled with sleeping students (a scenario I know from personal experience). If you are not careful and step on someone, they may scream, and even possibly wake up others around them, initiating a chain reaction. For the sake of argument, imagine that the energy it takes to wake up someone is inversely proportional to their size (younger kids "sleep like a baby"; for actual relativistic quantum particles, this energy is proportional to the speed of light divided by the particle's size). So, there are more small kids you may accidentally step on, but they are less likely to wake up. Can you make it across the room and to the bathroom? If you are careful, especially in not stepping on smaller kids (since there are many more of them), then the answer is yes. You may wake up a few kids here or there, but they'll fall back asleep, and you won't start a chain reaction. In other words, the infinities can be ignored, and we can get sensible predictions from our calculations. These sorts of interactions in quantum physics are called renormalizable. Gravity, however, is different. The higher the energy, the bigger its gravitational interaction, so the infinities can't be ignored and blow up your equations. In our Helli Hotel analogy, your midnight attempt to visit the bathroom will result in the absolute chaos of a room full of screaming, sleepy students. In physics, we call this breakdown of our mathematics "loss of predictability," like when you divide any number by zero in your calculator. These infinities remain in our equations and, with few exceptions (like the standard model of particle physics), plague all quantum theories, including the quantum theory of Einstein's relativity. Removing these infinities and regaining predictability is one of the main challenges to constructing a quantum theory of gravity. Particle theories can be renormalizable, the infinities of its theory tamed; gravity, not so much.

But what if the energy it takes to wake up smaller students (analogous to making particles out of the vacuum) were much bigger? What Hořava proposed was that if, rather than energy scaling as one over a particle's size, it varied as the inverse of the cube of a particle's size—that is, if smaller students are much harder to wake up—then we have a renormalizable interaction. And having a renormalizable interaction means that we remove the infinities, taming the mathematical chaos that often haunts quantum calculations. This nontraditional dependence of energy on size, a proposition that deviates from relativity, is known as a Lifshitz scaling (named after the Russian physicist Evgeny Lifshitz, who first noticed the effect near phase transitions of quantum material). And here's the kicker: particles that satisfy this Lifshitz scaling will have a speed that increases as the inverse of the square of their size, so that small (or energetic) particles can move arbitrarily quickly. This state of affairs, of course, violates the fundamental assumption of special relativity: that nothing can travel faster than the speed of light.

The first few months after Hořava's paper came out were a deluge of confusion. Physicists were not sure if the theory was mathematically consistent or which version (out of many) made sense. Nor was it clear whether it could match astronomical observations and therefore compete with general relativity as a guide to how objects move through the cosmos. However, I noticed something profound. I could show that Hořava's theory on large scales was equivalent to Cuscuton. Thus VSL, Cuscuton, and Hořava gravity were all different expressions of the same idea. The pieces of the puzzle were merging into one concept that could challenge inflation and rewrite our story of the Big Bang.

That summer, my colleagues at the Perimeter Institute organized a workshop to discuss all the confusing aspects of these developments; at this workshop, I also presented the surprising connections that I had come up with. It was an exciting time, as for the first time in my life, I could see a glimmer of hope for convergence and respect amongst different tribes of our most foundational physical endeavor. In closing, I remember Hořava, the esteemed string theorist, thanking loop quantum gravity founder Smolin for all his leadership in quantum gravity research, only eight years after the publication of *The Trouble with Physics*, Smolin's scathing critique of string theory. Alas, this détente did not last long, and people soon retreated back to their old tribes.

For me though, the taboo had been broken. There was a single framework—call it Cuscuton, Hořava gravity, or VSL—that could address our deepest problems in physics: the horizon problem of the Big Bang, the origin of cosmic structure, the infinities of quantum gravity, and much more. For example, in 1964 John Bell proved that predictions that come out of quantum physics cannot be consistent with local statistical behavior (what physicists call "local hidden variables") unless we allow for faster-than-light communication. For example, one may think that randomness in quantum mechanics is like a coin toss: a chef may decide to serve her customers pudding or pie, depending on whether the coin comes up heads or tails. However, it turns out that you cannot reproduce the kind of randomness we observe in quantum mechanics across space with this kind of local coin toss (a point confirmed by Alain Aspect, John Clauser, and Anton Zeilinger, who won the 2022 Nobel in physics for this work). Simply throwing a coin and signaling the outcome of the toss at the speed of light is not enough to achieve the sort of outcomes we see in quantum experiments; the random coin toss happens across space, all at once. While physicists and philosophers discuss the true meaning of Bell's theorem ad infinitum, a simple way of removing metaphysical magic from quantum physics is simply to allow subtle, instantaneous random signals. So, even though you cannot use such signals to send messages to your friends, they can solve paradoxes of quantum mechanics. I was finally convinced that VSL had promise and was now a cosmologist with an agenda: to revive the ether that Einstein left for dead, across space and time, from black holes and the Big Bang to the outermost expanses of our cosmos. Now that I no longer shunned VSL, it was time to team up with Magueijo to probe the Big Bang itself.

THE BI-THERMAL BIG BANG

The one thing that Magueijo had mentioned in passing in his 2008 "speedy sound" paper was a potentially thermal origin of structures. Recall that the problem for any model of the Big Bang is to explain why the CMB has almost the same temperature everywhere but with tiny fluctuations that are nearly scale-invariant at large scales. Returning to the pizza analogy, just as dough gets warm because it comes into thermal equilibrium with the surrounding hot air molecules in the oven, so does hot cosmic plasma harmonize its temperature easily, given the gargantuan speed of light VSL proposes in the very early universe. But like a speck of dust undergoing random Brownian motion due to frequent kicks from water molecules, the sound waves in the primordial plasma fluctuate due to the inherent disorder in a thermal bath, leading to small fluctuations in density. As the universe expands, it exits the VSL regime, inducing a sudden drop (or phase transition) in sound speed. This drop would transform a thermal black body into a pattern of nearly scale-invariant hot and cold regions, matching the CMB fluctuations that were so spectacularly confirmed by WMAP and Planck observations.

VSL received a lot of press attention back in the early years of the first decade of the 2000s for being able to solve the horizon problem without inflation. But it was the ability to account for cosmic structures that excited Magueijo. In fact, he admitted to Phil that when the press got enthusiastic about VSL over twenty years ago, the model couldn't really compete with inflation in explaining the origin of structures, but now things were different. In 2016, I knew of a few models of cosmic structures that featured a thermal origin. For example, nearly all inflationary models cool down the universe extremely fast via exponential expansion and resulting redshift. Then, you need to tag on a "reheating phase" by hand to turn all the energy of the supercooled inflaton field into a hot plasma. This mysterious reheating phase introduces an inherent uncertainty to all inflationary models. However, a different theory (such as a thermal VSL), wherein the universe remains hot all the way through, will not have this problem. But creating a model with an expansion history that matched observations was a daunting task.

Magueijo and I started bouncing ideas off each other. He had a history of taking people's concepts (in this case, Mukhanov's) and turning those concepts against them. Now, it was my turn to do the same thing. Recall that the branes in string theory, lower-three-dimensional membranes moving in a higher-dimensional space, played a significant role in making models of our universe, including Ekpyrosis and brane inflation. One of the features of these brane models is what is known as brane-bending mode, something that may appear like an ordinary scalar field (these are the sorts of fields used to model the inflaton or the Higgs) to us three-dimensional, sentient beings but that in fact represents that motion of our membrane in the higher dimensions. Like the inflaton, this scalar field fills the universe, and we can see its fluctuations propagate and gravitate. Interestingly, the more quickly the brane moves in the invisible dimensions, the more slowly it propagates in our three dimensions. But if the extra dimension is not a dimension of space but rather one of time, this construction produces our superluminal waves for free. Since the model had two time dimensions and was thermal, we ended up calling it the bi-thermal Big Bang.

What does a bi-thermal early universe look like? It turns out that things in it are pretty similar to those we see in the standard hot Big Bang—until we get to extreme energies close to the quantum gravity era. As we get to higher densities or temperatures, we notice that sound waves become noticeably faster, and their speed rapidly blows up as we go back in time so that anyone (who can survive these hellish temperatures) can contact anyone else infinitely many times, no matter where they are in the universe. This faster sound speed is indeed how we solve the horizon problem. But this bizarre acoustic history also guarantees the peculiar spectrum of sound waves. Recall that in the early universe, these sound waves ripple through the primordial plasma that is the hot early universe and leave their traces as the hot and cold spots we see in the CMB today. Understanding their propagation is essential for any model of the Big Bang. In our model, we start with a "black-body" spectrum of highly excited sound waves in thermal equilibrium. As the universe expands and cools, we exit the VSL era, when the speed of light relaxes down to its current value of 300,000 kilometers per second (which, although it sounds big, is tiny by Big Bang standards). This slowdown leads to a deep freeze in which sound waves stop moving, reminiscent of what happened to the jubilant city of Arendelle in Disney's *Frozen*, following Princess Elsa's escapade.

In inflation, the universe expands so fast that regions of space that start in close proximity rapidly lose contact. Imagine siblings, one ashore, bidding adieu to the other on a ship disappearing from view as it sails into the high seas and beyond the horizon. As we described above, a VSL era would achieve that same effect by slowing sound to a halt, so that the siblings can't communicate with each other. As the universe relaxes down to its normal physics (normal expansion after inflation ends, or the normal speed of light after the VSL phase), the horizon continues to push further outward, eventually encompassing the original region that disappeared from view. We can think of this era as one in which sound waves unfreeze as regions inside the horizon become reunited with their once-separated partners, just as Elsa rescued Arendelle from its icy fate when she rediscovered the love of her sister Anna. Like inflation, this whole process takes a few hundred thousand years and finally imprints sound waves in the CMB fluctuations that lead to the structures of galaxies we see today. Unlike the Disney version, all it takes for the universe to come back to life is time. True love must endure through time.

Is this cosmology an eternal or beginning one? Interestingly, it's both! If you were a sound wave, you'd think time stretches back forever, as you'd be able to go arbitrarily fast and thus arbitrarily far. In other words, if you could build clocks out of sound waves, there would be no beginning of time. Nonetheless, not much is happening during this "infinite" time, as sound waves remain in thermal equilibrium. On the other hand, according to gravity, time has a beginning; gravitational waves cannot travel very far. Thus, whether time has a beginning is ambiguous and depends on the observer. What of the singularity theorems, then? While you may naively think there is a singularity at zero time since the scale factor goes to zero, we know that a quantum theory like Hořava gravity can become scale-free (or conformal) on small scales (remember, Penrose envisaged as much for his CCC model). So, the vanishing of the scale factor has no physical significance; a change of coordinates can remove it. Incidentally, this feature is basically what allows the model to solve the flatness problem (that is, the question of why the cosmic space starts out so flat). It turns out that the scale-free equations require zero curvature from inception. When Penrose first began his study of black-hole singularities, theorists believed black holes actually contained not one but two singularities: the first at the event horizon and the second at the black hole's center. But David Finkelstein soon proved that the infinities at the event horizon existed because of a poor choice of coordinate and thus had no physical significance. We were making the same claim for the Big Bang. Our model removed the justification for calling the bang the beginning of time, just as that same justification had disappeared in other quantum gravity models. It seems this point is something all our models share.

Some physicists don't like conjecturing extra dimensions of time, as doing so destroys our classical notion of causality. We usually think that an effect is preceded by its cause in time, but if two different time coordinates exist, we will not be able to establish what is cause and what is effect. But then maybe causality is not a fundamental property of the universe and is only emergent for our large-scale, low-energy observations. Incidentally, this possibility also means that if the universe began, we might not be justified in demanding that it necessarily have a cause (we shall return to this point later). Maybe having two time dimensions is not even that radical. Itzhak Bars, the string theorist on sabbatical at the Perimeter Institute and working with Turok and Steinhardt, had a passion for two-time physics. He even co-wrote a book on the subject in 2010.[17] So, I thought he would be amused during one of his Perimeter Institute visits when I explained the idea of getting faster-than-light propagation via branes moving in extra time dimensions. I think that idea crossed a line for him, and we never discussed it again. Maybe he was not as radical as I thought.

Only eight days after João and I submitted our paper to *arXiv* in 2016, our favorite Iranian string theorist at Harvard, Cumrun Vafa, along with Robbert Dijkgraaf, then director of Princeton's IAS (Institute for Advanced Studies; Einstein's old home), now Dutch minister of Science, submitted a paper in which they said: "Negative branes are intimately connected with the possibility of timelike compactification and exotic spacetime signatures."[18] "Exotic space-time signatures" is physics lingo for *more time dimensions*, so while I still didn't exactly know what "negative branes" were, it looked like they were what we needed to get superluminal propagation at the Big Bang. I then wrote to Vafa, bringing our paper to his attention and suggesting that we should work together to look at the cosmology of their model. He immediately responded (with João and Dijkgraaf copied in): "Yes, indeed our setup is natural for cosmology, and we have already begun thinking about that. I will read your paper." And that was the last I heard from him on that topic.

At the end of the day, the true measure of a physical theory's success is how many data points it can fit, not how many string theorists you can convince to embrace it. Assuming that we (and indeed all matter) exist on a membrane that moves on the geometry we had used in our model and requires consistency with Einsteinian equations of general relativity. At this juncture, we had only one free parameter in our theory that could be fixed by comparison with CMB observations. And to me, that development was crucial; the more flexibility a model has, the harder it is to evaluate, as you can always change things to get your theory to agree with the data. But at this point, our theory was completely rigid, and so it had a hard set of predictions that could be tested against astronomical observations. In particular, we had precise predictions for the patterns of temperature fluctuations seen in the CMB sky (notably the "red scalar tilt" that we introduced in chapter 2).[19]

Similar to loop quantum cosmology, another smoking gun for the bi-thermal Big Bang comes from primordial non-Gaussianity, a kind of distribution roughly translating to slightly more very hot CMB spots than very cold ones. A perfectly Gaussian distribution will have as many hot spots as cold spots. The story of calculating this effect is also interesting, as it involved computing some very complicated integrals; basically, João and I were stuck for almost five years. But then pandemic lockdowns happened, and I met Maria Mylova, a Greek cosmology postdoc living in Korea, on Twitter of all places. She and a mathematician friend of hers, Mary Mochoscou, managed to do those integrals, and voila, we had another set of sharp predictions to be tested by those upcoming experiments.[20]

What about gravitational waves? I am of two minds on this question. On the one hand, in the paper with João, our model couldn't generate them (it was only the speed of sound that blew up at the Big Bang, not that of gravitational waves). On the other hand, in a paper with my former master's student Abhineet Agarwal, we argued that Hořava gravity (which is presumably more fundamental than general relativity) might produce gravitational waves with an energy density 0.1–0.01 percent of that of the sound waves in the primordial plasma. Thus the situation is still uncertain. Recall from chapter 2 that the BICEP team first thought they had measured this number to be about 20 percent, a result that later turned into dust. But as we shall see in our final chapter, the energy density of any primordial gravitational waves will be measured more precisely by future experiments. So, on the issue of gravitational waves, we are worse off than some of our competitors; many cyclic models have a falsifiable prediction that no primordial gravitational waves exist. Inflation is compatible with both detectable and non-detectable gravitational waves. I have to admit that our model is, too, and thus this flaw of inflation is one we share.

At the end of the day, as elegant (or contrived) as it may sound, the bi-thermal story of our cosmic creation is just another story on par with others we have told in this book (albeit a more testable story). Elements of the story may find application elsewhere. For example, Ghazal and her students showed recently that, with the correct Cuscuton theory, it's possible to model a well-behaved bouncing cosmology, one that avoids all the singularity theorems or quantum instabilities that bedevil other scenarios, like Ekpyrosis. In his book, João discusses a cyclic cosmology based upon VSL, the idea being that a large drop in light speed converts the cosmological constant into matter, kick-starting the Big Bang. Things then settle down, but as the cosmological constant resurfaces, it will eventually reignite the VSL era and reset the cosmic clock.

ECHOES FROM THE ABYSS

Remarkably, VSL theory can also be used to construct solutions to the black-hole information paradox. Recall that quantum mechanics insists that information can never be destroyed. But relativity implies that a black hole will mercilessly annihilate any infalling information. Thus, our most cherished theories of physics contradict each other. If, however, the extreme conditions of black holes generate a quantum structure as described by Hořava gravity, this structure could allow for faster-than-light propagation, implying that information can get out, resolving the paradox. As we saw in chapter 4, the world of astronomy was forever transformed by LIGO's detection of black holes spiraling into each other, merging to form a single entity. This spectacular discovery opened up the world of astronomy to an entirely new domain: that of gravitational waves, ripples in the fabric of space-time. It also presented us with another possible way of testing our ideas.

At the time that gravitational waves were discovered, I had two students visiting me at the Perimeter Institute who needed new projects to work on: Jahed Abedi, a PhD student visiting me from Sharif University, and Hannah Dykaar, an undergraduate student visiting from McGill University. So, I asked them to help me discover whether it was possible to see a reflection from gravitational waves. None of us had ever worked on gravitational waves, yet Jahed and Hannah were persistent, and thanks to the LIGO team, the data was publicly available to analyze. But what exactly were we looking for? Relativity's black holes are like bottomless pits; things that fall into them (including light or gravitational waves) will never return. However, giving black holes quantum structure near their horizons is like providing a bottom to the pit—and if the pit has a bottom, it might serve as a surface for reflections or echoes to bounce off. However, the deeper the pit, the longer the wait to hear those echoes. Luckily, the waiting time for LIGO black holes is not very long, only a fraction of a second. So, we looked at the data a fraction of a second after black holes had collided with each other, and—lo and behold—we found something peculiar: a faint echo. In the summer of 2016, I showed our preliminary results at a conference at Caltech, in Pasadena.[21] There was a lot of excitement and press coverage when we finally published our paper in December. I think this potential to see "echoes from the abyss" is one of the most exciting things in cosmology. My enthusiasm for this line of investigation has led my mentor, David Spergel, and I to a double wager. He thinks that the most likely window to new physics in our field is the Hubble tension (recall that this tension arises from the fact that different methods of measuring the expansion rate of the universe do not agree). Which of these approaches turns out to lead to scientific breakthrough will determine the victor. The winner gets dinner in NYC or Toronto (not the sleepy academic enclaves of Princeton or Waterloo, Spergel has emphasized).

The thing, however, with faint signals like black-hole echoes is that they are hard to see, especially if you don't know exactly what you are looking for. Hundreds of researchers have studied echoes in the years since, and dozens of groups have looked for them, most using different methods and models. A handful have found faint signals from some events, as we did. Most, however, haven't found anything significant, which is not too surprising given that methods with many free (often unphysical) parameters find it much harder to distinguish faint signals from noise. Our bi-thermal Big Bang model, by contrast, has no free parameters. In the meantime, the LIGO team won the Nobel Prize in physics in 2017 for their groundbreaking discovery of gravitational waves. In a historical coincidence, John Michell's two significant intellectual achievements, seismology and black holes, would reunite after over two-and-a-half centuries—in what we called quantum black hole seismology—through gravitational wave echoes that peer into the quantum abyss. In 2018, Hannah's mother, my colleague Donna Strickland, won the Nobel Prize in physics for her discovery of chirped pulse amplification in lasers, only the third woman in history to do so. Whether Hannah (now an astronomy PhD student at the University of Toronto) will follow in her mother's footsteps and make discoveries of her own, time will tell.

From a bigger perspective, Hořava gravity is a full-blown proposal for quantum gravity, with all the trials and tribulations that such a proposal entails. For example, if the ether that Einstein vanquished a century ago is alive and well today, why don't we see evidence for it in laboratories or in the sky? Take the collision of two neutron stars that happened 130 million years ago that was seen by the LIGO gravitational-wave observatory and the Fermi gamma-ray observatory within two seconds of each other in August 2017. In other words, whereas many would expect a Hořava-gravity proposal to hold otherwise, observational evidence shows that light and gravity travel at the same speed, to a precision of one part in two quadrillion. Such observational findings are why most physicists today remain skeptical—as George Ellis was back in 2003—of VSL or Hořava gravity. But then, string theory predicts six extra dimensions of space that we cannot see. In the same way that extreme energies may be necessary if we are to probe hidden, extra dimensions of string theory, extreme conditions may be required if we are to observe the presence of ether. This state of affairs is a drawback for both approaches, as neither string theory nor Hořava gravity gives precise predictions about when these effects shall arise. A marriage between quantum physics and gravity, like any other marriage, is a work in progress and requires compromise and patience.

When seeds are randomly scattered across a field, it may be hard to predict which ones will grow into huge trees in a few decades and which will simply fail. While this book exhibits some seeds of possible Big Bang theories, they are clearly at different stages of growth (or failure). The tree that is faster-than-light cosmology may show little above the ground but, in my view, has the deepest and most branched roots, extending well into the rest of physics. But then, not everyone cares to look deep into the ground; ultimately, it is the fruit of the tree, the empirical verification, that counts.

✷ 9 ✷
HOLOGRAMS AND MISSING DIMENSIONS

Once, my son asked me why people who live in different places on Earth speak different languages, even though they all came from Africa. While I gave him a very sensible answer about the evolution of language, there is a biblical response to this question that, in some sense, might be much more profound. According to the book of Genesis, mankind used to speak a single language. They decided to build the tower of Babel—a structure that would reach the heavens. God, who observed this transgression, confounded their speech so that they could no longer understand each other, forever separating humanity from the divine realm and from each other. This story might very well be that of physicists, trying to reach out to unravel the mystery of the Big Bang and being confused by their divergent disciplines. On the one hand, we have the language of geometry and causality, perfected by Einstein and cherished by relativists. On the other hand, we have the language of quantum mechanics, represented by gigantic matrices that embody the entangled uncertainty of quantum measurements, cherished by string theorists and particle physicists. Here, we shall tell the story of holography, the wonderful Rosetta stone that promises to bridge this divide. It starts with our old friend Jacob Bekenstein, and our journey will lead us all the way back to the Big Bang.

One day in 2006 (if I remember correctly), I took a stroll down from my office to Harvard Square to attend a physics seminar by the Israeli physicist Jacob Bekenstein. The observatory is co-located with the Smithsonian Astrophysical Observatory, together making the Harvard-Smithsonian Center for Astrophysics; it's the biggest astronomy institute in the world, housing close to a thousand astronomers in residence. The observatory is nestled within a beautiful, suburban residential neighborhood, typical of many across New England. Getting to the rest of the historical Harvard campus required a twenty-minute leisurely walk along Garden Street to get to the always-bustling Harvard Square. It used to mystify me why the astronomy department, despite being part of the Faculty of Arts and Sciences, was so far away from the rest of the campus. The distance was certainly an inconvenience, as I often wanted to attend seminars, like Bekenstein's, at the physics department; getting to and from their building required an extra forty minutes.

Later, I heard a rumor that this distance from campus was not entirely an accident. Right next to the administrative office at the observatory hung a large picture of a woman whom I had not heard about before coming to Harvard. Her name, printed in large letters, was Cecilia Payne-Gaposchkin, who I learned was the first person to obtain a PhD in astronomy from Harvard in 1925 (later, she became the first woman to chair the Harvard astronomy department).[1] Her doctoral research, said to be "undoubtedly the most brilliant PhD thesis ever written in astronomy" by Otto Struve, a preeminent astronomer of the time, concluded with the groundbreaking discovery that the sun was mostly made out of hydrogen and helium.[2] But the Graduate School of Arts and Sciences at Harvard would not even admit women until much later, in 1962. Prior to this point, women could only study at the Radcliffe, which was next door to the observatory. The rumor was that Harlow Shapley, then director of the observatory, established the astronomy department in order to award Cecilia Payne (her maiden name) a doctoral degree. As discussed, women had worked as "computers" at the observatory as early as the 1870s, doing the bulk of the astronomical data analysis and making lasting contributions, despite lack of official titles or degrees. Among their ranks, the names of Annie Jump Cannon (mother of stellar classification) and Henrietta Leavitt (pioneer of extragalactic astronomy, whom we met in chapter 1) can these days be found in any introductory astronomy textbook. If you have read Margot Lee Shetterly's *Hidden Figures* or watched the Hollywood film adaptation—about Katherine Johnson, Dorothy Vaughan, and Mary Jackson, all of whom made crucial contributions to NASA's human spaceflight program—you might consider the story of "Harvard's computers" as a true-world historical prequel.

While it is easy for some to think that we have progressed well beyond the gross inequities of the past, old habits die hard and often resurface. Remember my colleague, Donna Strickland, who in 2018 was only the third woman to win the physics Nobel Prize? Her win came after 206 of her male peers had received the same recognition. During my time at Harvard, the president of the university, Larry Summers, was pressured to resign (through a vote of no confidence by Harvard professors), in part because he had suggested that the underrepresentation of women at highest positions might be due to "systematic differences in variability" between the sexes, in addition to social factors.[3] But the hierarchies embedded in our academic institutions (and in few places more than at Harvard), are in part created through historical inequities, but they feed into future inequities that may be less explicit but are just as powerful: women versus men, astronomy versus physics, prosaic versus promising, old guard versus new guard. These supposed oppositions are all examples of hierarchies across academia (and beyond) that are created by our prejudices, yet survive by feeding into each other.

What stuck with me from Bekenstein's seminar, now almost two decades ago, was the extent to which these hierarchies were in plain sight. As we have seen throughout the book, by this point, Bekenstein was already a towering figure in theoretical physics; his discovery that black-hole horizons have entropy had even stumped Hawking, before the latter turned around and proved that Bekenstein was indeed right. Down in the seminar room at Harvard's Jefferson Lab, I found a short man in a yarmulke, diligently describing his steps in constructing a modification to Einstein's general relativity that could explain the evidence for dark matter, without requiring actual dark matter or violating other experimental constraints (John Moffat has been trying to do the same thing for decades). In essence, the existence of dark matter is inferred, as astronomers witness a discrepancy between the rotation of galaxies and the dictates of gravity's laws for their stars. Modified gravity, instead of invoking unseen stuff (that is dark matter) to fix the discrepancy, alters the very laws themselves. Bekenstein was one of the first to make such a modification not just to Newton's equations but to Einstein's. At the conclusion of his presentation, the audience—who were mostly Harvard physics postdocs—turned away from the speaker and started to direct their questions about the talk to another member of the audience, Nima Arkani-Hamed (more on him, soon); the audience then left together, leaving the speaker alone at the podium. At this point, I actually felt bad for Bekenstein and walked up to him so that he wouldn't feel ignored after an hour of speaking about the past decade of his research.

Looking back, I feel this talk was a watershed moment in research into foundations of gravity. The revolution that Bekenstein had started was about to leave him behind, and a new guard was taking over. While Arkani-Hamed, who was a physics professor at Harvard at the time, had been exploring similar modifications of gravity to Bekenstein's, he would soon move to Princeton's IAS and disavow his old ways, famously saying: "Don't modify gravity. Understand it!"[4] As if decreed by the high priest in the grand temple, those who worked on modified gravity to replace dark matter (like Moffat or Bekenstein) would turn into pariahs. In the famous words of Jacques Mallet du Pan: "Like Saturn, the Revolution devours its children."

THE BIRTH OF HOLOGRAPHY

As we noted before, in the early 1970s, Bekenstein insisted that black holes obey the laws of thermodynamics and thus have entropy, much to the indignation of Hawking and other physicists. Why the indignation? If black hole holes have entropy, they must give off heat—but as nothing can escape a black hole, they *cannot* do so. Hawking himself showed the way out of this paradox; he demonstrated that quantum mechanics allows an escape clause that permits black holes to radiate. The equations for this process perfectly matched the formula for entropy. Thus, a new sub-branch of physics was born: black-hole thermodynamics. But this solution had a strange consequence. The formula describes the entropy of the black hole as proportional to its two-dimensional surface area, whereas for all other objects we know of, entropy is proportional to the object's three-dimensional volume. To get an idea for how strange this consequence is, recall that entropy measures how many ways there are to arrange something such that it looks similar. There are fewer ways to arrange sand into a sandcastle than there are ways to arrange sand into a pile. The latter then has higher entropy than the former. Intuitively, then, the space of possible arrangements of particles should depend on the volume of space in which particles can move around. But that isn't what happens in Hawking and Bekenstein's formula. A black hole, then, can be understood by ignoring what's inside and focusing purely on its surface. Just as with a hologram, the appearance of depth can be an illusion. What a hologram does is record an object's light patterns and then recreate them with another light source, typically a laser. Deceived by this two-dimensional masquerade, your mind interprets these patterns as a three-dimensional object. Similarly, Hawking and Bekenstein's formula was interpreted to imply that black holes might be playing a similar trick on us, with what we think of as a three-dimensional black hole being a mere projection of a two-dimensional reality. At first this conjecture did not appear to have anything to do with cosmology, but the principle is now being used by physicists in a host of applications, including my own attempts to describe the Big Bang itself. It may be that, as with a hologram, we can ignore everything about the Big Bang but its surface.

The true birth of holography came in a dramatic face-off in 1981, in the attic of the San Francisco mansion of Jack Rosenberg (a.k.a. Werner Erhard), self-help guru and millionaire, who liked to entertain leading theoretical physicists of his time. There, Stephen Hawking dropped a bomb: black holes destroy information!

Relativists (those who specialize in studying Einstein's theory of gravity) had known of this finding for a while, as Hawking had published it five years earlier. But this meeting may have been the first time particle theorists learned of it. Indeed, Susskind described the confrontation as "the first shot" in his book *The Black Hole War*. Relations between the two denominations of physicists had been strained for some time. Famed quantum theorist Richard Feynman described leading relativists as a "bunch of dopes" doing nothing of interest.[5] So it's not surprising that particle theorists were ignorant of Hawking's groundbreaking paper. When they finally heard about it in Rosenberg's mansion, all hell broke loose, for destroying information in quantum mechanics is sacrilege. You can shred, burn, or dissolve a document, but all that does is scramble information into an unreadable form. But fundamentally, quantum theory says that information is still there, hiding in the remnants of your sabotage. Hawking realized that the only thing that can be emitted from a black hole is the radiation that bears his name and that is truly random. It has no information. After eons of time, the black hole evaporates, leaving nothing. Black-hole thermodynamics, while saving the universe from an entropy problem, opened a new quagmire that we call the information paradox. One set of laws (general relativity) says information is destroyed; the other set (quantum mechanics) says it is not. Two of the attendees didn't take this blatant violation of a basic tenet of quantum (and classical) mechanics lightly. One of them was Susskind; the other was Gerard 't Hooft, the Dutch theoretical physicist who would win the 1999 Nobel Prize in physics. Both 't Hooft and Susskind would go on to develop versions of what they called the holographic principle, generalizing Bekenstein and Hawking's black-hole entropy formula to hypothesize that all the quantum bits of information existing in any volume of space are in fact encoded in its outer boundary. As such, the information that falls upon a black hole is not lost but rather preserved on its surface in some regurgitated, scrambled way; it is then recycled back into the universe in the form of Hawking radiation. Thus, holography could preserve information that would otherwise be lost and so save physics from Hawking's challenge. But though the idea was tantalizing, no one knew how to realize it in detail, nor understood how it might impact cosmology.[6]

HARVARD PROFESSORS

During my time at Harvard, there were two Iranian American professors in the physics department: Vafa and Arkani-Hamed. We have already talked about Vafa. He left Iran right after graduating from high school and just before the Islamic revolution, finishing his PhD under Ed Witten in 1985. Vafa has always been proud of his Iranian heritage, maintaining ties with the Iranian diaspora and Iranian scientists back home. Arkani-Hamed was born in Houston, Texas, to Iranian parents (where his father, a planetary scientist, worked on the Apollo program), splitting time between the US and Iran. The Islamic revolution, which deposed the Shah of Iran, happened in 1979, when Nima was only eight years old (and I was a mere babe in arms). Like many other expatriates, Nima's parents moved back home, hoping to help establish a more democratic civil society. Unfortunately, most of them had their hopes dashed, as an even more autocratic regime slowly emerged out of the chaos of the revolution. In particular, the "cultural revolution" (from 1980 to 1983) led to a closure of universities followed by a purge of liberal academics. Fearing for their lives, Nima's parents had to pay human smugglers $50,000 to take them out of the country on horseback through the Western border, a journey that ten-year-old Nima barely survived, having come down with a 107° Fahrenheit (42° Celsius) fever. They would eventually settle in Canada, where Nima would finish his bachelor's degree in physics at the University of Toronto, and then his PhD at the University of California, Berkeley, south of the border.[7]

Just around the time that I left Harvard to come to the Perimeter Institute in Canada, Arkani-Hamed would also leave, but he traveled in the opposite direction, to Princeton, eventually turning into one of the most vocal advocates for the new holographic way to "understand gravity." There, at the venerable IAS, he would become colleagues with Juan Maldacena, another theoretical physicist, who had made the same journey from Harvard seven years earlier. Maldacena grew up in Argentina, but, like me, came to the US for graduate studies, finishing his PhD at Princeton in 1996. However, it was a 1999 paper he wrote during his brief stint as a Harvard professor, entitled "The Large-N Limit of Superconformal Field Theories and Supergravity," which revolutionized the world of theoretical physics as we know it.[8] Fittingly, it is the most cited high-energy physics paper of all time (with over twenty-three thousand citations at the time of this writing). Combining recent advances in understanding of string theory, branes, and the theory of quantum fields, Maldacena finally landed the idea Bekenstein had launched twenty-five years earlier, when he was a graduate student at Princeton. While Bekenstein and then Hawking gave convincing arguments (or educated guesses) that the entropy, or all possible information, of a black hole fits on the surface of its event horizon, they could not possibly know what this information was. In general relativity, black holes were classical and had no hair (or no information, other than their mass, spin, and charge). In order to show that quantum mechanics was consistent with black holes, one should ideally model the black hole as some arrangement of quantum objects. Although we often say that entropy is a measure of disorder, more technically it's a measure of how many microscopic states are equivalent to a certain macroscopic configuration. So, in the case of black holes, physicists hope to count these microscopic arrangements of quantum objects that make up the black hole and check whether their tally agrees with the Bekenstein-Hawking entropy formula. No physicist had ever achieved this task until 1996, when Vafa and Strominger at Harvard modeled a special type of black hole using tools from string theory.

Maldacena pushed this program one step further and provided a concrete example of a "holographic dictionary," showing that one type of physics in the interior of a space-time was precisely equivalent, or dual, to another on its outer surface. In other words, the theory of quantum gravity within a volume of space (also referred to as the bulk) is mathematically equivalent to ordinary quantum mechanics (without gravity) on the boundary of that volume, which naturally has one less dimension than the bulk. So, for example, if the bulk theory is three-dimensional, the boundary theory will be two-dimensional. This equivalence is known as the holographic principle, in analogy with the two-dimensional hologram that contains information about a three-dimensional object.

I am not sure about 't Hooft, but Susskind especially—being a pioneering string theorist—felt vindicated by Maldacena's discovery, which put his own qualitative conjectures and speculations about the holographic principle into a rigorous and precise mathematical framework. This precise holographic correspondence would become known as AdS/CFT, with *AdS* standing for *anti–de Sitter* space, which is a hypothetical space that has no matter and a negative vacuum energy. Unlike the positive vacuum energy that is currently speeding up cosmic expansion, a negative vacuum energy would pull things together. Such contraction is the opposite of what we see in our universe. *CFT* stands for *Conformal Field Theory*, a special class of quantum field theories in which the field looks the same whether you zoom in or out, a characteristic technically known as conformal rescaling (recall, this term was also the first C in Penrose's CCC). Maldacena's paper was seen as a tremendous breakthrough because it drew connections between otherwise disparate areas of physics; it meant we could understand gravity by studying particle physics. Calculations that were thought impossible on one side of the AdS/CFT correspondence are often easier on the other. Physicists then can use the duality to infer results that could never have been achieved without Maldacena's discovery. In the late '90s and early first decade of the 2000s, the AdS/CFT mania would spread across physics and math departments all over the world like wildfire. One of them, the Department of Applied Mathematics and Theoretical Physics at Cambridge, UK, was the place Hawking called home.

While many of Hawking's contributions and academic debates went above the heads of his popular-science audience, he still had a way of making them interesting by placing public and entertaining bets with other movers and shakers in their respective fields. To my knowledge, two of these bets concerned what happens to information as it falls into a black hole; Hawking maintained—as he had boldly declared in Jack Rosenberg's San Francisco mansion—that it was lost for good. In one of them, Hawking was joined by another famed relativist and future Nobel laureate, Kip Thorne. Together, they wagered that information would be forever lost in a black hole. On the opposite side was John Preskill, a former PhD student of Weinberg (who, as you may recall from chapter 2, beat Guth and Tye to the discovery of the monopole problem), now a leading quantum information scientist at Caltech who (not surprisingly) believed that quantum bits last forever. The other bet was against Don Page, Hawking's former Canadian PhD student, who is now a professor in Alberta. In my view, it was the AdS/CFT mania of the first decade of the 2000s, the belief that a unitary quantum dynamics preserves information on the boundary, that probably drove Hawking to concede his bets and give up a copy of *Total Baseball: The Ultimate Baseball Encyclopedia* to Preskill, and a novelty dollar bill featuring the face of Marilyn Monroe to Page. Hawking's concession was enough for Susskind (who was not party to either bet) to claim victory. Meanwhile, Thorne never conceded.

I met Susskind first as a graduate student at Princeton and then met him again at Perimeter, where he was a regular visitor while his son was doing a PhD at the University of Toronto. What made him such a powerful and charismatic character was his unique combination of eloquent oratory skills and authority. Most other leading scientists either avoid talking about their research to the public for the fear of oversimplification, or they have a hard time reaching their audience, as they have to navigate the minefield of caveats that prevents them from making categorical and succinct statements.

Yet, the combination of oratory eloquence and authority has a more ominous side. Think about the most powerful speeches that you have heard, those that inspire across and beyond generations, or that mobilize the masses into fighting for causes from justice to war. Those are the words that resonate in your gut and stay with you. When you think of those moving passages, I bet that they did not come with error bars, nor with a list of caveats and uncertainties. For we humans don't seem to be inspired by doubt, uncertainty, and humility but rather by confident and charismatic leaders and politicians who often speak well but rarely speak the truth. Millennia of history, through the Dark Ages and the Enlightenment, have shown us why the separation of church (or mosque) and state is a practical guideline for governance, as those who speak in terms of absolute heavenly perfection can seldom solve imperfect real-world problems.

Now, consider this. At the conclusion of a lecture to amateur astronomers in Silicon Valley in 2008, Susskind would declare: "This entire room and everything in it is really represented mathematically by a theory on the boundary which is very much like a hologram. That idea has now become a central pillar of physics. It is not any longer a conjecture. It is now a mathematical reality that the universe, the entire universe, can be represented as a boundary theory, with everything on the boundary, where the boundary is out near the horizon of the universe. Well, this is something that of course has never and probably will never be directly experimentally tested. It involves things which are far too small—Planck-area-sized things, far too small—but it is through the extraordinary ingenuity of physicists, not just string theorists, that the holographic principle is no longer a wild speculation. It has become the principle mathematical tool of theoretical physics."[9] By this point, all I could hear was Galileo rolling in his grave at the idea that, less than four centuries after his passing, something that had "become a central pillar of physics" could "never be directly experimentally tested." Galileo's Italian intellectual descendant, Rovelli, has been more vocal: "A physical theory that does not give predictions is not a good theory. We need definite predictions, like those that all good physical theories of the past have been able to produce."[10] And recall that Maldacena's masterpiece—which I assume propelled Susskind's confidence, and which put holography on firm mathematical foundations—only actually worked in an anti–de Sitter space, which requires negative dark energy density. In such a universe, the gravity of empty space would pull things inwards, counteracting the expansion of the universe and, if strong enough, causing it to recollapse. It's certainly a cosmic coincidence that within a year of the publication of Maldacena's paper, astronomers discovered just the opposite: that dark energy had positive density, which means that empty space speeds up cosmic expansion (or that we live in a world closer to de Sitter space). Of course, a mere flip of the sign cannot dissuade true believers. I remember Susskind giving a colloquium at the physics department at Princeton (while I was still a PhD student), drawing a cartoon showing a contiguous patch of "string theories in anti–de Sitter," surrounded by a sparse sprinkling of points surrounding it which represent "string theories in de Sitter." Then, he speculated that while the former are much more abundant (à la Maldacena), the anthropic principle (for some reason) would put us in the latter group, a thesis that he elaborates in his book *The Cosmic Landscape: String Theory and the Illusion of Intelligent Design*.[11]

Nonetheless, the Swampland conjecture (recall chapter 4) developed by Vafa and colleagues raised questions as to whether string theory can be compatible at all with a space with a positive cosmological constant. This line of thinking implies that either string theory is wrong, the conjecture is wrong, or dark energy is not a cosmological constant after all. It also spells problems for inflation, as the inflaton field is a form of dark energy in tension with the Swampland conjecture. I remember that Ghazal and I once shared a train ride with Maldacena on NJ Transit from Princeton to New York City. At the time, my PhD advisor, David Spergel, had suggested that I work on some flavor of cosmic strings as dark energy (remember that he had declared cosmic strings were dead after COBE, but I guess he had had second thoughts)—since we had been told that string theory allows neither for a positive cosmological constant nor for de Sitter space (that is, for constant dark energy). Maldacena, however, disagreed and said that in fact he and his team had just managed to construct such a model using branes in compact extra dimensions. Their 2003 paper, "Towards Inflation in String Theory," is now often referred to as KKLMMT construction (named after its authors: Shamit Kachru, Renata Kallosh, Andrei Linde, Juan Maldacena, Liam McAlister, and Sandip Trivedi; yes, the same wife-and-husband Kallosh and Linde).[12] Incidentally, Kachru is also married to brane inflation pioneer Eva Silverstein (both are string theorists at Stanford and former PhD students of Witten). At Stanford, inflation seems to be a family affair. Thus, yet again we see physics fracturing into warring parties, those that see holography as the new holy grail and those who are skeptical of the relevance that two hypothetical spaces, neither of which describes our universe, are equivalent to each other. We see, too, a sub-fight between those who think the Swampland conjectures are correct and threaten inflation and those who don't. With no way of probing the minuscule scales that are the domain of quantum gravity and holography, we seem to be stuck. But although we may never be able to build a particle collider so powerful it can probe the energy scales needed to see this new physics, nature has already done it for us. The Big Bang itself is a laboratory to study physics at the extreme, and it is here that the true secrets of nature lie hiding.

HOLOGRAPHIC COSMOLOGY

As should be clear by now, having been trained in astrophysics, I had my reservations about how and why theoretical physicists were enamored by the idea of holography, as I thought such ideas could never be tested and thus had little relevance for understanding the Big Bang. But my stance changed when I met Kostas Skenderis. Early in the first decade of the 2000s, when I was a graduate student at Princeton, Skenderis was an assistant professor in the physics department. Those days were the heyday of holography, and Skenderis was at its epicenter, writing some of the most influential papers on the topic. However, the first time I remember meeting him was in summer of 2009 at the Perimeter Institute, during a workshop entitled "Holographic Cosmology." Neil Turok had just started as the director, and was very excited about getting his cyclic cosmology from holography. After all, the main difficulty in cyclic models was how the universe could bounce from a crunching singularity, which was where quantum gravity was needed, and holography was the style du jour. However, my best memory of that workshop was of the time when I was talking to Robert Brandenberger and he had to rush to catch Skenderis's talk, which seemed to him to be "something genuinely new." Indeed, while other talks were making tenuous connections between cosmology and ideas in holography, Skenderis's talk, "Holography for Cosmology," was a serious proposal for getting cosmological predictions from AdS/CFT machinery, using a mathematical tool known as analytic continuation, which can effectively switch between time and space.

Recall that the central idea of holography is that all the information we need to describe the properties of three-dimensional space can be encoded on a two-dimensional surface, our world of height, depth, and length being a mere projection from a world with one fewer dimension. A three-dimensional black hole can be understood purely by looking at ordinary quantum mechanics (without gravity) on its two-dimensional boundary. Skenderis's holographic cosmology was even weirder, as it implied that the dimension we would be losing was not one of space but that of time. Just as the Grand Canyon captures millions of years of Earth history in a snapshot of its sedimentary layers, the entire history of the universe is captured within a single quantum snapshot.

One can tell a similar story about the fate of observers who fall into a black hole, what Susskind calls black-hole complementarity. This story is about two men named Jesus and Mohammad (not the religious figures, but), both real astrophysicists who worked at my institute around 2012, although I shall slightly exaggerate their roles and personalities. Jesus is more adventurous; let's imagine that he wants to jump into Gargantua (the supermassive black hole in the movie *Interstellar*) to discover the secrets of quantum gravity. However, Mohammad is more conservative, so let's say he opts to stay behind but promises to keep a detailed log of his observations. According to Susskind, what Mohammad will observe is that Jesus approaches the horizon (never actually reaching it) but is slowly dissolved into the Hawking radiation, which gets superhot near the horizon, becoming part of the entropy of the black hole's surface and eventually evaporating. However, Jesus (like Matthew McConaughey's character in *Interstellar*) may feel perfectly fine at crossing the horizon, eventually discovering the secrets of quantum gravity, and/or die a violent death near the singularity; we don't really know (recall, this question was the topic of the firewall paradox from chapter 6). Nonetheless, as far as Mohammad is concerned, Jesus died when he crashed into the black-hole horizon surface; all Mohammad's observations can be described in terms of what happens on the surface of and outside the black hole. These diverging experiences seem to lead to a contradiction (Jesus could not have died at both the event horizon and at the central singularity). But Susskind claims no contradiction exists, as long as Jesus and Mohammad never meet again to compare notes. This argument is analogous to Niels Bohr's notion of complementarity in quantum mechanics, which states that you can know precisely the position or speed of a particle, but not both. Of course, you should ignore the last ten minutes of *Interstellar*, in which McConaughey's character escapes the black hole and meets his now-much-older daughter. If such a thing were to happen, it would indeed lead to a contradiction.

In a similar sense to Susskind's black-hole complementarity, holographic cosmology does not tell us exactly what happened at the Big Bang. However, it suggests that we can understand cosmological observations today much in the same way that we can understand black-hole evaporation without knowing exactly what is going on inside black-hole event horizons. If you are not satisfied by this narrative (or lack of one), you are not alone. Holography is kind of like catching a killer but learning nothing about their motive or means.

Here, Skenderis was building upon a topic he had co-pioneered back at Princeton, which became known as the holographic renormalization group. In physics, a renormalization group is what emerges when one blurs (one's view of) a system to focus on information on progressively larger scales. Recall from the last chapter that renormalization was the technique used to deal with mathematical infinities in quantum field theories. Theories that are renormalizable have their infinities cancel out, yielding finite answers for calculations. A renormalization group builds upon that technique, tracking the scale dependence of these theories' properties. For example, imagine looking at an elaborate painting of an epic battle scene. You may focus your eyes on each soldier's face in order to read their expression, but you can also slightly blur your view to see the grand layout of the battle scene. In physics, the renormalization group is used to understand how properties of systems with a lot of chaotic, complex behavior may be effectively different on different scales. For example, the charge of an electron slowly weakens over larger distances, as it is screened by the virtual electrons and positrons that randomly pop in and out of the quantum vacuum. In the holographic version of renormalization, if the quantum system lives on a boundary of a gravitational system, then these properties on larger scales in fact tell us what's happening deeper in the interior, that is, farther away from the boundary where the quantum system lives. Now, if we rotate space into time, à la holographic cosmology, then looking at structures over large distances at the end of inflation tells us what was happening early during inflation, while smaller distances tell the story of how inflation ended.

At the time, Skenderis was a professor at the University of Amsterdam. While he was already a high-profile string theorist, he needed the expertise of a cosmologist to connect his ideas to cosmological observables. As luck had it, Paul McFadden, a former PhD student of Turok's, was also a postdoc at Amsterdam, so they worked together to develop the basics of the dictionary that connected cosmological observables (that is, what we see in the CMB or galaxy distributions) to the properties of a quantum theory on a holographic boundary. In the context of inflation, this proposal was actually not that radical. Given that inflation produces nearly scale-invariant fluctuations, which source structure formation after it ends, we might just be able to replace the whole inflationary story with its outcome, expressed as an almost-CFT at the moment inflation ends (that is, at our holographic boundary). I say "almost" because, for the same reason that inflation has to end at some point, the quantum system cannot be *exactly* scale-invariant, and thus not cannot be exactly a CFT.

However, the real radical leap—which may turn out to be holography's biggest service to cosmology—is in considering that we may not need the entire story of inflation. The point of any duality in physics (such as the T-duality that we discussed in chapter 4, or AdS/CFT) is to show that apparently different systems are actually equivalent. If I can formulate all the features that I need to explain observations from a timeless quantum theory at the onset of the hot Big Bang, then perhaps we can settle for a frozen picture without the inflationary story and all its baggage, like eternal inflation and the measure problem. We will lose an explanatory framework, but we will gain a more parsimonious description. In particular, I may use quantum theories that don't have a simple inflationary story (maybe because they are deep in the quantum gravity regime), but I can still make predictions using my holographic (almost) CFT.

I again met Skenderis, as well as his wife Marika Taylor (a former PhD student of Hawking) and Paul McFadden, in Amsterdam, during a visit in 2011. They happened to have rediscovered our Cuscuton fluid, which made me happy. In fact, that conversation inspired our work together on getting the Big Bang out of the collapse of a four-dimensional black hole (recall chapter 7) shortly afterward. Skenderis and Taylor would move to Southampton, while we would recruit McFadden a postdoc to the Perimeter Institute, where I turned out to be his mentor. While McFadden and I regularly met to discuss science over lunch at the Perimeter Institute, it was becoming clear that he was ready to move on from holographic cosmology to new topics, and yet I was getting more and more excited by it. I suspect my growing excitement as sparked by the endless bickering I had witnessed between pro- and anti-inflationary camps, each promoting their own story of what happened at the Big Bang, using esoteric and abstract criteria—bickering that prompted me to ask: "Do we really need a story, or yet another creation myth?" After all, a multitude of stories might yield the same outcome, and then we could argue ad infinitum about which story is more believable, but we clearly cannot converge if we cannot agree on a common set of criteria for what *believable* means. The mature thing to do, it seemed to me at the time, was to stop asking what happened and instead look for the simplest way to explain what we see now. (Then again, sometimes you long for a bedtime story.) It was at this time that my PhD student Beth Gould (whom you'll recall from chapter 7) was looking for a new project, so I suggested that we reach out to Skenderis.

Skenderis and McFadden had an idea for turning inflation on its head. Holography proposes a duality between two theories, which can be useful if one of the theories is hard to do calculations in, but the other is simple. In other words, when the boundary quantum theory is hard to understand, we may be able to use the familiar, classical general relativity and its holographic duality to do calculations. On the other hand, when we have hardcore quantum gravity effects in the bulk (such as when general relativity goes singular at the Big Bang or within black holes), then we can resort to simple calculations on the boundary. So, a holographic theory of ordinary inflation based on general relativity will not be particularly useful, as we can already do simple calculations in the bulk picture with gravity (what Mukhanov pioneered, if you recall chapter 2), while the boundary theory may be extremely complicated.

But what if the quantum fluctuations in the bulk are so large that we cannot even tell if the universe is inflating or not? Let's call this scenario "bizarro inflation." Then, the holographic cosmology of Skenderis and McFadden may be the only practical way to discuss what may come out of this colossal mess. In order to use it, we just need to specify some simple quantum system on the holographic boundary at the end of our bizarro inflation. So, what is an example of a simple quantum system? Skenderis and McFadden picked something like the standard model of particle physics. However, while the standard model has only sixty-one types of particles, in order to be consistent with CMB data (basically 1 in 100,0000 temperature fluctuations), holographic cosmology needs thousands of distinct particles to exist in the timeless quantum system. Incidentally, roughly speaking, this numerical requirement was what "large-N" in the title of Maldacena's 1999 seminal paper referred to.

So, at first pass, it looks like holographic cosmology transforms bizarro inflation into a bizarro, timeless particle physics: it looks weird, but it turns out to be essentially the normal particle physics we have known how to deal with for half a century. However, the plot thickens when we try to compare the theory's predictions with detailed CMB maps. Recall from the last chapter that Planck found a small deviation from the scale-invariant spectrum of sound waves during the Big Bang. Can holographic cosmology explain this deviation?

This point was where Beth came in. Thanks to calculations by Skenderis and his collaborators, Luigi Delle Rose and Claudio Coriano, we knew that the bizarro, timeless standard model, while nice and well-behaved on small scales, becomes unwieldy on large scales. This difference leads to a characteristic scale dependence in CMB fluctuations, but it also tells us that at some large scale, we can no longer trust our calculations. Beth found that on scales smaller than roughly six degrees in the sky, the model works beautifully and is consistent with the CMB fluctuations seen by Planck. However, on larger scales, where we don't trust the calculations anymore, it begins to fail. Now, recall from chapter 5 that the CMB on large scales shows weird anomalies that loop quantum cosmology claims to explain. Could it be that the large quantum effects in the bizarro, timeless particle physics may actually resolve these anomalies? To find the answer, we will have to wait to develop specialized large-scale computer simulations, as pen-and-paper techniques are simply not capable of getting the job done. Skenderis and his collaborators in the UK are slowly working on developing this machinery. I personally can't wait.

If you have made it this far, it is probably safe to assume that you are interested in finding out the story of what happened at the Big Bang. After all, we have made a point of telling the stories of (both the people and the science of) various Big Bang scenarios. You may thus wonder what happens if you run the movie of the universe backward in a holographic cosmos. This, the model does not tell. It may be that the concept of time, which is the key ingredient of storytelling, is an illusion for classical observers like us, and only emerges after the Planck era. This possibility brings us back full circle to the spirit of the decades-old Hartle-Hawking proposal: that clocks would start ticking only after an initial, timeless quantum era. If observers cannot survive through those conditions, then there is really no story to tell. Yet, it also may be that there is a complex, even breathtaking, story of what came before the Big Bang.

However, all we can observe today is the final page of that story, and like a reader in a rush, that final page is all that holographic cosmology will ever find. What you may need is not always what you may want, when it comes to the scientific method. However, if you're in search of a story, there are other narratives we shall find that explore the world of holography from different angles. Nonetheless, one principle that all approaches have in common is that space (and even possibly time) is not fundamental but is an emergent property of quantum systems. Just as the wetness of water is not found in individual atoms of hydrogen and oxygen but emerges when they are combined at certain temperatures, so distance and time might not be found at the most fundamental scales of our existence. As the Penrose-Hawking singularity theorems relied on our traditional notion of space and time, this principle can provide us with another path on which to journey beyond them.

THE FUTURE OF HOLOGRAPHY AND THE EMERGENCE OF SPACE-TIME

In 2015, our old friends Maldacena and Arkani-Hamed, now colleagues at the Institute for Advanced Studies in Princeton, wrote "Cosmological Collider Physics," a paper that provided a prescription for finding new particles during inflation in subtle patterns of correlation (or non-Gaussianity) of the CMB. This approach followed the same philosophy as holographic cosmology, that is, it held that all observable information lies in an object's final surface, so we should try to define the rules of the game over there. Not surprisingly, their paper launched literally hundreds of studies that further developed the technology for computing highly detailed relationships between subtle statistical properties of observations within general classes of inflationary models. At the end of the day, though, these subtle signals are unlikely to be seen at least for decades (if ever), and even if they are seen, we will still be faced with the same problems that Ijjas, Steinhardt, and Loeb complained about: that there are many more potential models than there are observations that could possibly constrain them.

But to really understand the essence of holography, we shouldn't think of it as a variant of string theory or even as a theory of quantum gravity. We have two different languages for describing nature on different scales and in different situations: geometry and quantum mechanics. Holography is the dictionary that allows us to translate between these two languages. In this sense, arguably, holography as a concept cannot be right or wrong but rather is custom designed to translate a geometric (or gravitational) theory into a quantum theory. We may choose to use one or the other (depending on which one is easier in a given situation), but we still have to decide on the rules of the game we want to play (which quantum gravity theory to choose) and the story we want to tell (what happened at the Big Bang), in either language.

In February 2006, two Japanese physicists, Shinsei Ryu and Tadashi Takayanagi, noticed something peculiar about the holographic dictionary. The amount of quantum information that could be shared between two regions of space in a quantum theory on the boundary of anti–de Sitter space appeared to be proportional to the area of the minimal surface (like a soap film) that connects them in the bulk. You may think that you have seen this relationship before, and you'd be right—it's just a generalization of Bekenstein's entropy formula for black holes. This generalization pointed to something that could be very profound: that shared quantum information (technically known as quantum entanglement) is intimately related to geometric quantities such as area and distance.

Susskind and Maldacena pushed this idea one step further. They suggested that every set of entangled quantum particles in the universe has a wormhole connecting them. More concretely, what we perceive of as the geometry of space-time might be an emergent phenomenon (like sound waves) that's a specific manifestation of quantum entanglement between space-time points. This suggestion has become known as EP = EPR, the EP referring to Einstein and Podolsky, inventors of the wormhole, and the EPR referring to quantum entanglement (as Einstein, Podolsky, and Rosen had pioneered thinking about that subject, too). Sean Carroll (whose model with Jennifer Chen—of the baby universe budding off from a mother cosmos—we saw in chapter 6) was one of the people inspired by this proposal, and he embarked upon the program of constructing a concrete geometry that obeys Einstein's equations of gravity from entanglements within a quantum system.

We started this chapter with the story of Cecilia Payne-Gaposchkin, the Harvard astronomy department's first PhD recipient. Carroll was the 234th recipient of the same degree. Just as Sheldon in the TV show *The Big Bang Theory* became disillusioned with string theory and switched to dark-matter research, so life reflects art, as real-life Caltech theorist Carroll gave up on dark matter to pursue quantum gravity. Later, he recalled to us his volte-face. "I got frustrated; I can literally pinpoint the paper I wrote which is a model of dark matter," he told us. "We killed ourselves writing that paper because there's so many constraints now from astrophysics and nobody cared, because it's just one of dozens or hundreds of models of dark matter. Who cares about that? What are the chances it's actually going to be right? . . . I'm getting too old for that shit." Carroll quickly switched his focus to the foundational problems, becoming an evangelist for Hugh Everett's many-worlds interpretation of quantum mechanics. The plethora of universes implied by this system is a horrifying extravagance for some. Still, to others, its mathematical austerity, needing no hidden variables or other baggage, is the height of parsimony.

Conventionally, physicists think of quantum fields as the fundamental building block of reality, but to a "mad dog Everettian" like Carroll, it's the wave function of the universe that should be placed front and center. Everything else, it is conjectured, can be derived—not least gravity itself, creating a path to a quantum theory of gravity and a deeper understanding of our true origins. The problem of developing such a theory, as Carroll sees it, almost lies with the inner working of a physics classroom: "When you teach quantum mechanics, you start with a classical theory and quantize it, because your students already know classical physics . . . It's not that we shouldn't do it, it's that nature doesn't do it." An alternative viewpoint is to start with the wave function and reverse-engineer Einstein's theory from this purely quantum domain. As Carroll put it, "the wave-function represents reality; it's not just the tool you use to make predictions—that's what the world is. . . . [E]veryone else starts with some classical system and gives it a wave function. I'm saying, 'No, the wave function should come first.'"

Carroll and his collaborators are using quantum entanglement to define distances; parts of the wave function that are highly entangled are nearby, and those that aren't are far away. This connection implies a formula for relating separation in space to the degree of entanglement; spatial geometry then emerges from quantum mechanics. Carroll calls this geometry "space from Hilbert space," the latter being a mathematical world describing all possible ways for the wave function to be. But for this proposal to work, one must assume that Hilbert space has only a finite number of possible states, an anathema to many physicists who assume that the total number of states is infinite. This assumption may seem like dogma, but Lorentz invariance, the principle at the heart of relativity, that all observers should measure the same speed of light, seems to depend on this symmetry. If Hilbert space is finite-dimensional, this sacred principle may have to go. And encouragingly, jettisoning it could lead to testable predictions—if only the new work could tell us precisely when the symmetry is broken. Alas, it does not (yet) do so.

While the proposal is still rudimentary, it may already hint at a way to describe the expansion of the universe and our Big Bang origins. The expansion of space can be mapped onto the entanglement of quantum information bits (or qubits). There are fewer entangled states as we follow evolution back in time. Imagine a network of adjoining nodes being disconnected. Eventually, we arrive at the lowest possible number of entangled qubits. As Carroll explains, what's startling about this picture is that "you can calculate when it happened, and when it happened was the beginning of inflation." This is the moment when space emerges from Hilbert space. As the wave function is the most fundamental element in this picture, the universe can be described by the Schrödinger equation, and according to Carroll, there is "an immediate consequence: namely, there's no beginning to the universe in the Schrödinger-equation time does not begin or end; there's no singularities; every moment in time is just as good as every other moment." Carroll is the first to admit that we are very far from building a testable model of cosmology from these ideas, though.

If you find it hard to grasp the concept that something as abstract as a wave function can be the fundamental building block of reality, you are not alone. When Phil and I talked to Carroll about it, we challenged him to convince us that this idea could make sense. So, he proceeded to tell a story of Ludwig Wittgenstein, who asked fellow philosopher Elizabeth Anscombe: why did anyone believe the sun orbited the Earth? "It looks that way," she replied. "But what would it have looked like if it were the other way around?" replied Wittgenstein. According to Carroll, we should only ask of a theory, "Can it match reality? What predictions does it make? What would we expect life to be like if that were reality, just a wave function evolving in time? . . . [W]hat it would look like is the quantum mechanics that we know and love, so why work harder? Why try to add more crap to the universe?"

From my own experience, I can appreciate that convincing others to take radical ideas seriously is an uphill struggle. That is how it should be, even for an eloquent communicator like Carroll. "When I talk about atheism to old religious people, they just think I'm an idiot," he says, but young people are more open-minded. "It's the same thing when I talk about quantum gravity to audiences. Old people . . . are like, 'We know it's string theory, why are you bothering us? Why are you wasting our time? . . .' But young people are like, 'Yeah, quantum mechanics, we don't understand that; maybe we should think about that and see how space-time emerges.'" Clearly, Carroll is not alone in getting "too old for that shit."

And that state of affairs well summarizes the status of holography, which promises to be the ultimate Rosetta stone of theoretical physics. There is the old school, battle-tested, string-theorist-approved AdS/CFT, where everything works, except for the wrong sign of vacuum energy (apologies to the 2011 physics Nobel Prize committee), the six extra dimensions, and the missing supersymmetry of string theory. There is the timeless holographic cosmology of Skenderis and McFadden, which according to my former student Gould may be enough to fit observations but doesn't give you a good bedtime story to tell your kids. And finally, there is Maldacena and Susskind's ER = EPR that has mesmerized Carroll, where quantum is the rule and geometry is an illusion. But here, the rubber is still far from meeting the road. What many of these ideas have in common, though, is the notion that space and time, once deemed scaffoldings of our existence, are mere veils of a deeper, hidden reality—so that if we go back in time to the Big Bang, we will eventually encounter a world that is alien to our everyday experience. Cosmology enters a quantum domain where our intuitions, built to evolve in a classical world, fail spectacularly. Traditional notions of before and after, cause and effect may simply disappear in this strange quantum realm.

Another bizarre consequence of the quantum-entanglement wormhole connection was that, while wormholes used to be only stuff of science fiction, they suddenly became normalized in mainstream physics. In 2022, some physicists even claimed to have made them in their lab in Santa Barbara (they had not).[13] Although Maldacena and Susskind's original idea didn't require wormholes that allowed for time travel (so-called traversable wormholes), such wormholes required only a small leap, and even Maldacena himself later cooked up solutions with normal ingredients in standard physics.[14] But if time travel is possible, what could it mean for the origins of the universe?

✱ 10 ✱

CAN THE UNIVERSE CREATE ITSELF?

هر کسی کو دور ماند از اصل خویش
باز جوید روزگار وصل خویش

> From their roots, those cast away
> Shall seek out their union day
> RUMI

In 1958, science fiction writer Robert Heinlein wrote a groundbreaking short story called "All You Zombies," which was made into the highly acclaimed 2014 Hollywood movie *Predestination*. (Warning: spoilers ahead.) The protagonists are really a single protagonist, as the character Jane, her child, lover, mother, and a mysterious man are all the same person. Time travel and gender reassignment have allowed them to be their own father and mother. Reflecting on their unique perspective on personal origins, they think, "I know where I came from—but where did all you zombies come from?" Some of the most influential thinkers of our times, including philosopher David Lewis and astronomer Carl Sagan, have cited the story. It also has echoes in the Egyptian God Atum, who created himself from the primordial waters known as Nu. But could there be parallels in this story to the world of the Big Bang? Could physics similarly allow the universe to be its own parent?

Time travel has become a familiar trope ever since Mark Twain wrote *A Connecticut Yankee in King Arthur's Court* and H. G. Wells penned *The Time Machine*. But it's not just science fiction writers who ponder the possibilities of traveling to the past. When, in 1915, Einstein gave us general relativity, he showed that the fabric of space-time warps in the presence of matter. Two decades later, the Dutch mathematician Willem Jacob van Stockum discovered perhaps the strangest solution to Einstein equations imaginable. He found that it was possible for space-time to curl into a loop. Anyone moving along the circle would travel back in time to their starting point. This loop has become known as a closed time-like curve (CTC). Van Stockum would go on to become a test pilot in the Canadian air force and volunteered to fly missions over Europe during World War II. Just a few days after D-day, his aircraft was shot down, killing everyone aboard. But before his untimely death, he had the privilege of working at the Princeton IAS, hoping to study under Einstein. Kurt Gödel was another European refugee who joined the great master in Princeton. After revolutionizing mathematics by showing that statements existed that were true but couldn't be proved, Gödel became friends with Einstein. The two would often take long walks together, but what they discussed remained mysterious. It may well have involved van Stockum's strange CTCs, as Gödel found that a rotating universe could generate such a phenomenon and allow time travel into the past. He presented the model to Einstein as a present for his seventieth birthday. Unfortunately, Gödel suffered from mental-health problems and had a deep fear of being poisoned.

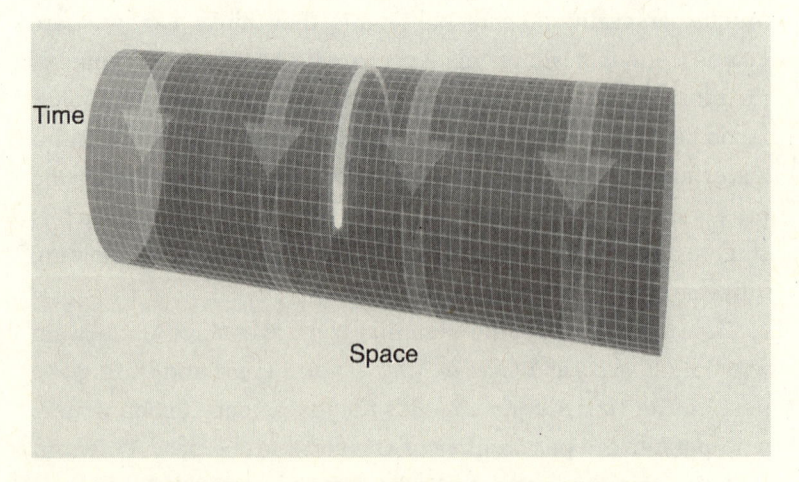

A closed time-like curve with a particle moving in a circle in time. Credit: David Yates.

He died from malnutrition, when his wife, the only person he trusted to inspect his food, was hospitalized.[1]

Since Gödel's death, others have found similar solutions that might allow us to travel back in time. Frank Tipler suggested that a vast, spinning cylinder could so distort the space-time around it that it could create a CTC. Kip Thorne showed that wormholes can act as doorways to the past. And my old Princeton professor Rich Gott demonstrated that two massive cosmic strings moving past each other can similarly warp space-time into a loop. Rich looks like he's auditioning for *Doctor Who*, although his conspicuous Kentucky accent may prevent him from ever getting the part. But we shouldn't be fooled by his wide-eyed stare and passion for time travel; Rich is an authority on Einstein's theory of relativity.

THE TIME TRAVELLING PRINCETON PROF WHO WEARS THE COAT OF THE FUTURE

When I arrived at Princeton in September 2000, I was eager to learn more about astrophysics, so I walked around corridors and into professors' offices to ask them what they did. It was easy to identify everyone with a topic: Spergel worked on the CMB, Strauss on galaxy surveys, Bahcall on galaxy clusters ... and then there was Gott. For one thing, he was kind of hard to find, for reasons that I cannot remember. When you did find him, the conversation would last for hours and touch upon all sorts of bizarre things. One day it would be about time travel and cosmic strings. Another day it would be about inflation and the Big Bang. As a young man, Gott had worked with Penzias and Wilson, operating the telescope that detected the CMB, but the paper of his that I used most in my thesis had to do with the formation of spherical clusters of galaxies. Another day, the conversation would center on history, statistics, and mapmaking. In fact, later he made a logarithmic map of the universe as a carpet for one of the corridors in the astrophysics department: you start from the solar system and get to the Big Bang after ten seconds of walking. More recently, his flattened map of Earth made *Time* magazine's list of the top one hundred best inventions of 2021.

Rich's mindset was all too confusing. Looking back at our interactions after two decades (and two children of my own), I can recognize that he has been doing, all along, what we all start doing as toddlers: exploring our surroundings and having fun doing it. Most of us grow out of this kind of joyful exploration, starting to set goals and agendas for our lives and define boxes that we should fit into, because that's what's expected of us. Gott did not.

When I was writing this chapter, I decided to have another chat with him to see how his thoughts might have changed in the intervening two decades since I'd last seen him. I asked him whether he thought his Princeton colleagues approved of him working on speculative things like time travel. To my surprise, he said that everyone was happy, as jumping from topic to topic was common, recounting examples from early years of some of my other Princeton professors. "It was all about having fun," he said.

My closest interaction with Gott came when I was the teaching assistant for his undergraduate general relativity course early in the first decade of the 2000s. Remember, I had left Iran only a couple years earlier, so my English was not very good, and trying to understand (what I now know was) his thick Kentucky accent was a challenge. Moreover, I was there to prove myself as an astrophysicist, and he was just trying to have fun. The class itself was also a lot of fun, mostly thanks to Rich's entertaining style. It is kind of ironic that when I look back, my current style of doing science and topics of interest are closest to Gott's, compared to those of all the other (highly accomplished) Princeton professors whom I ended up working with or talking to. Even the fact that he was a professor of astrophysics at Princeton (though he retired in 2016) might also have been a relic of a bygone era, when a sense of fun and a genuine curiosity to understand nature were valued over the political savvy to manage huge collaborations, grants, or armies of students and postdocs.

And what could be more fun than time travel? In his book *Welcome to the Universe*, co-authored with Neil deGrasse Tyson and Michael Strauss, Gott describes an elaborate show that he would put on during his seminars on time travel, involving a change of dress from a shirt to the "coat of the future," and a briefcase that travels on a time loop. (I truly regret having missed this show.) But it turns out that, as you may imagine, time travel is not quite as straightforward in reality. For one thing, if you decide to gradually turn around on your path in space-time and go back in time, you will need a vehicle that travels infinitely fast, thus faster than the speed of light (and definitely faster than the DeLorean's 88 miles per hour, from *Back to the Future*), which is not allowed by Einstein's special relativity. Now, having read chapter 8, you might think that that's OK, but it turns out that none of those scenarios would allow for time loops; although we talked about faster-than-light travel, we were referring to a speed faster than *today's* light-speed limit—as the speed of light increases in the early universe. Think about what happens when you get on a highway: the speed limit increases, but you'll still get a ticket if you go past that higher speed. VSL just increases the speed limit near the Big Bang and so still prohibits time travel.

The plot thickens when we add quantum mechanics to the mix. Recall that virtual particle and antiparticle pairs pop out and then back into the quantum vacuum all the time. Now, you also could think of these virtual pairs as a particle moving on a time loop, similar to Rich's briefcase: A positron (that is, an anti-electron) moving forward in time is like an electron moving backward in time. Creating briefcase/anti-briefcase virtual pairs is an incredibly unlikely event but could theoretically happen. Does quantum mechanics therefore allow for time travel? I think the answer can be both yes and no, depending on what you think of the nature of causality in quantum mechanics. We often say that correlation is not causation, but the fact is that correlation is all that we can measure, while causation is something we infer given a set of assumptions. For example, I may think that whenever I flip a switch, the light will turn on, and thus that the on switch is the cause of the light turning on. But reaching this conclusion requires assuming that I have free will to turn the switch on and off, whenever I want, independent of my environment. This assumption, of course, is not true. For example, it could be that I decide to flip a fake switch whenever it gets dark, but also that the light has a sensor that turns it on when it notices it is getting dark. In this scenario, the fake switch is not really the cause of the light. Now, if we cannot clearly identify what is a cause and what is an effect, how can we tell whether we are moving forward or backward in time?

THE JINN جن

In 1992, the Russian physicists Andrei Lossev and Igor Novikov named particles that run on time loops "jinn particles."[2] The word *jinn* (anglicized as *genie* and spelled جن in its original Arabic) comes from pre-Islamic Arabic folklore, which was adopted by Islam and even appears in the Quran. Jinns are mythical creatures, like spirits, who live in a parallel world to ours and can be good or evil, depending on whether they accept the Good Book. They can suddenly appear or disappear to us humans (committing all sorts of mischief), which is why Lossev and Novikov adopted the term. Having now split my life equally between North America and the Middle East, I can say that "jinn stories" are culturally comparable to Western ghost stories that are supposed to give you chills when told around campfires (I recall stories of jinns that had human bodies but with horse's hooves in place of feet). When I was in middle school, we had a Quranic-studies textbook that quoted a verse mentioning jinns. Our teacher used the opportunity to dedicate two weeks of the class to various myths and Islamic folklore about jinns, arguably the most exciting two weeks of middle school (imagine if *The Exorcist* had been on your school curriculum). Just as our teacher was done with the lessons, we received a new textbook from the Ministry of Education that made no mention of jinns at all. Quite literally—like Lossev and Novikov's jinn particles—jinns disappeared from our class, and our teacher told us to forget everything he had taught us in the preceding two weeks.

Interestingly, we can think of the entire corpus of quantum mechanics as the culmination of a human world where time runs forward and a jinn world where time runs backward (for aficionados, this description is known as the double-path integral and is based on Richard Feynman's formulation of quantum mechanics). When humans and jinns agree on a story, the story is likely to come to pass, but if they disagree, then either version has only a low probability of happening.

For a theory with probabilistic predictions such as quantum mechanics, it is hard to pin down causal order and thus the direction of time. In contrast, it is incredibly easy to come up with solutions to Einstein equations that allow time travel. The simplest solution imaginable posits a flat space with a periodic time coordinate, kind of like (everything other than Bill Murray's character in) the 1993 movie *Groundhog Day*, in which Murray keeps endlessly reliving the same day, February 2, only to wake back up to the song "I Got You Babe" playing on the radio. In this universe, events happen in the same order, over and over again, with whatever you're doing at 11:59 p.m. seamlessly blending into whatever you're doing at 12:00 a.m. (although the period of the cycle doesn't have to be one day).

For better or worse, we don't seem to live in the *Groundhog Day* universe (though I shall revisit it shortly). But could it be that this fact is only a shortcoming of where we happen to live in the cosmos, while CTCs and time travel are commonplace in other corners of our universe? As I mentioned earlier, Rich had exact solutions for space-times of moving cosmic strings. The problem there is that we still haven't found cosmic strings, and even if we do find them, they'd have to move really, really fast (more than 99.99999999996 percent of the speed of light), and we'd have to orbit around them even faster to go back in time. More organic time travel might be possible if we plunge into a spinning, supermassive black hole (as Jesus did in the last chapter), such as the one at the center of our galaxy or M87, the behemoth in the middle of the Virgo cluster that was imaged by the Event Horizon Telescope. Within the heart of a spinning black hole and long after we pass the event horizon (assuming that we survive the tidal forces stretching us into spaghetti, or quantum firewalls), we'll encounter what's known as a Cauchy horizon, the border between a world with time travel and a world without it. Just as Alice found—through her descent through the rabbit hole and out into Wonderland—the mathematics of general relativity suggests a strange world of ringlike singularities and time travel beyond the Cauchy horizon, where we could even escape through a white hole and into a different corner of the universe (or multiverse), if we managed to maneuver around the singularity. A similar Cauchy horizon exists for Rich Gott's cosmic string CTCs. These idealized mathematical solutions make great ingredients for sci-fi movies, but can they really happen in nature? The images that telescopes can make of space-time around black holes stop well before hitting the event horizon, and thus what lies beyond is a matter of theoretical speculation. While idealized space-times may predict fantastic phenomena, like a pencil resting on its tip, they may not be realized in the real world due to classical or quantum instabilities. For example, in 1989 my colleague Eric Poisson, along with Werner Israel, showed that the Cauchy horizon of a spinning black hole becomes unstable and then singular as matter falls into it.[3] In 1992, Hawking studied the behavior of classical and quantum fields in the vicinity of Cauchy horizons and suggested that they all might suffer from the pathology of an unstable vacuum that would blow the closed time-like curve apart. Based on this suggestion, he then proposed the "chronology protection conjecture," which held that "the laws of physics do not allow the appearance of closed time-like curves."[4] To me (and most physicists), this conjecture makes a lot of sense: if you have predictive laws of physics that determine how things evolve forward in time from some initial conditions, you cannot suddenly fall into a time loop without having some sort of catastrophic breakdown. But what if you start within a time loop (say at the Big Bang) and then cross a Cauchy horizon into standard cosmology (say inflation)?

"I GOT YOU BABE"

In 1996, an unknown student from China published a paper entitled "Must Time Machines Be Unstable against Vacuum Fluctuations?" in the journal *Classical and Quantum Gravity*, questioning the great Stephen Hawking's chronology protection conjecture.[5] The student's name was Li-Xin Li, and it so happened that he was also applying to graduate schools in the United States. At the time, Rich Gott at

Princeton's department of Astrophysical Sciences happened to be reviewing the applications—and he discovered his perfect collaborator. As Rich recalled to Phil, "I felt this [Li's paper] was a fantastic paper, and I said, 'Wow, this is a student who would like to come'—so I got him to come to Princeton." Together, Gott's expertise in general relativity combined with Li's mastery of quantum mechanics make for the ideal defense against Hawking's affront to time travel.

When Li arrived as a PhD student in Princeton in 1997, they didn't waste any time. Within three months, Li submitted a paper with Gott (later published in *Physical Review Letters*), providing a simple and beautiful counterexample to the chronology protection conjecture.[6] To understand this counterexample, let us go back to our favorite Punxsutawney, Pennsylvania, and the famous never-ending (fictional) events of *Groundhog Day*. The residents of Punxsutawney experience a gravitational acceleration of 9.8 meters per square second, due to Earth's gravity. However, according to Einstein, you can also imagine that Punxsutawney is a spaceship accelerating at 9.8 meters per square second in outer space. As long as they don't look up at the sky (or, say, try to travel to New Jersey), the residents cannot tell the difference between the two possible scenarios just based on the gravity that they experience. Now, imagine that in the Punxsutawney spaceship, the events of *Groundhog Day* happen over and over again every day (to the frustration of the astronaut Bill Murray). But then, according to Hawking's calculations, there will be a singular Cauchy horizon that prevents anyone from entering this region.

Time loops can have different durations. In the movie *Groundhog Day*, the loop lasts for a single day; in the Netflix show *Dark*, the loop lasts for decades. For almost every duration you pick, a singularity will be present at the Cauchy horizon separating the time travel region from the regular universe. And that singularity motivated Hawking's chronology protection conjecture—as it would mean that the time travel region would only exist in the place where our mathematics breaks down and we can't trust our results. The duration of the loop can be thought of as the circumference of a cone. But Gott and Li found that, just like in the Hartle-Hawking model, it is mathematically possible to construct a space-time where the tip of the cone is smoothed out, removing the pointy tip that represents the singularity. Unlike Hartle-Hawking, Gott and Li's model found no need to invoke imaginary time; they simply needed to change the duration of the time loop so that it would correspond to a cone with a smooth surface like a badminton shuttlecock. In this way, Gott and Li created a path that would allow a closed, time-like curve to be smoothly connected to our past.

In this picture, the acceleration represents the inverse of the distance to the tip of the cone; thus, the smaller the acceleration, the longer it takes to go around the circumference. For example, given the 1g acceleration that we experience on Earth, and if events repeated not daily but only every six years, we'd get Gott and Li's space-time that removes the singularity.[7] So, the residents of Punxsutawney (or Bill Murray) could choose to live the rest of their lives on a spaceship in a never-ending time loop stuck in the past, or they could decide to depart on a ballistic rocket, and—like an Amish teenager going on Rumspringa—leave the CTC world behind without the fear of being crushed by a singularity.

Only a month after submitting their rebuke of Hawking's chronology protection conjecture, Gott and Li were ready to publish their comprehensive forty-three-page treatise entitled "Can the Universe Create Itself?"[8] These days, normally when the title of an article ends with a question mark, the answer is "No!"—an adage informally known as Betteridge's law of headlines. Apparently, this principle was not quite so established in the 1990s, when authors were more wary of making extraordinary claims. So, in spite of their title, Gott and Li went on to construct a model of the Big Bang in which a region with CTCs tunnels out of an inflationary vacuum and is subsequently patched, through a Cauchy horizon, onto a more ordinary inflationary space out of which we are born. As Gott describes it, "the great thing about inflation is that a little tiny inflating piece grows it up to be an enormous inflating piece, each little bit of which is exactly like the little bit that it started with. So, what if one of those little bits was actually the bit that it started with? Then it could give birth to itself. Something funny had to happen at the beginning of the universe, and time-travel solutions to general relativity seem supremely suited for the purpose." Gott's comment that something funny had to happen at the Big Bang reminds me of a recent talk by Penrose in which he described inflation as fantasy, and although Penrose is critical of inflation, this comment was not supposed to be pejorative.[9] Rather, he was saying that something fantastical had to have happened 13.8 billion years ago; the only question was, what?

An even more fantastic story that Gott and Li tell is one wherein a supercivilization, somewhere across the vast cosmos, attempts to create an inflationary universe in the lab (which is what Farhi, Guth, and Guven were imagining in chapter 2) and accidentally manages to create its own inflationary beginning. Thus, Gott and Li conclude, "the laws of physics may allow the Universe to be its own mother," much as the character Jane is her own mother, in the Robert Heinlein short story that we discussed at the start of this chapter. But William Hiscock disagreed, arguing that the vacuum state associated with Gott and Li's time machine was unstable and would blow up the CTC.[10] Rich describes Hiscock as his "brother" because they both independently arrived at the cosmic string time machine. However, brothers bicker, and the two dueled over whether Hiscock

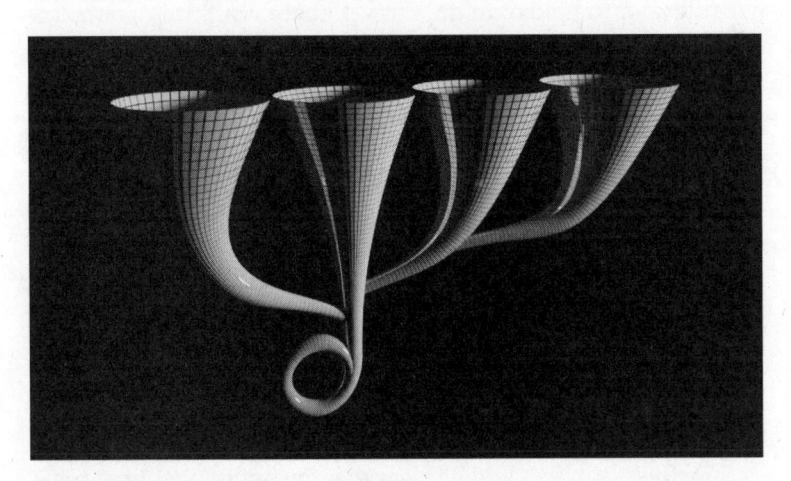

In the Gott and Li model, multiple inflating universes emerge from an origin region that allows for closed time-like curves. Credit: David Yates.

was using the right vacuum, Gott insisting that if you are careful about using the right vacuum states and the right period (as we discussed above), "all the terms canceled out beautifully; it does not blow up at the Cauchy horizon . . . [I]t does not prevent you from crossing through there and entering the time machine or leaving the time machine." Sadly, Hiscock died at just fifty-seven years old, of a rare blood disease.

The most obvious objection to CTCs is the so-called grandfather paradox, in which a time traveler travels back in time and murders their own ancestor, preventing their own birth and thus the very events that killed their ancestor in the first place. Or we could think of Marty McFly in *Back to the Future*, preventing his parents from becoming romantically involved and thus jeopardizing his own existence. If time travel is impossible, then these paradoxes are easily resolved. But they are not the only solution. As Gott put it, "You know why the Titanic sank? It was because of the extra weight of all the time travelers. . . . If you went back and tried to warn the captain of the Titanic, you know he'd ignore you like he ignored all the other iceberg warnings, because we know the ship sank. It's one four-dimensional sculpture; it does not change." In other words, relativity implies that the future and the past are fixed; neither can be altered. If you find this idea an affront to your sense of free will but still want to hold onto time travel, there is a third option. The many-worlds interpretation of quantum mechanics, championed by Sean Carroll (who we met in the last chapter) and others, implies a dizzying multiplicity of worlds wherein quantum events lead to a branching structure of reality. In this scenario, if you travel back in time and prevent your parents from falling in love, you would simply find yourself in a new branch of reality where you are wiped from existence. But the branch you came from would be unaffected, and so there would be no paradox. There is one more option. It may be possible for the entire universe to travel back in time, even though no conscious agent could do so. We shall return to this idea toward the end of the chapter.

Nonetheless, we should be clear that the notion that time travel (into the past) is possible still remains a minority point of view. For example, in 2022, Roberto Emparan and Marija Tomašević from the University of Barcelona studied time machines in holography and showed that—in the dual quantum theory—regions with time machines completely decouple from regions without time machines.[11] Yet science is not a democracy, and as we discussed before, the holographic principle (as popular as it may be) has not been tested by experiments. But then, speaking of experiments, no obvious means exists to test Gott and Li's fantastical ideas, if regions with CTC have now disappeared.

After I finished my teaching assistantship for Gott, he told me that he wanted to buy me lunch, as he had done for all his previous teaching assistants. Even though I spent four years at Princeton, somehow, we never found the right time to do it (which, in hindsight, I find kind of strange). Maybe if I had, I could have found a chance to ask him more about these ideas—for example, how he had reconciled his self-creating universe with his views of a creator. Ironically, many critics (like William Lane Craig, whom we'll meet in the next chapter) have described this idea as an act of desperation by atheist scientists to avoid God. But this take is a poor rush to judgment. Gott said, in a 2007 interview, "I'm a Presbyterian. I believe in God; I always thought that was the humble position to take. I like what Einstein said: 'God is subtle but not malicious.' I think if you want to know how the universe started, that's a legitimate question for physics. But if you want to know why it's here, then you may have to know—to borrow Stephen Hawking's phrase—the mind of God."[12]

DOES HISTORY REPEAT ITSELF?

When I was a high school student in Tehran from 1992 to 1995, two books mysteriously became must-reads amongst my peers: Stephen Hawking's *A Brief History of Time* and a compilation of essays entitled *From Lenin to Gorbachev*, on the rise and demise of communism in Russia, arguably one of the most significant drivers of the history in the twentieth century. But why are we so obsessed with history? Why do humans celebrate or grieve events that happened centuries, even millennia, ago, every year? Back in high school, during the times that we were not playing soccer or discussing wormholes and proletariat in the schoolyard, one of the topics that we learnt about in class was the (allegedly shameful) history of the Russo-Persian wars, bungled by the Iranian kingdoms of the past. During the course of five distinct wars from 1651 to 1828, four different northern provinces of Iran (then known as Persia) were taken over by Russia. Then, in 1909, Russian troops again invaded northern Iran, taking over the city of Tabriz and other (remaining) northern provinces, not leaving until the Russian communist revolution of 1917 was in full swing. If you thought these southerly incursions would stop with communism, you would be proven wrong: in 1979, the Red Army took over another Persian-speaking country for a decade, this time Afghanistan (as poignantly depicted in the Hollywood movie *The Kite Runner*, based on the highly acclaimed novel by Khaled Hosseini), to protect Soviet interests against American influence (through mujahideen guerrilla groups) in the midst of a civil war. Patterns of history seem to repeat over and over. As much as we would like to think that we learn from the past, there are natural forces greater than our best intentions and wisest intellects that pull us back into the same vicious cycles. Maybe we would take this lesson from history, if we entered one of Rich Gott's time machines. Or maybe not.

After working with Skenderis to constrain holographic cosmology with cosmological observations (as we discussed in the last chapter), Beth Gould was in search of a new topic to cap off her PhD thesis. So, I suggested that she come up with an idea that she'd be excited to work on. Starting from her undergraduate and master's theses, Beth had always been interested in the foundations of quantum mechanics. At the same time, like many young people who get into physics, she was also a big fan of science fiction and time travel. As we saw before, specifically when I discussed the human-versus-jinn world, causality and the direction of time become ambiguous in the quantum microscopic world, where time loops (such as virtual quantum particle/antiparticle pairs, what Lossev and Novikov called jinn particles) are commonplace.[13] Around the same time, Penrose had been stirring a lot of controversy with the conformal cyclic cosmology (CCC) model and his elusive circles in the CMB sky.

Now, one thing about gravity is that it can couple very small and very large scales. For example, if you take a wave packet and make it smaller and smaller, its momentum and energy increase, per Heisenberg's uncertainty principle, which means it can make a bigger and bigger black hole. In fact, as we learned in chapter 4, the T-duality of string theory can map the physics of the very small to the physics of the very big. And now, here is the kicker question: Could the whole universe, much like tiny jinn particles, be on a giant time loop? Could the entirety of cosmic history, from the Big Bang to eternity, exactly repeat itself? As Beth recalled to Phil, "One of the questions I had is whether or not anyone had ever looked at this possibility of having the universe come back around like this . . . I should see if I can put something together and see . . . if it's plausible or ruled out." Nothing she found ruled out the possibility that the universe might be cycling through the ultimate time loop, one joining the end of time to its beginning.

At this point in the book, you may wonder whether you are having a déjà vu moment. Didn't we already talk about cyclic cosmologies in chapter 6? Yes and no. The cyclic cosmologies we have talked about so far (à la Penrose, Baum-Frampton, Turok-Steinhardt, and others) envisaged a repeating, grand pattern of cosmic expansion (and possibly contraction) history but were still faithful to traditional notions of causality. The entire history of past eons is wiped out via late-time inflation driven by dark energy (with the possible exception of Penrose's ghostly circles), and you start over from a clean slate, with different random processes that form stars and galaxies anew in each eon.

But now let us imagine that our entire universe is running on a time loop, a cosmic Punxsutawney, so to speak. The duration of this day, however, may be trillions of years, even infinite as per Penrose's CCC. In fact, unlike in Gott and Li's model, the loop's exact duration does not matter, as cosmic structures expand and freeze (or stop evolving) during late-time inflation.[14] However, when the universe

The timeline of the universe according to periodic time cosmology. Contrast this model with the Gott and Li model, where the closed time-like curve is only present in the early universe. Here the entire history of the cosmos is one giant time loop. Credit: Christian Unterdechler.

unfreezes again at the Big Bang (and the cosmic clock hits 6:00 a.m. as Sonny and Cher's "I Got You Babe" starts playing), structures need to be in place (with Bill Murray waking in his cosmic bed) to replicate the exact same events, down to the last details of the formation of the Milky Way, our solar system, Earth, and beyond.

This point is quite nontrivial. For starters, you may wonder how the structures that are expanding into infinite size via late-time inflation could suddenly shrink back to tiny scales at the Big Bang. Here, Beth and I decided to follow Penrose's lead to adopt his secret sauce to CCC: conformal rescaling.

As you recall, Penrose suggested that you need mass to make any kind of clock or a ruler. Without mass, the universe would lose its sense of scale, and so the small beginning of the universe and the large end of the universe become almost equivalent. With this insight, we should expect to observe the large-scale structure of the universe to be nearly scale-invariant. The largest structures in the universe (on scales of sixty-one billion light-years and bigger) remain always frozen in our standard cosmological model (as light cannot travel any farther), and so if they are statistically scale-invariant (which means there is no preferred scale), they still look the same under conformal rescaling. I'm not sure if Penrose realizes how important his idea is in understanding this key feature of the early universe. It turns out that conformal rescaling and scale-invariant CMB are (yet another) match made in heaven. Just as Penrose could use this ingredient in his cyclic scheme, we could use it to ensure that the fiery start of the universe and its cold death would merge and form a loop. The end of time would become the beginning, and the beginning would become the end.

But why nearly but not perfectly, scale-invariant? On smaller scales, things do evolve; there are sound waves, galaxies, clusters, superclusters, and even you and I that form and die. But then, if conformal rescaling shrinks these fossils to small scales, the endless supply of scale-invariant fluctuations on larger scales can seed the exact same structures as in previous eons.

You may now worry about violating the second law of thermodynamics, requiring an ever-increasing entropy, which could be a concern for any cyclic cosmological model. However, if space is infinite and we have an endless supply of scale-invariant fluctuations on large scales, then entropy is infinite, implying that the second law cannot be applied.

In our paper titled "Does History Repeat Itself?," Beth and I outlined the mathematical structure of this model. But the crucial hurdle would have been to put it on a computer and compare it with the CMB observations. Having tested many other ideas, like our black-hole model (chapter 7) and holographic cosmology (chapter 9), it was now time for Beth to examine her own. The first thing she had to do was to match the late-time evolution of the universe to our initial conditions. We know that conformal rescaling shrinks the universe at each eon, but by how much? Beth ran the model against the current Planck data and found that the answer to be almost 100 times. So, for example, the gigantic Laniakea (Hawaiian for "immeasurable heaven") supercluster of galaxies is currently 520 million light-years across. While its atoms and stars disappear at the crossover into the new eon, the large-scale structure of Laniakea will shrink a hundredfold to 5.2 million light-years at the next eon, eventually collapsing and forming stars anew, this time as a measly galaxy. But then you may wonder how history repeats itself, if a supercluster turns into a galaxy. The answer is that there must be another super-Laniakea, at a whopping 52 billion light-years across, somewhere in the universe, which would then become the Laniakea supercluster when it shrinks a hundredfold in the next eon. Repeat the pattern, and the universe becomes a fractal, with infinite copies of any structure, on different scales (although most of them will be outside our observable horizon).

But Beth discovered something even more surprising. Fitting CMB data required dark energy to increase by just a few percent every few billion years. I find this discovery to be profound: While it is quite bizarre that the future evolution of dark energy is required to match the fossils of our past, we should expect exactly that if the universe is on a time loop. Moreover, it gives us a testable prediction: while most of my colleagues think dark energy is a constant, our model predicts it increases over time, at a precise rate.

What about other cosmological problems that inflation was historically credited with having solved, namely the flatness problem, the horizon problem, and the monopole problem? The conventional strategy is to imagine a set of generic initial conditions that then turns to what we observe through an inflationary scenario. However, if time is periodic, we do not get to pick our initial conditions, as our past is our future; all we have is self-consistency. In particular, space must be infinite and flat, because a curved space has a size and so cannot be conformal at the crossover into the new eon. Similarly, we need uniform space and scale-invariant fluctuations (on large scales), so that structures can match under rescaling. These uniformities address the flatness and horizon problems, as logical requirements for the consistency of the model. As we discussed in chapter 2, the monopole problem, one might argue, is only a problem for the particle theories, popular in the 1970s, that produced stable monopoles. While Preskill, Guth, and Tye's monopole problem is almost as old as I am, no theoretical consensus or experimental hint yet exists as to whether such theories, or their monopoles, should exist in nature.

What about other possible future tests of periodic time cosmology? While it would be really cool to find multiple copies of the same structures (as in our red-rose picture; see color plate section), each one shrunk a hundredfold, signals are, realistically, likely to be more subtle. For example, as we mentioned above, the spectrum of CMB fluctuations is fixed by dark energy's history, so better measurements of cosmic expansion (as promised, for example, by the upcoming Euclid satellite and the Rubin Observatory), combined with higher-resolution CMB maps (as we shall discuss in chapter 12), can further test this consistency. What about primordial gravitational waves? Here again, our only constraint is self-consistency. So, it is possible to have a scale-invariant spectrum of gravitational waves on large scales (turning red on smaller scales, just like sound waves), or none at all. At minimum, you have gravitational waves produced by astrophysical objects (like LIGO observed), or even Penrose's infamous circles, but they would be from previous eons. However, unlike in Penrose's model, we know the signal would be very much weakened by cosmological redshift during late-time inflation, and thus it would be very hard to see. And, finally, self-consistency might impose some weird non-Gaussian features, but that point is yet to be worked out. To summarize, what I love about periodic time cosmology is how minimalistic it is. You don't have to invent new processes to generate structures in the universe, as what you see is what you get. But then, the physics is so exotic that I cannot see how (even in principle) it can be tested in the lab, which makes the model a bit too fantastical for my physicist taste. So, thank God for cosmology.

After finishing her PhD, Beth did a postdoc at Southampton, UK, with Skenderis (with whom, as you may recall from the last chapter, she worked on holographic cosmology) and another at Queen's University back in Canada. But the COVID-19 pandemic hit her really hard, just as it impacted so many across the world, so she decided to take a break from physics. For as long as I can remember, Beth had been enamored with Russian culture, so she decided to move to Russia, which certainly was not easy due to pandemic travel restrictions. Nonetheless, in the fall of 2021 she managed to travel to Russia, where she began searching for a place to live and a job. But in the preceding decade, while Beth had been busy devising her "New Views on the Cosmological Big Bang" (the title of her PhD thesis), another war had been brewing to the south of Russia. Just as they had four decades earlier in Afghanistan, Russians claimed to be threatened by Western influence at their doorstep, and on February 24, 2022, tens of thousands of Russian troops poured across the Ukrainian border. Beth managed to leave just in time, as most airlines stopped flying to Russia and most other countries banned Russian airlines. But eventually she returned to Russia, where she now lives.

It is often said that those who cannot learn from history are doomed to repeat it. Yet, as Friedrich Nietzsche prophesied in his "Eternal Return" doctrine, the real lesson may be that history will repeat itself; every event that has ever been will occur again. You will get a second chance at life, and a third, and so on ad infinitum, even though you may not remember them. And it's not only religion that promises a hereafter—talking of religion . . .

✣ 11 ✣

SCIENCE OF RELIGION, RELIGION OF SCIENCE

أَلَا بِذِكْرِ ٱللَّهِ تَطْمَئِنُّ ٱلْقُلُوبُ.

Lo, the name of the Lord shall bring confidence to your heart.
QURAN 13:28 (الرعد)

We have come to the end of our survey for competing models of the Big Bang, but our journey is not complete. Having encountered different narratives about the Big Bang, you may wonder whether it is time to decide which one really happened and which are false memories. Alas, no; we must be wary of the misplaced confidence that fills our contemporary discourse. Certainty is seductive, showing confidence and projecting strength; but it is not appropriate at the frontiers of science, in the realms where physics is still trying to uncover the secrets of nature. Yet when it comes to the Big Bang, brash convictions are ubiquitous, whether they be from scientists, philosophers, or theologians. The phrase *follow the science* should really be amended to *follow the science and embrace uncertainty*. But there is uncertainty, and there is uncertainty. I am sitting in my bedroom, closing the doors to avoid getting distracted by kids playing outside. When I look around me, I see a dresser and a bed. While I am not sure exactly how long each piece of furniture is, I can guess their measurements within a foot of uncertainty. But whether my kid is just outside the door, about to burst into the room, I don't know. That possibility has a much higher level of uncertainty, what quantum physicists call a "cutoff." While you are within the cutoff of a theory, you have a rough understanding of what's happening, within some uncertainty. When you go outside the cutoff, all bets are off. When we talk about the Big Bang, that outside-the-cutoff uncertainty is the uncertainty that we talk about. One motive for writing this book was to give a glimpse of science at the edge of the knowledge, past the cutoff; to show that it involves not just theoretical conjectures, equations, and experiments, but is also a human process, full of conflict and emotion: bitterness, hope, doubt, and belief. But another motive for the book was to show readers that the public have been misled. Controversies supposedly resolved by modern cosmology are very much alive. The Big Bang does not prove that the universe had a beginning, and dark energy does not rule out cyclic behavior. These debates have been laid out in this book, and it is always tempting to pick a side prematurely. But that's what we don't want to do. Until the data has spoken, it's better to be an explorer than a believer.

But while we can't tell you exactly what happened at the Big Bang, we can at least line up the narratives and assess their plausibility. In our introduction, we mentioned a number of criteria that Big Bang imagineers hold dear. First, they seek theories that have explanatory power. Inflation succeeds here, because it can explain why we live in a flat, uniform, vast universe filled with innumerable galaxies. But it doesn't answer the question of why the entropy of the universe was so low at the Big Bang, how the singularity is resolved, or why we live in three dimensions of space and one of time. Other models, like VSL or cyclic universes, can answer some of the same mysteries. Yet, some frameworks have explanatory virtues that inflation lacks. Loop quantum cosmology can resolve the singularity, while string gas cosmology can tackle why we live in three space dimensions. Cyclic and hourglass cosmologies may remove the need to describe the universe's beginning.

If explanations are our revenue, then assumptions are our costs. I'm sure that many of my colleagues will think that VSL is making an outrageous assumption—that the speed of light can become arbitrarily fast—confronting a century of triumph of general relativity. But as this varying light speed is a prediction of a more fundamental theory, namely Hořava gravity, then its cost will rise and fall as confidence in that theory expands or diminishes. Penrose's CCC assumes that information is destroyed in black holes. The community widely rejects this idea, but as long as the information paradox is unresolved (notwithstanding the weekly clickbait articles to the contrary), then I'm happy to entertain such views. We might define *explanatory power* as the ability to assume a little and explain a lot. Inflation at first glance requires only general relativity and quantum mechanics and so looks attractive; I think that's why it seduces so many cosmologists. However, an inflation model has to assume an inflaton field, the field's myriad variants, and its illusive measure. That's where problems arise. Our assumptions, then, don't come at a fixed cost; rather, they fluctuate, hopefully as our understanding of nature develops, but all too often, I fear, at the whims of fashion.

Any model of the Big Bang should arise from a deeper theory of physics. We do not know the underlying particle physics of the inflaton. Attempts to model inflation in string theory have been problematic, and while loop theorists regularly assume that exponential expansion happened, exponential expansion is not a prediction of that theory, either. Models that come more directly from some of the quantum gravity proposals we have looked at may have a clear advantage here. But as none of these ideas has been confirmed by experiments, and as particular models all have their own assumptions, seeking more fundamental theories is hardly decisive. Steinhardt and collaborators' cyclic cosmology is not doing much better. It also assumes an arbitrary field to get things working. Of course, assuming extra fields is not a fatal flaw; I have used them myself. One of my more memorable moments of joy was in seeing the connection between our Cuscuton field, Hořava gravity, and VSL. Sometimes these hypothetical fields even end up being real. Recall the great fanfare that greeted the discovery of the Higgs field in 2012—fifty years after its existence was postulated. Finding more direct evidence for any of the postulated fields that populate our models could similarly be a watershed moment. Given that we proposed Cuscuton in 2007, I am keeping my schedule for 2057 free.

Our Big Bang suspects must be compatible with all known facts about nature. To show this compatibility, we need calculations, which often challenge even the best of us. In particular, the detailed properties of the CMB and the distribution of galaxies (amongst others) should be computed and compared to observations. I'm immensely proud of the work done by my students and collaborators to make such calculations possible, for a number of the proposals this book describes.

And we likewise applaud the string gas and loop quantum cosmologists who have attempted to carry out their own calculations of this kind. Inflation is compatible with the data, but that compatibility is hardly surprising because there are so many versions of inflation. My friends tell me that simple inflationary models have survived contact with experiment, and thus that strong evidence exists for the theory. They say they would have given up on inflation otherwise. But exactly what counts as *simple* is debatable. What about those proposals—like Baum-Frampton, de Sitter equilibrium, or Carroll-Chen—that haven't made detailed CMB calculations? I'm not ready to write them off. I'd rather consider them as being like string theory: works in progress. But until they get to the stage of making CMB calculations, they cannot compete as candidates for our prime suspect.

Lastly, our models need to make predictions for new observations. This requirement is really the ultimate test for any scientific theory; it is what can turn speculation into fact. For example, while relativity beautifully explained gravity as the curvature of spacetime, it was not until Eddington's eclipse expedition that we could fully embrace relativity as our trusted description of the world. Amazingly, scientists still seek to challenge the theory with a variety of experimental probes, from atomic clocks to gravitational-wave detectors. Likewise, for the Big Bang, we are looking for models that make concrete predictions for the strength of gravitational waves, the statistical properties of the CMB, or the distribution of galaxies. In the predictions game, inflation is a tricky case, because every individual model of inflation makes predictions—but does that mean the paradigm as a whole does? Or maybe thinking so is a fallacy of composition; just because every member of a flock has a mother, it doesn't follow that the flock itself has one. Inflation started out as a theory and became a class of many theories, each theory making different assumptions and resulting in different predictions. This transmutation is not taught in standard textbooks. But inflation isn't the only paradigm with this issue; just saying that the universe is cyclic or that the speed of light changes doesn't give us firm predictions, either. One needs more guiding principles, a heck load of data, or possibly both.

In my opinion, no model scores perfectly on all the criteria we have laid out. That conclusion may seem like a disappointment, but it's what makes the field so exciting to me. We are on the edge of knowledge, not knowing if glories or ruin await. But while we currently cannot conclude in favor of any one model, the fact that the debate is still ongoing can have major implications for the way scientists understand cosmology. One thing we hope you appreciate from this book is that the certainty that the universe was born a finite time ago in a state of infinite density is now gone, replaced by a lineup of fascinating candidates, some that might restore an ultimate beginning and others that promise an eternal cosmos. This is the quiet revolution of twenty-first-century cosmology. It's a dramatic new landscape, one that thrills some but alarms others. Many theologians claim that the "fact" that the universe had a beginning confirms their religion and provides evidence of a creator. Meanwhile, some scientists sell speculations of naturalistic origins as facts, creating a real danger that theories may calcify into dogmas. Our task here is not to preach to scientists or theologians how to think about the Big Bang, but rather to demonstrate where the roads that they have taken may lead. And this task brings us to an apparent face-off: where does science stop and religion start?

THE RELIGION WITHIN COSMOLOGY

Looking back on my own past, my memories are filled with the imagery of my youth, from waking up to sounds of prayer calls at dawn to the terrifying encounter with the morality police at my college campus, when Ghazal and I were escorted to the security headquarters for the "crime" of speaking together in a classroom. This background may give me a rare perspective on the topic of science and religion. The first half of my life was lived under an Islamic theocracy, where religion was invoked to justify, quite literally, every aspect of public and private life. But the second half was driven by science, not because the Canadian and United States governments are guided by it (they aren't), but rather because I left my home country in pursuit of science, to uncover the deepest and darkest mysteries of the cosmos.

When I came to the United States for graduate school in 1999, I brought with me my religious upbringing, from kindergarten through high school and then college. While living religiously was the norm back home, keeping the faith as a stranger in a strange land did take its toll. Despite its bad reputation, the United States is actually very inclusive and welcoming to different cultures compared to where I came from, and so it didn't take long for me to find people of common outlooks. With liberal presidents in power (Clinton in Washington, DC, and Khatami in Tehran), the US-Iran animosity of the 1980s, fueled by the Iranian Islamic revolution and the hostage crisis, was in apparent decline, and there was reason for optimism. But as we discussed in the last chapter, history has a way of repeating itself. Shortly after I arrived in Princeton, nineteen Muslim men (out of seven hundred million worldwide) hijacked US airliners and struck them against the Twin Towers and the Pentagon. Even though the 9/11 hijackers were from Saudi Arabia, UAE, Egypt, and Lebanon, four months later, George W. Bush would declare Iran, Iraq, and North Korea to be an axis of evil that needed to be defeated to right all the wrongs of the world. For me, a faith that used to be a source of comfort, connection, and community had suddenly become a liability.

But this is a book about the Big Bang, where brave (some may say strongheaded) cosmologists take what they learn in particle accelerators on Earth, and from galaxies in distant corners of the cosmos, to infer what happened 13.8 billion years ago, at crushing densities and unbelievable temperatures we have never seen before. In the same way that the Big Bang, as a physical process, is informed by our knowledge of physics, our approach to the science of Big Bang is a human process and thus informed by our understanding of humanity. Put differently, you can take the cosmologist out of religion, but can you take religion out of the cosmologist?

Religion matters because some of our leaders, those at the very top of the food chain, believe that "God told [them] to end the tyranny in Iraq," while others think they are God's representatives on Earth and that all those who oppose them are cronies of "the Great Satan."[1] It is not the absurdity of these dogmas that is most remarkable, but rather the fact that for millennia we have established our hierarchies, social mobility, and reward systems around them. So, could it be that suddenly—*poof!*—enlightenment, and we are free? Or do we just replace our dogmas with samgod?

Speaking of dogmas, consider the doctrines of original sin in Christianity or the infallibility of the prophet in Shia Islam. These tenets are taken as fundamental elements of our religious belief systems, to the extent that even the faithful can be shunned, excommunicated, or burnt at stake if they do not subscribe to them. But in the theoretical physics community, you may be exiled into the wilderness if you entertain the idea of faster-than-light signals (Magueijo), time loops (Gott), violating null energy conditions (Steinhardt and Ijjas), modified gravity to replace dark matter (Moffat and Bekenstein), or basically anything other than inflation and string theory. While exile may be better than being burnt at the stake, consider that if anonymous reviewers decide your starting assumptions are "not well-motivated," "plausible," or "consistent with known physical principles," you cannot get your papers published, get funding, or hire students. But perhaps we shouldn't take this analogy too far. Phil thinks I'm being too pessimistic. He pointed out that the rebels he interviewed for this book seem to be doing fine, their anonymous reviewers and employers have been more open minded, their papers get published, and they have good academic jobs. *Vive la résistance!* But I'm not fully persuaded. Perhaps he is being too optimistic? We should also be wary of survivorship bias, the idea that the famous faces and voices we get to read and hear in top journals and TV interviews don't necessarily represent the struggles of the broader community. So, take this caution as a warning about what can and (speaking from the trenches) all too often does go wrong.

While I wonder whether my battle scars are special to my journey, Phil thinks they are much more widespread. It is a near universal trait for humans to organize into in and out groups. Christian versus Muslims, Shia versus Sunni, urban versus rural, and of course most of us have our favorite sports teams whom we support, however embarrassing their performance. That is why, here in this book, we have tried to present the strengths and weaknesses of the many competing ideas about the Big Bang. Of course, one can't blame scientists for getting excited about their own models, and I'm no exception. But I hope I have convinced you that my proposals are at best, on good days, decent candidates for pieces of our cosmic puzzle. But I don't really know what happened at the Big Bang. Nobody does.

Now, take the string-versus-loop-quantum-gravity debate or the inflation-versus-cyclic-cosmology debate, each of which we have described in graphic detail throughout this book. In both these debates, proponents on each side accuse the other of being charlatans, populists, elitists, liars, etc. (some less diplomatically than others, though we are not at liberty to disclose all the details). This state of affairs parallels the debates amongst people of differing faiths (or opposing political parties), where proponents face off about the characters of their leaders and not about those leaders' messages. There is then the argument that many more physicists work on strings or inflation than work in loop quantum gravity or cyclic models, and that therefore the former must be correct. But then would you say that the New Testament must be more correct than the Torah, since there are over a hundred times more Christians than there are Jews? Popularity and scientific consensus are distinctly and critically different.

Of course, I am aware that some parties in all four camps mentioned in the aforementioned arguments may be seriously offended by my comparison of their venerated scientific disciplines with old religions. (And some religious readers might be offended by the comparison too.) But we are all cut from the same cloth and must guard against elitism. The forces and instincts that have guided our social interactions through the ages may persist, albeit

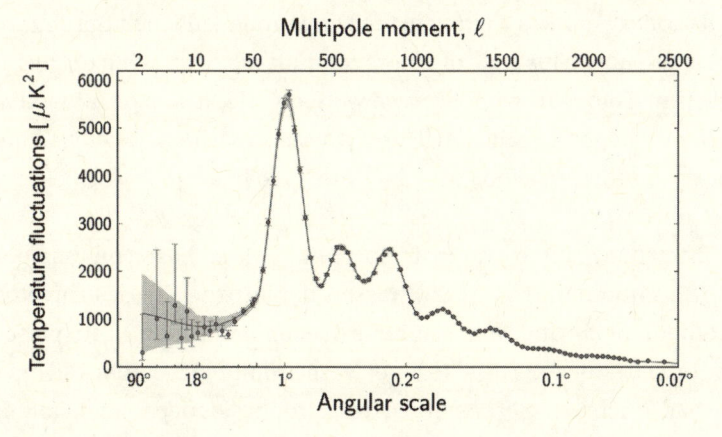

To verify a scientific theory (as opposed to a religious belief), the data must match expectations. Here, the Planck spectrum of cosmic microwave background (CMB) temperature fluctuation in the sky matches successful Big Bang models. Examples include Starobinsky's inflation or my bi-thermal model with João Magueijo. Copyright ESA and the Planck Collaboration.

under different guises and in spite of our intentions. My theorist friends point me to the amazing mathematical successes of their frameworks: canceling anomalies, dualities, or exact solutions of complex systems of equations. These frameworks offer so much to revel in, yet when I close my eyes, I see the wondrous St. Peter's Basilica in Rome, or the breathtaking Shah Mosque in Isfahan, constructed to draw awe and inspire the faithful. But then, is the grandeur of one's cathedral a measure of the truth in one's theory? My cosmologist friends recount the amazing observational success of inflation, solving unexplainable mysteries of the Big Bang and predicting the spectrum of sound waves that match the Planck satellite's observations across the sky with unprecedented accuracy. But again, I close my eyes and hear the voice of my middle school teacher: "How could there be so much order, so much beauty in the world, without an all-powerful creator?"—and I fear that neither my cosmologist friends nor my teacher can be proven wrong. For I remember well when the data seemed to suggest an open universe (that is, a space with negative curvature), or large non-Gaussianity, or primordial gravitational waves (recall the BICEP debacle), and there were always many perfectly "well-motivated and natural" inflationary models waiting in the wings and ready to fit the data, as there are now. This state of affairs reminds me of growing up during the Iran–Iraq war; whenever we won a battle, it was positioned as a victory of good against evil, and whenever we lost, we envied our martyred soldiers who got so lucky to go to heaven. A true believer can never lose.

The 2001 Hollywood movie *A Beautiful Mind*, based on the book of the same name by Sylvia Nassar, depicts the life of Princeton mathematician and Nobel laureate in economics John Nash (played by Russell Crowe).[2] I used to meet Nash around Princeton, and once we had lunch in the basement of the astrophysics department (along with a few other graduate students, as per tradition). In the movie—spoiler alert—Nash has to battle his schizophrenia, which manifests as hallucinations in the form of government agents who recruit him for anti-Soviet intelligence operations. However, through many ups and downs, he eventually learns to live with these hallucinations, without any medication. What if, contrary to what Karl Marx imagined, religion is not the "opiate of the masses" but is, rather, akin to John Nash's hallucinations—built into our DNA? We could call such hallucinations faith and let them guide our lives; or we could declare ourselves cured, yet find ourselves still following the same illogical ideologies and groupthink. Or maybe we could accept their inevitable existence and absurdity but decide to keep them in check and plan our lives in spite of our human realities and blessings. It is no wonder, then, that we are filled with misplaced confidence, mistaking our speculations for facts, nor should it be surprising that so many see in the heart of the mystery of the Big Bang their favorite deity staring back. Lemaître, the Catholic priest, cosmologist, and father of the Big Bang, warned against exactly this tendency. But, as we shall see, the temptation is often too strong to resist.

THE COSMOLOGY WITHIN RELIGION

Chief amongst those who ignore Lemaître's advice not to mix science and religion is the Christian philosopher William Lane Craig, who has managed to get his claims of a link between the Big Bang and God into the philosophical literature and, at least on one occasion, a scientific journal.[3] He is a master debater. Writer Sam Harris said that Craig has "put the fear of God into many of my fellow atheists."[4] Cosmological arguments were recently named in a survey of philosophers as the best reasons to believe in God, and Craig has championed the most popular version from among these contentions, which is known as the Kalam cosmological argument.[5] It's often expressed as a syllogism; you can even buy it printed on a T-shirt:

1. Whatever begins to exist has a cause.
2. The universe began to exist.
3. Therefore, the universe has a cause.

Growing up in Iran, we also were taught this same syllogism—there called the causality argument—in our high school's religion class. (The Arabic word *kalam* translates to *argument*, so calling something the "argument cosmological argument" sounds kind of ridiculous.) Needless to say, I'm not convinced by this argument, and neither is Phil, who critiqued it in a journal and in a YouTube series in which I was a talking head, along with colleagues including Roger Penrose, Carlo Rovelli, and Sean Carroll.[6] What Phil pointed out was that the argument seemed to be based on the old consensus that the Big Bang singularity proves that there was a beginning to time. Craig's argument, then, is the antithesis of this book. We claim that we don't know if the universe had a beginning. But Craig insists that "this cosmological singularity, from which the universe sprang, marked the beginning, not only of all matter and energy in the universe, but of physical space and time themselves. The Big Bang model thus dramatically and unexpectedly supported the biblical doctrine of creation ex-nihilo. Indeed, given the truth of the maxim ex-nihilo nihil fit (out of nothing comes nothing), the Big Bang requires a supernatural cause."[7]

Again, almost no cosmologists believe in the singularity; it's an artifact of pushing Einstein's theory of gravity beyond its limits. The singularity marks a state of infinite density, curvature, temperature, and pressure and so should be denied, not embraced, by Craig—who also tells us infinity is impossible. I'm also skeptical that the notion of causality can be applied at a fundamental level. In my own models where, for example, there are multiple dimensions of time, time loops, or no time at all, causality may not be an appropriate concept. It's plausible that our perception of cause and effect, like the wetness of water, is an emergent concept, absent from fundamental physics but present in our day-to-day lives. So Big Bang cosmology does not show that the universe had a beginning; and even if it did, we have no right to demand that such a beginning must have had a cause.

When Craig first published the Kalam cosmological argument in 1979 (the year I was born), the Penrose-Hawking theorem was often interpreted to assert an absolute beginning.[8] Ironically, 1979 was the very year that Guth stumbled upon inflation, which violates one key assumption of the Penrose-Hawking theorem: that gravity is always attractive. So, Craig now focuses on the Borde-Guth-Vilenkin theorem we met in chapter 2—but all that theorem really shows is that inflation in general relativity cannot by itself remove the singularity. But this limitation means very little; as we have seen in previous chapters, there are other ways to resolve the initial singularity from quantum or classical bounces to conformal or Euclidean space-times, spin torsion effects, closed time-like curves, etc.

My wife Ghazal and collaborators have recently shown several ways in which space-time can extend beyond the boundary set by Borde, Guth, and Vilenkin.[9] As she told Phil, "Nothing dramatic happens . . . In some of our solutions, it looks like they're going into a static universe; some of them seem to be going into contracting universes." Theoretical physicists Lesnefsku, Easson, and Davies have gone further, arguing that there is a mathematical flaw in the theorem, allowing inflation to be eternal into the past. Meanwhile, Will Kinney and colleagues have tried to make Borde-Guth-Vilenkin more general and claim they can apply it to some of the cyclic cosmologies we covered in chapter 6. This back-and-forth clearly shows that the science regarding the universe's ultimate beginning (or lack thereof) is far from settled.

In 2005, Vilenkin did say that the theorem proves a beginning to the universe, and Guth and Borde were mostly quiet about its implications.[10] So, for years Craig and his followers (not all Christian) would parrot the same quote from Vilenkin, ecstatic that their faith had seeming support from science. But they might have been wiser to follow Lemaître's advice not to mix science and religion, lest changes in the former should embarrass the latter. Sure enough, when Craig stepped into the ring with Sean Carroll to debate "God and cosmology," his usual knockout blows missed their target. Carroll, an advisor to the Marvel movie empire, was able to deliver a superhero move that shook the core of the argument when he quoted his former postdoc supervisor, Guth, saying, "I don't know whether the universe had a beginning. I suspect the universe didn't have a beginning. It's very likely eternal but nobody knows."[11] The revelation that Guth was a past-eternal-universe advocate was a devastating blow, as Craig had staked so much on the Borde-Guth-Vilenkin theorem (BGV) as proving an absolute beginning. Ignoring Lemaître's advice proved to be a costly mistake. Kalam advocates might have held out hope that Vilenkin would throw them a lifeline; after all, it was his quote they had used for so many years. But scientists are so unreliable. When Phil asked Vilenkin about the BGV theorem afterward, he replied that it only proved the expansion of the universe had a beginning, whereas the question of whether the whole universe had a beginning, the theorem doesn't answer. One alternative to claiming that BGV proves a beginning is to propose that there was a past, contracting universe, as the theorem only rules out an expansion extending into the eternal past. However, it is often claimed that such a crunching universe would be messy and unstable and hence unable to expand into our smooth universe. But as we have seen in previous chapters, many models and simulations exist that show that our universe could make a smooth transition from contraction to expansion. Of course, we don't know whether they are correct, and Craig is keen to emphasize that "there's no evidence at all that the universe is beginningless"—which is true. But the contrary argument has no evidence to support it either. It cannot be repeated enough: observations only confirm the hot Big Bang, not the Big Bang singularity.

But even if every cosmological singularity is shown to be erased by quantum effects, like a sand sculpture at high tide; Kalam advocates believe they can use thermodynamics to prove a beginning anyway. The argument says that as entropy increases, after an infinite age, we should find ourselves in a maximal-entropy state. In such a state, the universe would be a totally disordered random mess, with no structures of any kind. Since we don't live in such a universe, it can't be a past-eternal entity. That argument is related to what puzzled Roger Penrose, namely why was the entropy so low at the Big Bang when high entropy seems more probable than low entropy? But a low-entropy state is improbable anywhere on the timeline of the universe, whether it happens to be at the start of a universe with a beginning or in the middle of one without. So, the puzzle does not provide evidence for one class of models over another. My own point of view is somewhat more nuanced: that growing entropy, like the sunrise and sunset, is an observer-dependent effect. Simply put, the second law of thermodynamics says that if you start from a special state, you are likely to evolve to more and more random states, but the state that I find special may be different from yours: a boring day in my life may be the day that you win the lottery. Recall, from chapter 6, that we described a deck of cards being shuffled as a process that increases entropy. But in some sense the original order is just as unlikely as any other sequence, and it's our arbitrary choice to call the starting state one of low entropy. Similarly, it's not that the entropy was low at the Big Bang; that's just how it appears to us. A different approach to the problem is to assume an eternal, infinite universe. Many models, such as de Sitter equilibrium, Carroll-Chen, Baum-Frampton, CCC, and loop quantum cosmology can be thought of as having a mechanism that operated before the Big Bang that put the entropy of the universe into its low state. Unlike the singularity, such a mechanism could explain the puzzle. This principle actually unites the warring tribes that circle around the Big Bang. When Phil asked Guth if he had in common with Penrose the idea that an eternal universe could solve the low-entropy paradox, he reluctantly said he agreed "with that phrase, but no further." So, ironically, the very issue that supposedly shows a beginning to the universe might imply the opposite.

GOD OF THE FINE-TUNING

Another aspect of modern cosmology has drawn the attention of philosophers and scientists alike. That aspect is the so-called fine-tuning of the universe for life. According to this argument, the universe was born with perfectly set physical parameters; a tiny change in any of roughly two dozen constants of nature would lead to a barren universe, one devoid of complexity and life. How, then, did we get so lucky? Are life-permitting constants so unlikely that we need God to intervene and set the dials just right? Craig also championed this argument in his debate with Carroll with yet another syllogism:

1. The fine-tuning of the universe is due to either physical necessity, chance, or design.
2. It is not due to physical necessity or chance.
3. Therefore, it is due to design.

Craig ruled out necessity by appealing to string theory and its huge landscape of possible worlds. The obvious problem here is that, as we saw in chapter 5, string theory is not the only game in town and may simply be wrong. Even if it is right, we don't know if the landscape is a real feature of the theory. Recall from chapter 4 that Vafa's swampland conjecture suggests many of the imagined universes in the landscape are not real. So, necessity cannot be ruled out; it's perfectly conceivable that the constants of nature have been called "constant" for good reason; they simply couldn't have any values other than what we observe, and therefore they are not unlikely.

But what if we grant that Craig is correct in his summation of fundamental physics? It would certainly seem then that necessity is ruled out, but then chance seems less daunting. One can overcome any low odds with high frequency; if enough lottery tickets are sold, someone is bound to win. The multiverse of eternal inflation and the very string theory landscape Craig appealed to to rule out necessity seem to accomplish this task. One can, of course, dismiss the multiverse if one is prepared to embrace alternatives to inflation, but as many of its competitors cast doubt on the beginning of the universe, that solution may be hard to stomach for Big Bang theologians.

My friend Stephon Alexander and collaborators have even constructed a cyclic universe where the constants of nature vary from cycle to cycle.[12] Perhaps we should consider this model as another type of multiverse.

Craig, though, argues that even if a multiverse exists, it cannot solve fine-tuning, as such a multiverse would be dominated by the bizarre Boltzmann brains, fluctuating out of the vacuum, that we encountered in chapter 6. But as we saw in chapter 2, probabilities in eternal inflation depend on a measure, a method of regularizing the infinite. Cosmologists have found measures where these fluctuating brains do not outnumber normal observers. In fact, in the presence of a cosmological constant, our own patch of space-time will expand forever—and in infinite time, even the most unlikely events, including Boltzmann brain formation, are bound to happen eventually. So whatever reasons one has to prefer a single universe over many, Boltzmann-brain domination shouldn't be one of them.

A better reason to reject the multiverse is that models of eternal inflation might not be right. What then for the skeptic of fine-tuning? As we saw in chapter 7, other alternatives exist that are not listed by Craig. For example, Smolin's cosmological natural selection creates a selection pressure analogous to Darwinian evolution. Thus, we don't have to choose between chance and design; natural selection is neither.

But I think a simpler solution to the problem exists, one that doesn't require us to invoke a multiverse. We can simply deny that fine-tuning exists as a distinct phenomenon altogether. Getting all aces in a poker game requires an explanation because we know that the outcome not only could have been different but that it was likely to have been different. If the constants never change, then they can't ever be different to what they *are*—so we can't say they are unlikely. Even the flatness problem, one of the first motivations for taking inflation seriously, may not actually be a problem after all—since we don't really know that the probability of starting with an extremely flat space is low.

It may be that fine-tuning is a pseudo problem. But many theologians will quote cosmologists with greater reputations than mine to persuade you in the other direction. Fortunately, Phil's interviews show that many of these giants of the field who are often quoted as supporting fine-tuning are far more doubtful of it than advertised. Roger Penrose, Alan Guth, and Steven Weinberg have all spoken out against the claim that multiple constants of nature are fine-tuned for life, even though they are some of the scientists quoted to support the argument.[13] One area of confusion is that physicists do say theories are fine-tuned, but that kind of fine-tuning indicates a theory's deficiency and does not necessarily say anything about the probability of a life-permitting universe. Penrose, for example, thinks that the low entropy of the universe is fine-tuned in the sense that it's not what we should naively expect. But it's not fine-tuned *for life*, as the entropy could be much larger and life would still be here. Both Weinberg and Guth do think the cosmological constant is fine-tuned in the sense that quantum theory predicts a much larger value than we observe. But Weinberg conceded that he wasn't even sure about that point, and Guth caveats it by assuming that dark energy is a cosmological constant, which might not be the case. Moreover, almost all the scientists who do entertain the idea that the universe is fine-tuned for life *also* think that a multiverse is a plausible solution to this fine-tuning—a fact very rarely mentioned in presentations of the argument. Those who live by the appeal to authority must also die by it.

Surprisingly, one thinker who believes that God offers a poor explanation for fine-tuning is himself a believer. Philosopher of physics Hans Halvorson's take on the issue is startling, turning the whole argument on its head.[14] As a Christian, Halvorson believes that God is the ultimate author of all reality and that he is thereby responsible for setting probabilities for life. So, then the fine-tuning argument backfires—because if God loves life, the probabilities for life should be high. Fine-tuning is the claim that the probabilities for life are low, potentially undermining theism. In reality, theism makes no clear prediction about what the probabilities of life should be. But if arguments in favor of fine-tuning undermine arguments for God's existence, why is Halvorson still a Christian? It seems that he's sympathetic to my view that probabilities for the constants of nature are totally ambiguous, making fine-tuning a mirage; he tells us that "most of our intuitions about probability have been homed in very simple cases where you have a finite number of possibilities." Think of a dice roll: there are normally six possible outcomes, each with ⅙ odds, so the total probability adds up to one. We have what's called a measure over the probability space. But no known finite space of possibilities exists for the constants of nature; they could literally take on any value. Therefore, as Halvorson puts it, "in infinite space there actually is no such thing in general as the invariant measure, the measure that it's just sort of the default. So already you have to think, *wait, where are we getting these numbers from?*" Lydia and Tim McGrew are Christian philosophers who have also critiqued the fine-tuning argument for very similar reasons.[15] Carrying on with the dice analogy, imagine we want to know the probability of getting a 6. We can confidently calculate this probability only if we know certain things: How many numbers are on the die? Are they equally weighted? And how many times do we get to roll? With one die, rolling a 6 is just as likely as rolling a 2, but with two dice 6 is the more likely roll, as there are many ways to get a 6 (5 + 1, 4 + 2, 3 + 3) but only one way to get a 2 (1 + 1). We understand how dice work and have seen them do so on multiple occasions, so we can be justified in our expectations of certain outcomes. We have no such knowledge for the constants of nature.

Another thing that puzzles me about fine-tuning advocates is that many of them also claim that life could not have come about via natural means and thus requires divine intervention. But then why would God have needed to fine-tune the constants? As Halvorson says, theists "need to make their mind up. Are they going to try to find beauty in the laws of nature and then maybe at the end of the day say, 'To God's credit' . . . or are they going to try to find holes in the laws of nature and say, 'To God's credit'? . . . That's a fundamental tension." But then, I guess, the Lord works in mysterious ways.

DOES COSMOLOGY DISPROVE THE CREATOR?

If one wants to believe in God, cosmology is not a good reason to do so. But that does not mean it proves God does not exist. The New Atheists have also been guilty of misleading the public as to cosmology's findings and implications. In 2009, Lawrence Krauss gave a talk entitled "A Universe from Nothing," summarizing some of the ideas we highlighted in chapter 3.[16] It soon went viral, garnering more than two million views, and Krauss turned it into a popular book.[17] Much of the criticism of Krauss was badly misplaced, as commentators mistakenly claimed that he had assumed the universe came from a preexisting vacuum—which is not "nothing." In chapter 3, we saw that Vilenkin's model postulates that the vacuum itself tunneled into existence from a state with no space or time, so this criticism is dubious. The real problem I see with Krauss's talk was never picked up by his detractors, who seemed to miss that the evidence he used to support it, is quite weak. For Krauss claimed that only a flat universe can begin from nothing and that we should be worried if the universe wasn't perfectly flat. He then proceeded to reveal observations that show the universe is flat. I hope you'd agree that this line of argument creates a false impression that a universe from nothing has been confirmed by observations. All the models that give rise to the universe from nothing, like Vilenkin's tunneling from nothing and, potentially, Hartle and Hawking's no-boundary proposal, are closed (with positive curvature). Inflation will make the universe appear flat even if it did not come from nothing. Recall that loop quantum cosmologists assume that there was a bounce and then inflation, as did the Baum-Frampton model. De Sitter equilibrium and the Carroll-Chen model are eternal to the past and future and include inflation; even the father of inflation himself, Alan Guth, asserts that the universe may have existed forever into the past. Alternatives to inflation such as VSL or slow contraction can also make the universe flat. Thus, the flatness of the universe says next to nothing about whether the universe came from nothing. In his afterword to Krauss's book on the subject, well-known atheist Richard Dawkins said, "The spontaneous genesis of something out of nothing happened in a big way at the beginning of space and time, in the singularity known as the Big Bang . . . If 'On the Origin of Species' was biology's deadliest blow to supernaturalism, we may yet come to see 'A Universe from Nothing' as the equivalent from cosmology." But Darwin's theory of evolution, as Dawkins has so eloquently explained in his own books, has been confirmed by extraordinary evidence from multiple fields; not so for a universe from nothing. Recall from chapter 3 that Vilenkin admitted that so far, no one has thought of a way to test his model. It's one thing for Vilenkin to propose the idea as an interesting and stimulating conjecture. I have no problem with that. But for Dawkins to claim that Vilenkin's scenario is actually what happened is for him to take the very same faith-based position he abhors in his religious opponents. Here, we see another sign of speculation becoming dogma, of science turning into a creation myth. The problem is not with the hypotheses themselves but with the hype that surrounds them.

Unfortunately, Dawkins and Krauss are not alone in making overblown claims for a naturalistic creation ex nihilo. Stephen Hawking was without doubt a scientific hero, a genius at science and publicity; but we deify him at our peril. And one flaw in the great man was his own tendency to exaggerate the case for speculative ideas. When his book *The Grand Design* was released in 2010, it was accompanied by front-page headlines from the *London Times* to CNN, from the *Daily Mail* to Reuters, that read "Hawking: God Did Not Create the Universe."[18] The book appeals to similar concepts as those put forward by Krauss and the no-boundary proposal; in it, Hawking and his co-author, Leonard Mlodinow, claim that "spontaneous creation is the reason there is something rather than nothing, why the universe exists, why we exist."

Ironically, such a strong statement is not needed for these atheists to refute their opponents. What we hope you learn from this book is that spontaneous-creation models remain simply one variety of more than a dozen different proposals for what happened at the Big Bang. Many competitors we have explored assert that the universe is eternal. Carroll has published what has become known as the quantum eternity theorem, which suggests that a universe with nonzero energy must be eternal to the past and future.[19] Why this notion should worry the atheist Krauss is mysterious. Models without a beginning could just as equally be used to justify Hawking's assertion that God is not needed to start the universe. What we have seen is that there is no reason to think the universe ever *got* going; it's perfectly conceivable that it has always *been* going. But the temptation to inflate one's claims and project a misplaced confidence seems too seductive for theist and atheist alike; both are guilty of misleading the public. For while we cosmologists dabble in hyperbole and quantum uncertainty, what truly has no boundary is our self-confidence in our speculative hypotheses.

The quest to understand our ultimate origins burns deeply within us all, so much so that it is almost impossible to find a culture lacking a creation myth. And that fact should make us reflect on our own research; are the stories we have told here nothing more than secular creation myths? Can science become a religion? There are certainly similarities. Science has its holy books and its high priests; mathematics is its esoteric language. As we said at the start of this chapter, there are danger signs that we are walking into a world of dogma where "truths" are accepted uncritically, where the masses follow the bandwagon and refuse to think outside the box. But the debates we have seen show that accepted wisdom can be challenged. There are no fatwas sentencing critics to death, just strongly worded replies in magazines. "Science is self-correcting," we say. It's when we mistake conjectures for facts that science can become like religion. We must practice humility at the frontiers of knowledge, something that is lacking in many popular accounts of cosmology. But can we go too far in either replacing or surpassing existing paradigms? When are we guilty of dressing up speculation as physics? Where is the line between thinking outside the box and going beyond science altogether? When does science communication move beyond education, entertainment, and storytelling, and turn into propaganda? In order to check which of our models are flights of fancy and which are reasonable conjectures, we could simply apply what we know about the scientific method and determine if the relevant research programs match our criteria. But just as we must not avoid the conflict at the heart of cosmology, we cannot shy away from the battle over how to define science itself. For this question brings us into another fog of war, to what philosophers of science call the demarcation problem. Not only is science messy, so is the philosophy of science.

WHAT IS PSEUDOSCIENCE?

Exactly what distinguishes science from pseudoscience? You might agree with us that astronomy is science and astrology isn't, that chemistry is a respectable discipline while alchemy is hocus pocus, that vaccines and masks can prevent disease whereas homeopathic remedies are nothing but placebos, and that intelligent design has no place in our classrooms. In these cases, philosophers concur. It would be great if we had a simple formula to test which of our Big Bang candidates are science and which are folly. But alas, things aren't so straightforward. Remember, "follow the science and embrace uncertainty."

Many physicists think that the formula to distinguish science from bunk was found by British Austrian philosopher Sir Karl Popper, who suggested that a scientific theory is one that can be falsified by some imaginable experiment. From this perspective, some of the models we have surveyed, such as cyclic models that predict no gravitational waves, are in good shape; they are clearly falsifiable. But inflation is compatible with any level of gravitational waves, detectable or not. Does this compatibility mean it is not science? Some critics think so, but Popper's criteria might be a bit too simplistic to apply to what is a difficult and subtle problem.

A survey by Brian J. Alters found that only 11 percent of philosophers thought that a single criterion could be used to demarcate science and pseudoscience.[20] Perhaps this figure is so low because philosophers claim they've found Popper's Achilles' heel in the Duhem-Quine thesis. Pierre Duhem was a French theoretical physicist with a passion for the history of science, and Willard Quine was a hugely influential logician. Their thesis states that theories live in an interconnected network of scientific knowledge and so cannot be falsified by a single experiment. If a theory is apparently contradicted by data, it might be because the theory is false or it could be that some "auxiliary hypothesis" our inferences depend on is wrong.[21] The most obvious example of an auxiliary hypothesis is that the data is reliable. One only has to recall the BICEP incident we covered in chapter 2 to remind ourselves that this isn't always true. You may think that as more data piles up, such an auxiliary hypothesis would become gradually less tenable. That conclusion, however, assumes you are measuring the right data. For example, most polls prior to the 2016 US election predicted that Hillary Clinton would win the presidency. However, it turned out that the people being polled were not a representative sample of voters. This kind of mistake is known as a systematic error, which is incidentally what eluded the BICEP collaboration. They had great statistics (more than 5 sigma) but the wrong auxiliary hypothesis (underestimating the dust in the Milky Way). The Duhem-Quine thesis claims that it's always possible to add an auxiliary hypothesis to save a theory, and so the falsification criterion is too simplistic to reflect how science really works. Indeed, to the satisfaction of Turok and Steinhardt, the cyclic model was not falsified, as the auxiliary hypothesis turned out to be true.

Many philosophers of science take inspiration from a different British Austrian of the same era, Ludwig Wittgenstein. He was a logician, not particularly interested in the demarcation problem; instead, he struggled over the simple question of how to define a game. One might think that a game provides pleasure, but so can eating ice cream. A game has rules, but so does driving. A game is competitive, but so is business. In the end, Wittgenstein concluded that there is no single criterion for defining a game; rather, it should be understood as possessing what he called a "family resemblance." In the absence of genetics, we might look at someone's appearance, blood type, and mannerisms to try and ascertain if they are kin. Similarly, we must examine multiple criteria to decide if something qualifies as science.[22] Falsification may be one such criterion; we shouldn't throw the baby out with the bathwater—but there will be many others. That is why we picked several tools to interrogate our suspects.

Perhaps the physicist's ultimate tool is Occam's razor. Its author, William of Occam, described it with the comment, "Plurality should not be posited without necessity." It's often interpreted as meaning that the simplest theories are the ones we should prefer. One way to quantify this idea is to say that the fewer the free parameters in a theory, the better. What especially excited me about the bi-thermal Big Bang model we met in chapter 8 was that it has only one free parameter. However, the number of free parameters is not the only consideration. One might also ask: are the underlying assumptions of the theory parsimonious or not? Is the proposition that the universe expanded at a gargantuan rate simpler than the proposal that the speed of light was much faster in the past? Both involve phenomena we have no experience of. The defenders of inflation point out that it requires no new physics and so better satisfies our condition of parsimony. But personally, I think that as our usual tools of relativity and quantum mechanics come to a head at the Big Bang, then we should embrace new physics, not avoid it. Judgment calls then become unavoidable, and that necessity for judgment is really what lies at the root of the conflicts we have seen over the Big Bang. The knife that cuts between science and pseudoscience does not have a sharp edge. Or perhaps we can compare it to a modern razor that has multiple blades.

Science is a multifaceted concept, like a Hindu god with many arms. It can be defined as a method merging mathematical theory and experiment, as a subject of study distinct from the arts and humanities, and as a body of knowledge. While an idea's simplicity, mathematical elegance, or explanatory power might make it plausible, elevating it beyond mere speculation, only contact with experiment should allow it to be admitted into the cathedral of established scientific theories. This fearsome guard at the gate ensures that our conjectures don't turn into dogmas; a few confirming facts should not be confused for overwhelming evidence. But patience may be required. When Aristarchus, more than two thousand years ago, first dared to imagine the Earth orbiting the sun, he could never have dreamt that such a hypothesis would become testable with a telescope thousands of years later. Some of the models we have described have predictions for experiments; those without such predictions might develop them in the future. These proposals may be seeds of ideas that lead to testable models, or they might just fade away like a midnight dream. Or they might persist for millennia, as do some of the myths we live with today, like a nightmare you can't shake. Time will tell.

WHAT IS PROTO-SCIENCE?

We may then focus our attention not on the distinction between science and pseudoscience but between established science and its young seedlings, or proto-science. My friend João subtitled his book on VSL "the story of a scientific speculation."[23] But then you might wonder what makes a speculation scientific? I think Feynman answered this question best when he said that science is "imagination in a tight straitjacket," a garment made of data and existing knowledge.[24] We can see scientists struggling with that straitjacket in many of the stories we have told. Inflation theorists puzzled over how to solve the graceful-exit problem before settling on the slow-roll process; early models of loop quantum cosmology floundered because they predicted the universe would collapse with a positive cosmological constant; VSL appeared implausible to me until I noticed its connection to Cuscuton and Hořava gravity. These cases are all examples of cosmologists struggling with Feynman's straitjacket. That such contortions exist doesn't necessarily mean they're all valid, but it does mark a key difference between scientific speculation and creation myths in which anything goes.

Proto-science is essential if we are to uncover the truth about the Big Bang, and we should welcome novel attempts to solve the mystery. Curiosity is the engine of progress. But we must not mistake what is at best a plausible hypothesis for a fact. If proto-science is to be worthy of its name, it must confine its conjectures to compatibility with well-established knowledge. Received wisdom can be challenged, but not without carefully working out the consequences and checking that they are consistent with observations. If science is working properly, the best theory will outcompete its rivals. But we need an open playing field. Inflation should be challenged, as should string theory. If our pretenders to the throne succeed, that success will mark a major revolution in our understanding of nature. And if they fail, our confidence in the ruling elite may be reaffirmed. Science must not become a one-party state prematurely; for this reason, it is essential to develop the models we have covered and understand their strengths and weaknesses, developing them until they can be tested against the data. We then should follow Popper, perhaps not in his focus on falsification, but in his other great work, *The Open Society and Its Enemies*.[25] Just as our story began in ancient Greece, so Popper's account of liberal democracy pays homage to Athenian democracy and its encouragement of criticism. This encouragement is Popper's most enduring legacy, which we hope will be embraced by all.

Science is a dance between theory and experiment. Sometimes theorists lead and data follows; at other times, experiment tightens Feynman's straitjacket, constraining the space for theorists to play. Data without interpretation and theory without experimental confirmation are hollow enterprises. Only when the two dance as one can we be confident in our conclusions. So far, we have seen theorists leap high into the air; in order to bring them to the ground, we need the gravity of empiricism. In our final chapter, we will see how we can gather more data that might finally end the game and allow us to declare a winner. Astronomers have one advantage over other seekers of forgotten time: we don't have to rely on dusty heirlooms or contested stories of the past. For a telescope is a window into the past that can directly observe our most ancient history. In our final chapter, we shall see how we might be able to set the dial on our time machine to 13.8 billion years BC and get real data that can reveal the true nature of this most mysterious event.

✳ 12 ✳

THE END OF THE BEGINNING

In the summer of 2022, the James Webb Space Telescope (JWST) unveiled for the first time its unparalleled views of the heavens. The ten-billion-dollar instrument revealed the atmosphere of a distant world, showed us a galaxy born only a few hundred million years after the Big Bang, and displayed gorgeous nebulae in detail never seen before. But the hype for such extravagant science projects is not always as honest as the science. I noticed that the Google Doodle described the new pictures as the deepest ever taken of the universe, and I quickly tweeted that they were not. Nor can JWST see back to the dawn of time or the origins of the universe. Similar marketing was used by the Large Hadron Collider at CERN, which was described as a Big Bang machine. Both do fantastic science, but neither instrument is designed to determine exactly what happened at the Big Bang.

The cosmic microwave background (CMB), the afterglow of the Big Bang, is the oldest light in the cosmos. It's described as a baby picture of the universe, but this infant is 380,000 years old. Before its light was set free from the primordial plasma, the cosmos was so dense it was opaque; it was only after 380,000 years of expansion that the universe became dilute enough to set free the CMB photons from the thick fog of the dense early universe. Just as we can't see the sun's interior, we cannot look into the era before the emission of the CMB. It seems we have hit a barrier for discovery. If you've ever wondered if we could build a telescope so powerful it could see back to the Big Bang itself, the answer is a clear no, at least not with light. To peek into earlier times, we will need a messenger that can travel through the wall of the CMB. So far, like a spy listening in on a conversation next door, our knowledge of what lies beyond the wall has relied on the sound waves that propagate through the primal plasma and wiggle the CMB temperature fluctuations seen by WMAP and Planck. But what if we want to learn more than the mere hum of the Big Bang?

GEOMETRIC RIPPLES FROM THE BIG BANG

Just as a rock thrown into a pond creates water waves, so moving masses in the cosmos can generate gravitational waves. These geometric ripples in the fabric of space-time cause it to deform, such that it's stretched in one direction and compressed in another. When two black holes collide, they send out a burst of gravitational radiation so cataclysmic that, for a fraction of a second, they can release more energy than all the stars in the entire observable universe. But for distant observers, this mother of all explosions creates a gravitational wave so feeble that the relative change in lengths caused by the squeezing and stretching of space-time (known as the strain) is typically one in a billion trillion. This degree of change is equivalent to changing the Earth's distance from the sun by the length of a single atom. The violent era of the Big Bang can also generate primordial gravitational waves. These waves can travel unimpeded through the opaque primordial plasma, potentially sending us a message from the Big Bang itself. So, while light cannot escape from the dense early universe, gravitational waves can—and with them ride the hopes of early-universe cosmologists to recover the memory of the universe's birth.

But any signal that is ghostly enough to travel through the plasma of the early universe will also be frustratingly difficult to detect. For decades it was thought that finding any gravitational waves would be next to impossible. Only someone with considerable chutzpah would think they could do such a thing. Joseph Weber was a microwave engineer who had, in the 1940s, approached George Gamow, looking for a project. But strangely, the scientist who was one of the fathers of the Big Bang and who predicted the existence of the CMB turned him away, saying he couldn't think of anything for Weber to do. Weber eventually built suspended aluminum bars that he thought would ring when a gravitational wave passed through, but his claims of successful detection were met with skepticism. At one meeting in 1974, a fistfight almost broke out between Weber and his critic Dick Garwin; the two were separated only by the moderator's cane.[1]

Weber's singing bars became his white whale, sinking his reputation as he clung to discredited claims of detection. But Rai Weiss was inspired by his quest and quickly realized that lasers (a technology that Weber had helped nurture) held the key. By splitting and bouncing beams of light between mirrors placed at great distance, their waveforms can then be combined. Such a device is known as an interferometer. Recall that in the late nineteenth century, Michaelson and Morley used an interferometer to try and measure the Earth's motion through the hypothesized ether. Their inability to find any trace of the ether inspired Einstein to develop relativity. More than a hundred years later, this technology continues to foster revolutions in science. A gravitational wave that passes through an interferometer gets stretched in one direction and squeezed in the other, causing the waveforms to go momentarily out of phase. If the interferometer's arms were long enough, we might just be able to measure the impossibly small changes in length that the gravitational waves generated. Using this method, Weiss (along with Kip Thorne and Barry Barish) finally won the Nobel Prize in 2017 for discovering gravitational waves at LIGO (the Laser Interferometer Gravitational-Wave Observatory) fifty years after the idea was first proposed. LIGO's arms are 4 kilometers long. In the mid-2030s, ESA plans to launch a space-based detector called LISA (the Laser Interferometer Space Antenna), which will have three satellites creating a triangle of laser light with arms a whopping 2.5 million kilometers in length. This length will enable it to see sources—such as supermassive black holes colliding—that are invisible to LIGO. I'm hoping LISA may confirm my hypothesis of black-hole echoes, and we'll be looking closely at the data for other deviations from Einstein's relativity that may reveal subtle effects from quantum gravity. LISA will also resolve the Hubble tension (which, you will recall, was the striking mismatch between measurements of the age of the universe using supernova data and those using CMB data) by precisely probing the cosmic geometry in an unprecedented way, testing the exotic models of dark energy that we explored in chapters 6 and 10. And finally, LISA may settle my double bet with David Spergel. But neither LIGO nor LISA is optimized to see gravitational waves from the Big Bang. To target these, we will need something far grander.

BIG BANG OBSERVER

In the closing years of the twentieth century, Neil Cornish was doing a postdoc with David Spergel, working on ways to test whether the universe was spatially finite. If the cosmos is bounded, then ancient light from the primordial plasma could circumnavigate the entire universe, creating matching patterns in the CMB.[2] Cornish and Spergel worked to determine precisely how these patterns might appear to the WMAP spacecraft. None were found, implying that the universe is either spatially infinite or bigger than we can ever see. But between writing the code and waiting for the data, the two started to look far into the future. NASA was encouraging scientists to think big and dream of how we might probe the Big Bang itself. Around the same time, Alberto Vecchio was doing a postdoc at the Max Planck Institute with Carlo Ungarelli; one of their first papers together explored empirical tests of the pre–Big Bang model we examined in chapter 4.[3] But soon, they had their gaze on the same target Cornish and Spergel were focusing on. Both teams independently realized that LISA would likely be insufficient to go after the ultimate prize: primordial gravitational waves. Collaborating with Chuck Bennet, the principal investigator of WMAP, Spergel and Cornish came up with a design for what is, in my opinion, the ultimate telescope: the BBO, or the Big Bang Observer. This project is hugely ambitious, perhaps the moon shot for early-universe cosmology. BBO requires not three spacecraft, as in the case of LISA, but twelve. The seemingly excessive number of satellites are there so that BBO can keep track of all the different sources of gravitational waves ringing through the universe. These other waves must then be subtracted from the data to finally reveal whether any faint murmur of gravitational waves remains, left over from the Big Bang itself.

Cornish was happy to admit to Phil that the BBO will be "stupidly expensive." Each satellite has two mirrors roughly the same size as the Hubble Telescope and requires 300 watts of laser power. By comparison, LIGO at peak capacity runs at 200 watts, and the sort of laser Phil uses to entertain his cats runs at a mere five-thousandths of a watt. One way to think of the BBO is that, while LISA doesn't need to be as sensitive as LIGO, given that its arms are so much longer, BBO does. As Rai Weiss put it, "It's just a wild idea. It's way, way far from everything that's being done right now." Indeed, BBO will require the sensitivity of LIGO (which has 4-kilometer-long arms) at the scales of LISA (which has arm length in the millions of kilometers). But Bernard Shutz, a leading gravitational-wave expert, describes the BBO, or something like it, as the ultimate goal of the field. One way to get around the BBO's insatiable demand for laser power is to deploy lower-watt beams that are then boosted by a device called a master oscillator-power amplifier. LIGO uses this technique, and Japanese scientists hope to implement it on their own versions of BBO, called DECIGO (which stands for *Deci-Hertz Interferometer Gravitational Wave Observatory*, or just *decide and go*—my friend Eiichiro Komatsu's advice to prospective funding agencies). Only 10 watts of initial laser power are needed. The problem, however, is that the spacecraft's relative position has to be maintained within the wavelength of light being used (515 nanometers). It's been said that BBO is LISA on steroids, but DECIGO is LIGO in space. The Japanese intend to launch a smaller-scale pathfinder mission in the mid-2030s in order to test the required technology for the full mission.

Beyond BBO and DECIGO, scientists are investigating an entirely distinct technology, not laser interferometry but atom interferometry. This alternative technology exploits the wavelike nature of matter. As in laser interferometry, particles are sent through a beam splitter, energized by a laser, and separated into two halves, which are then recombined. If a gravitational wave passes through the interferometer, the wave will go out of phase. Neutral atoms don't react to electromagnetic fields, and so they can be more easily shielded from noise sources like cosmic rays and solar wind. An atom interferometer needs just two spacecraft making a single line. But it's too early to say whether it can really outperform the tried and tested. Even more outlandishly, physicists have proposed using asteroids as test masses; these test asteroids could be equipped with atomic clocks, transmitters, and receivers. Pulses would be sent between them, and by measuring the time the pulses take to make the round trip, we could conceivably detect gravitational waves. A similar proposal that I have worked on with my colleagues is to consider how sound waves in a supercooled Bose-Einstein condensate (a special quantum state of matter) can be sensitive to gravitational waves. In this setup, we basically use sound waves to do what lasers do in LISA. The advantage is that since the speed of sound is much slower than the speed of light, the apparatus can be much smaller. Unfortunately, the noise sources are also much bigger, and we need more time to see if this technique can be competitive with current technology. These ideas, then, are the dreams of Big Bang explorers; they are where we hope to be in some distant future. They are what we might do if time and money were no object. I don't know if I will live long enough to see these projects fly, but one can hope.

In my view, BBO and DECIGO are our most ambitious proposals for seeing deep into the Big Bang using gravitational waves. But we must learn to walk before we can run. Currently, the pinnacle of CMB detectors is Planck, but before that was WMAP and before that COBE. And even COBE might not have flown had Penzias and Wilson not recognized that first glimmer of the CMB three decades earlier. For primordial gravitational waves, it's as if we are still in the era before Penzias and Wilson, not knowing if our signal is even there to discover. Fortunately, there is a way we might probe for their existence without incurring the huge costs needed to see their complete form.

A giant Laser Interferometer Gravitational-Wave Observatory (LIGO) detector, located in Livingston, Louisiana, USA. LIGO detects minuscule changes in the lengths of its two 4-kilometer (2.5-mile) arms as gravitational waves ripple through our planet. This new type of astronomy remains our best hope for testing some of the models of the Big Bang by detecting primordial gravitational waves. Although LIGO's design sensitivity may not be enough to detect these waves, they are a prime target for future detectors. Credit: Caltech/MIT/LIGO Laboratory.

THE FUTURE OF THE CMB ON THE GROUND

In chapter 2, we saw that the BICEP observatory created a false alarm, claiming a detection of a type of polarization known as B-modes. If of primordial origin, these B-modes would have been the imprints of Big Bang gravitational waves on the CMB. The discovery turned to dust, literally and figuratively, as further investigation showed that dust was indeed all that the team had likely measured. But the quest for these swirling signs in the primordial light goes on. Just as we were finishing this manuscript, a tantalizing hint of a sea of very-long-wavelength gravitational waves, seen through their subtle imprints in the radio wave signals of rapidly rotating neutron stars, was published by a consortium of international observatories.[4] Whether this new finding will prove to offer another frontier on which to probe what came out of the Big Bang, as opposed to something more mundane, akin to BICEP's dust, is the billion-dollar question.

Back on Earth, until recently, searching for CMB polarization was a wild frontier full of rivalry and competition. But now unity is the new theme as diverse projects are beginning to merge. David Spergel, who was one of the first to raise questions about BICEP's "discovery," acknowledges that that debacle partly drove this newfound spirit of collaboration. I recently got to catch up with my former supervisor to ask him about the state of the field post-BICEP. He said, "All of us were unhappy with how the BICEP team had operated . . . [T]he fact they went to the press with such a strong claim without sharing data, the fact that they don't make data available, the fact that they were really pretty careless in what they did. I think one of the things we [the CMB community] accomplished, first with COBE and then with WMAP and now with Planck, is we developed a reputation across the physics community for doing robust and careful analyses, and I think that the BICEP incident somewhat undermined it." As you may recall, Spergel had worked on cosmic strings, a theory that was a rival to inflation for explaining the scale-invariant nature of the CMB, but the results weren't positive. I asked him to describe his experience. "When the COBE results came out, did I want them to fit the model I was working on?" he asked rhetorically, then answered: "Yes. Was I initially depressed when they didn't? Yes. To me this was not at all a waste of time; I was very glad I did the work because we took an idea, we developed it, we showed what the predictions are, and I think we can say that that's not the origin of fluctuations." Such a trajectory offers a good model for all new Big Bang scenarios to pursue. It's essential to realize that most of these scenarios will fail but that through those failures, we learn.

After COBE, David became one of the key scientists on the follow-up mission, WMAP, which reported additional evidence for inflation. David and I do not see eye to eye on this issue. I challenged him that there were so many different inflationary models that one of them was bound to be confirmed by the satellite's measurements and that thus the paradigm had become unfalsifiable. At first, he replied that inflation is falsifiable. We could have seen observations that were highly anisotropic (different from one side of the sky to another), which would have been "highly problematic."

He also claimed that we could have seen a statistical pattern in the CMB called non-Gaussianity, but that again, we didn't. (Recall that a Gaussian distribution follows a bell curve where there are roughly as many hot spots as there are cold spots.) But David, echoing the Duhem-Quine thesis we discussed in the previous chapter, eventually conceded that inflation "could fit non-Gaussian fluctuations, but they would get increasingly baroque. So, theories often aren't falsified; they're forced into a more and more baroque corner, and inflation could have been forced into a baroque corner." But however the WMAP results are interpreted, the project unquestionably marked a revolution in precision cosmology. On the heels of these achievements, David came to the attention of one of the world's most important science philanthropists.

After a distinguished career as a professor of mathematics (notably, co-inventing what is now known as Chern-Simons quantum field theories) and code breaker for the NSA, Jim Simons came into some funds from his father that he decided to invest.[5] The project would later spiral into perhaps the most successful program-trading hedge fund of all time: Renaissance Capital. And now Simons has decided to return his billions into the science that made him so successful in the first place. According to David, Simons is "a remarkable guy. He made his money by being incredibly smart, and he's an incredibly inquisitive person." The philanthropist is investing in an observatory that may see the elusive B-modes. Curiously, he seems to be hoping that his new telescopes won't achieve such a goal. In a promotional video, Jim Simons appeared to back the model of Steinhardt and Ijjas that produce no B-modes, saying, "Most people think inflation is valid . . . I like the other theories better. It's hard for me to believe that time began. Everyone imagines that time would go on forever, but why wouldn't it have been going on forever? Why should time have started? It doesn't seem reasonable to me. It's not very aesthetic, whereas something that oscillated between big and small and big and small and was always there—[the idea that] there was always a universe—that, aesthetically, is much more appealing."[6] Nevertheless, Simons says, "either way we win." Winning here means finding out the answer, not confirming one's own pet theory. There's nothing wrong with having a preferred horse in the race, but as good Big Bang explorers we need to accept whatever the evidence reveals. And Simons is someone who wants to know the answers. He's fascinated with origins and also has funded attempts to shed light on the genesis of life and consciousness. In the Atacama Desert, the observatory that bears his name is building new telescopes to examine different frequency ranges. The sheer number of cryogenically cooled detectors being assembled is particularly astonishing. For WMAP, David told us that "at our highest frequency, where we had the most detectors, we had four, and I knew each one of those four detectors." The Simons observatory will have tens of thousands of detectors. But even that number will be eclipsed by what has become known as CMB S4. This project will lead to a gargantuan half-million detectors. But the pain and embarrassment of the BICEP incident will cast a long shadow over any potential discovery. As David said, "How do you convince yourself that what you see is a signal? It's not dust, it's not some foreground, and I have my own wish list for what I want to see." This list includes ensuring that the signal is seen in multiple directions, accounting for different foregrounds and multiple frequencies in different detectors. But after all those hurdles have been surpassed, once "we [have] convince[d] ourselves we see the fluctuations," then, David says, "the very next thing, perhaps the next day or that day, you say, 'I want to measure the tilt [for gravitational waves].'" Recall that the tilt is the dependency of the energy of waves on their wavelengths, and it is a crucial marker for determining which model of the early universe is correct. Cosmologists describe two tilts. The first, known as scalar tilt, is concerned with sound waves in the primordial plasma; the other, called tensor tilt, refers to the Big Bang's gravitational waves. Scalar tilt has already been measured with significant precision by WMAP and then Planck, but to see the tensor tilt clearly, we need to survey the entire sky, something practically impossible for ground-based detectors. If these primordial gravitational waves are ever detected, a follow-up, space-based mission will probably be needed to measure this all-important tensor tilt. Many cosmologists have pitched projects with acronyms like CORE, EPIC, and PRISM, but these proposals have been sidelined in favor of a spacecraft now being pursued by the Japanese. This spacecraft is called LiteBIRD, or the Lite Satellite for the Studies of B-Mode Polarization and Inflation from Cosmic Background Radiation Detection.

THE FUTURE OF THE CMB IN SPACE

At around the same time that I arrived in Princeton, Eiichiro Komatsu was a visiting PhD student from Japan, working with David Spergel. We were both studying clusters of galaxies at the time, so we used to talk a lot. When the time came to pick a PhD supervisor, Eiichiro was one of the people I talked to who strongly recommended working with David. I did heed his advice. Eiichiro is now one of the leading scientists on LiteBIRD. He described to us his experience working on WMAP as "the most intense but the most productive and exciting time of my life."

    The statement he and his colleagues produced on NASA's WMAP website—a statement that those of us skeptical of inflation found somewhat egregious—read: "The data provide compelling evidence that the large-scale fluctuations are slightly more intense than the small-scale ones, a subtle prediction of many inflation models."[7] This finding was the monumental discovery of the red scalar tilt we first mentioned in chapter 2. If these primordial sound waves had the same power at all wavelengths, the value of the scalar tilt would be 1. As an analogy, imagine music where the bass and treble are as loud as each other. This is the sort of harmony we might hear if we could listen to the sound waves that rippled through the early universe. A blue tilt would mean that there was more power at shorter wavelengths, as if—in our musical analogy—the treble were louder. A red tilt would mean that there was more power at longer wavelengths and would translate to a tilt value of less than 1; it would be akin to finding more power in the bass. This red tilt is thought to be generated in simple inflation models because the longer wavelength fluctuations exited the horizon and froze at earlier times, when cosmic expansion was more intense.

The detection of this scalar tilt, which we observe through temperature variations, was one of WMAP's most remarkable achievements, one that cosmologists hope to replicate for the tensor tilt arising from primordial gravitational waves. As Komatsu recalled, when preparing the WMAP data, he and his collaborators "paid attention to that quantity right from the beginning." After the team had collected measurements for several years, it slowly became clear that the scalar tilt was red, with a value around 0.96. However, the statistical significance of their finding wasn't yet high enough to allow them to be sure it was real. "But then," Komatsu continued, "by nine years finally we hit this mark of 5 sigma.[8] We're like, 'Wow, gosh.'" Still, they wondered: "Is this really true?" Fortunately, when ESA's Planck mission released its cosmology data, it found exactly the same number. Komatsu recalled, "I thought, 'Okay, this is very very strong evidence [for inflation]. To me, this is truly astonishing.'"

But whether inflation really does predict a scalar tilt of 0.96 remains controversial. On the one hand, Mukhanov predicted this exact number in 1981, for Starobinsky's model of inflation. On the other hand, critics point out that there is always some inflationary model that will give any value for the scalar tilt. As we said in chapter 8, my model with Magueijo, for a bi-thermal Big Bang, makes a precise prediction of 0.9648. Another test is therefore needed to help resolve the dispute. That is why so many eyes are focused on CMB polarization. Komatsu thinks that the satellite he and his colleagues are working on might do exactly that, namely give us a much higher-resolution readout of the crucial scalar tilt.

LiteBIRD began its journey when particle physicist Masashi Hazumi was hospitalized in 2005. According to Eiichiro in an interview, "He thought about his life. He decided to change the subject of his research to cosmology." Hazumi helped form the embryonic LiteBIRD community and turned it into the world's leading candidate for a new CMB mission. At a recent conference, Planck scientists literally presented a torch to the team, acknowledging LiteBIRD as the next great mission to probe the early universe. However, this accolade might have been premature. Eiichiro told us that when the email came selecting LiteBIRD, "I was relieved, but at the same time

I couldn't quite know what that meant." The mission is currently in phase A of selection, but phase B is "the moment of truth," when the bulk of the money is committed. LiteBIRD needs support from international partners, especially ESA and NASA. Eiichiro recalled that then the 2020 decadal survey (a list of priorities selected by US astronomers) was released, "they didn't recommend any funding for LiteBIRD, so that's kind of [a] bummer for us." They did, however, advocate that a space-based mission should be made a priority if B-modes are detected from the ground.

NASA is also considering a mission named PICO (Probe of Inflation and Cosmic Origins) that promises ten times better resolution than even LiteBIRD can achieve. But this greater sensitivity comes at the cost of more waiting and more money. My friend Tarun Souradeep is also leading CMB Bharat out of India, entering into the race with a spacecraft similar to LiteBIRD in sensitivity. Incidentally, Souradeep was the PhD student of Narlikar, who himself was a student of steady-state pioneer Fred Hoyle. Had Narlikar not won that position, it would have gone to Hawking, and the history of cosmology might have been very different.[9]

Clearly, the coming decades will see enormous improvements in our ability to detect primordial gravitational waves. If they are discovered, we will want to gather as much information as possible, particularly to learn the tensor tilt and to make detailed maps of the cosmic gravitational-wave background, just as we did for the cosmic microwave background. Such reconnaissance will most likely be carried out via space missions. First, a mission like LiteBIRD will be able to detect CMB polarization, giving us a probe of the Big Bang's gravitational waves on cosmic wavelengths of billions of light-years; finally, the reconnaissance efforts will shift to space-based observatories like BBO and DECIGO, which will be able to examine those waves on much shorter wavelengths of millions of kilometers. This multipronged strategy will be our best bet to get our grandest view of our Big Bang origins. As some of my colleagues and I have recently shown, with enough sensitivity, we may even be able to match the patterns in these (CMB and gravitational-wave) maps, as they both picture the same Big Bang but at different depths. But what will all this new information mean for the models we have already examined?

Before we answer that, it's worth noting that should we discover primordial gravitational waves from inflation, such a discovery would indirectly confirm Hawking's most famed claim: that black holes emit radiation. Now, we don't know for sure that there were any black holes in the early universe, but an inflationary universe itself is like a black hole turned inside-out. In the same way that light cannot escape from a black hole's event horizon, light from sources outside the cosmic horizon of an inflationary universe cannot reach us inside. Following the same logic as Hawking's original calculation, we can show that the cosmic horizon has a temperature and emits gravitational waves, which we could potentially see in our telescopes if inflation was rapid enough.

Now let's consider our first controversy, the rebellion against inflation that sought to establish a cyclic universe. Detecting primordial gravitational waves would rule out several cyclic models, including the Ekpyrotic model, its more recent variants, and, according to Penrose, CCC as well. But a non-detection would not rule out inflation, as inflationary models can make these waves so faint they will never be detected. However, the cyclic model proposed now by Steinhardt and Ijjas does predict much weaker secondary gravitational waves, with a blue spectrum, that we described in chapter 6. If such waves alone were observed, the discovery would surely mark a triumph for this conception of the universe. Inflation is not the only model to predict primordial gravitational waves; some string-inspired alternatives do, too. In the case of string gas cosmology, the waves should have a blue tensor tilt (that is, more power at shorter wavelengths); in inflation models, they typically have a red tensor tilt (more power at longer wavelengths). And for the pre–Big Bang, we should see blue primordial gravitational waves in laser interferometers but no B-modes in the CMB. If the tensor tilt showed primordial gravitational waves with a blue spectrum, would it be a moment when the rebels stormed the castle and overthrew inflation? Might Brandenberger and Vafa (or Gasperini and Veneziano) swipe the Nobel Prize many think is waiting for Guth and Linde? Would my mentor, David Spergel—who said, "We're dead" after COBE ruled out cosmic strings—make the same comment about inflation? Alas, it is possible to create an inflationary model with a blue-tilted spectrum, and so inflation would not be dead, but it might be forced into a particularly "baroque corner." What of string theory's rival, loop quantum gravity? Recall that loop quantum cosmology (the cosmology inspired from the mother theory) needs to make assumptions about what matter fields are present. We can modify its predictions by choosing a different inflationary model or even assuming inflation did not occur. One might think of it as a dictionary that translates a classical model's predictions into a quantum one. Nevertheless, loop theorists have been claiming that the anomalies in the CMB may be an effect of quantum gravity in the sky. These features are curious but don't have the statistical significance to mark them as a real discovery. New observations will help distinguish whether these oddities are smoking guns of a bouncing universe or nothing but a mirage in a noisy dataset.

ONE TRILLION GALAXIES

Looking out in the heavens over billions of light-years reveals networks of strangely shaped objects: celestial whirlpools, sombreros, cigars. All of these objects are, of course, galaxies, and when they congregate into clusters and superclusters, they form the largest structures we know of in the cosmos, sometimes called the cosmic web, which I also study as part of my research. One might imagine they have little to tell us about the extreme physics needed to probe the universe near the Big Bang. But the seeds of galaxy structures were planted at the very earliest moments of the expansion, and different proposals give contrasting predictions for their distribution. For inflation, simple models with just a single field predict that their distribution will follow a Gaussian pattern, but more complex, multi-field models predict a specific pattern that is non-Gaussian. So far, our best constraints on non-Gaussianity come from the CMB, which is like a two-dimensional snapshot, while the actual

Big Bang was three-dimensional. Can we map the Big Bang in its full three-dimensional glory? That is precisely what the distribution of galaxies provides. For some non-Gaussian universes, galaxies are more social, clumping together more frequently than Gaussian ones. There are, however, literally trillions of galaxies out there in the cosmos. But capturing a small sample, let's say a few hundred million, and mapping them out is not beyond the realm of experiment. NASA has an upcoming mission to do exactly this, led by Olivier Dore out of the Jet Propulsion Laboratory, another veteran of WMAP and a friend of mine from old Princeton days. SphereX has a main telescope that is actually smaller than the one Phil uses in his London backyard. But for wide-field infrared spectroscopy (needed to estimate the positions of galaxies), a large telescope is not needed—making the mission inexpensive at only a few hundred million dollars. Combined with other wide-field telescopes currently being built, we will have vastly improved our measurements of non-Gaussianities by the next decade. SphereX is due to launch in 2025. Some models, like Smolin's cosmological natural selection and Poplawski's torsion bounce, are arguably only compatible with certain inflationary models. Others, like loop quantum cosmology, create small but measurable differences in their predictions for the CMB once an inflation model is selected. Proponents of the no-boundary proposal say they can also make specific predictions from certain inflationary models. But which is the right model? By measuring non-Gaussianity and gravitational waves, these theories can be further constrained.

From my point of view, these predictions are not concrete, as they often leave enough wiggle room to allow us to adjust our expectations, no matter which way observations turn. For example, both the Ekpyrotic and inflationary models made predictions for the size of non-Gaussianities, with different degrees of certainty. In 2013, the Planck satellite reported values that were much closer to the inflationary prediction than the Ekpyrotic one; inflation stuck its neck out with a narrow range of predicted values and survived. Ekpyrosis, with a much wider possible range, played it safe. Statisticians would say this moment was a win for inflation. To quote Planck's original paper, "the Planck data put severe pressure on ekpyrotic/cyclic scenarios."[10] I doubt this assertion sat well with Lehners and Steinhardt, who published another paper saying that "Planck 2013 results support the cyclic universe."[11] As a result, the Planck collaboration softened their tone, changing "put severe pressure" to "significantly limit the viable parameter space."[12] Science is a messy business that involves more politics than meets the eye. Don't get me wrong, though! I am sure that if Planck 2013 had discovered large non-Gaussianity, a legion of inflationary cosmologists would have been celebrating, claiming that their more "realistic" models had already predicted just such a finding. Politics can go both ways.

But what does a concrete prediction look like? Here, I am biased. The bi-thermal VSL model, for example, predicts a precise value for the scalar tilt to four decimal places. Another example is Starobinsky's model of inflation, which predicts that the energy of gravitational waves should be fifty times weaker than what BICEP claimed to measure, with a small margin of error. Galaxy surveys such as SphereX will uncover a treasure trove of information about cosmic structures on small scales, testing precise predictions from models like Skenderis's holographic cosmology or Beth's periodic-time model. All models of the Big Bang should strive to meet this bar.

Despite the difficulties and ambiguities of testing inflation and its rivals, I see a possible path through the trenches that might lead to a real resolution of our conflict. Those "baroque" inflationary models that predict a blue tilt in the gravitational-wave spectrum should also produce non-Gaussianities.[13] By contrast, a model like string gas cosmology predicts a blue tilt in the gravitational-wave spectrum and—assuming that the scale of the fundamental string is close to the Planck scale—will produce a Gaussian spectrum.[14] Another surprising way to test inflation is by probing the fingerprints of "primordial standard clocks" (such as a uniformly oscillating massive scalar field, should it exist) on cosmic sound waves, which can give us a readout of how much the cosmic expansion was accelerating (if at all) when fluctuations were first seeded. Ironically, the proposal was by Avi Loeb (along with my friend Xingang Chen and Zhong-Zhi Xianyu), who—as you will recall from chapter

2—once suggested that inflation was untestable.[15] Loeb (working with Sunny Vagnozzi) has also since argued that the detection of primordial gravitons could be a similarly sharp tool. The difference between primordial gravitons and gravitational waves is like the difference between sunlight and radio waves that we use, say for communications. The former is a collection of random particles, while the latter is a lot of them oscillating together as a coherent wave. While producing primordial gravitational waves, inflation would erase any trace of relic gravitons, and thus observing them would rule out the whole paradigm. Loeb and colleagues argue that these relic gravitons, if present, could potentially be seen with futuristic gravitational-wave detectors, or through their subtle effect in the density of cosmic radiation. In fact, in an interview with Phil, Loeb made a dramatic volte-face, saying that he, Steinhardt, and Ijjas had been wrong to say that inflation was untestable.[16]

Decades earlier, Jacob Bekenstein faced skepticism when he claimed that black holes have entropy. Eventually it turned out to be his loudest critic, Stephen Hawking, who proved him right. Perhaps history will repeat itself in the decades ahead if inflation is confirmed by a test devised by one of its archnemeses, Loeb. Or perhaps relic gravitons could do the opposite. Either way we win, for uncovering the truth is the real goal of science.

Similarly satisfying will be confirming other tangential predictions of the models we have considered, such as phantom or decaying dark energy, or (my favorite) black-hole echoes, which provide nontrivial tests of aspects of our alternative Big Bang models. Lest we forget, recall that advocates of eternal inflation claim (in spite of my skepticism) that a detection of positive spatial curvature would rule out the framework. Current constraints on cosmic curvature will be improved by an order of magnitude by the missions we have described above, and so I fear for the lives of Andrei Linde and Martin Rees's dog.

Thus, by combining measurements of primordial gravitational and sound waves with non-Gaussianities, cosmic curvature, and expansion history, we may finally have a method to either dethrone inflation or strengthen its status, even possibly peeking into what happened before exponential expansion began. This achievement is our endgame. It may be a long way off. It was only in 2016 that LIGO reported the very first "direct" detection of gravitational waves. This finding was but a mere preview into a hidden realm using an entirely new branch of astronomy. And every time we have peered into the heavens using novel techniques, we have made revolutionary discoveries. We can only wonder at the magnificent picture of the universe that will be revealed when LIGO's successors pull back the curtain to reveal our cosmic origins. And when these new signals are joined on stage with more precise data from the CMB and galaxy surveys (many of which we didn't have space to cover here), a profoundly sharper view of the Big Bang (complete with non-Gaussianities, cosmic curvature, tilt, primordial gravitational waves, and so forth) shall emerge. Or so we hope.

THE FINAL PROBLEM

As we have seen, no one measurement can tell us what truly happened at the Big Bang; rather, the CMB polarization, gravitational-wave, and galaxy surveys will all be pieces of evidence helping us narrow down the suspects. There might also not be one answer; as in many great detective stories, sometimes there is more than one killer. The true description of the Big Bang may involve ideas from multiple models. It's possible to combine bouncing or even cyclic universes with inflation (a view Ghazal favored, at least as of 2021). My own periodic time cosmology fused ideas from my old friends Richard Gott and Li-Xin Li with the conformal rescaling proposed by Roger Penrose. So, we shouldn't rush to the conclusion that only one model we have examined will be relevant to the final picture. The real Bang may yet turn out to be a beautiful Frankenstein monster.

Here, I think it's worth taking stock of how far we have come. For thousands of years people have speculated on the origins of the cosmos. But thanks to tremendous advances in physics and astronomy, most notably Einstein's remarkable discovery of relativity, we possess the tools to model its evolution. This modeling led to the singularity theorems of Penrose and Hawking, which for decades were taken to show that the universe must have had a beginning, settling an age-old debate. But then came inflation, which claims to show that our Big Bang was one of many in a vast multiverse. We are told by some that inflation is a proven fact, but this claim hides a deep controversy about the validity of the theory and about how to interpret the data. The conflict may be resolved as we begin our journey to probe the gravitational-wave universe, but it is not a closed case yet. Inflation has its challengers, some of which I have introduced or helped develop; other challenges have been spearheaded by giants of physics like Penrose, Veneziano, Vafa, and others. And perhaps there is no better test for inflation than allowing it to compete against these rebels. We will see how this game plays out in the future. Even if inflation happened, many questions remain, chief amongst them that of how to resolve the singularity where our tools of relativity and quantum mechanics finally face off. String theory, armed with holography, is the community's prime suspect for a quantum theory of gravity that will supersede Einstein. But currently this model conjures confusion, with multiple models of the Big Bang and no consensus. Loop quantum cosmology gives a clearer picture of an hourglass universe and a Big Bounce but has a long way to go to convince a skeptical community. And other theories, like Hořava gravity and causal sets, wait in the wings with the potential to rewrite our cosmic history in their own idiosyncratic ways. Contrast this state of affairs to the situation when I stood on that rooftop in Tehran as a child, looking through a telescope, giddy with delight at the ghostly celestial objects in the night sky. There, I wondered where it all came from. Back then, the Big Bang beginning was gospel; asking what came before the bang was not allowed. Barely any of the proposals we have examined even existed. We have come a long way, and we should take a moment to pause and marvel at the grandeur of our playground, where hard science meets human imagination. The discovery of the Big Bang was thought to have ended our cosmic enquiries, but instead it has thrown them wide open. It is no longer true to say that we have no idea what happened at or even before the Big Bang. We now have a rich landscape of ideas, each with its own strengths and weaknesses. And I have a sense of awe in pondering that one of these ideas just might be right. But be wary of claims that we are already at the summit, when the fog has not yet cleared.

Our quest to uncover the origin of the universe requires theorists to put pen to paper risking their careers on models that may not bear fruit, and engineers to devote decades to perfecting required technologies that may never be put to use. But no discoveries are possible without taking risks. How might we persuade our colleagues to bet on the Big Bang? First, we will have to convince astronomers that projects like BBO or DECIGO are feasible. This certainly won't happen before LISA, humanity's first space-based gravitational-wave detector, launches (the launch is scheduled for the 2030s). If LISA is successful, scientists will look for the next big thing in gravitational-wave astronomy. If it fails, it would be bad, but perhaps we can recover with better ideas. The stakes are high. But many of my colleagues are confident in ESA's ability to pull this critical mission off, as it did with Planck. Then we have to convince the wider scientific community that our dreams are worthy. While it would be nice to see bases on the moon, a human being walk on Mars, or the next great particle collider, the reality is that funds are limited, as they should be. Other issues, like dealing with climate change, are far more urgent. So, we as society have to decide what to prioritize with the limited budget we have for curiosity-driven science. Do we want to know what matter is made of? Or what's it's like to stand on another planet? Or should we dare to discover how our expanding cosmos was born? For us humans, this last question may yet be the most intriguing of all. But someone needs to foot the bill. The cost of a project like BBO or DECIGO may be so large that only a multinational effort will make it feasible. There is precedent for such an effort: CERN's Large Hadron Collider has contributions from dozens of countries and cost $4.8 billion, while their proposed future collider is estimated to cost more than $20 billion. The International Space Station, with much less visible scientific results, cost $150 billion. And these outrageous price tags are often spread over decades and so are not as daunting as they first seem. Getting governments together to satisfy our hunger for knowledge is possible but will be a demanding undertaking.

My eldest son insists on keeping the closet lights on so that he can fall asleep. As children, many of us shared his fear of the terrors that lurk in the dark, a primal instinct that seems lost to grown-ups with their false sense of confidence. But for those who imagine the start of our cosmos, the dark excites us and the monsters are our guides. Living in the deepest corners of our vast cosmos or our mathematical constructs are dark matter and dark energy, black holes and Big Bangs, string theories and singularities—they all offer clues to understanding. To scientists—my friends and colleagues—the true terror is only a refusal to investigate these secrets. For, at its best, the scientific method—honed by solving the great mysteries of cosmos—remains our ultimate survival tool in an ever-changing world. I recall being my son's age, and how the fear of the dark slowly evolved into a love of the night sky.

This book asks how deeply we can all remember. Our ancestors' thirst for answers led to our creation narratives and scientific speculations. Curiosity about our origins is an eternal human quest. But the Big Bang is a genesis story whose authors believed it must be rewritten. Lemaître, Gamow, Dicke, and others all suspected it had a prequel. For years, many dismissed this idea, but now cosmologists have developed a multitude of candidates and tools to investigate our ultimate origins—from inflation to holograms, from cycles to hourglasses, from Big Rips to Big Bounces, from time loops to time loops again. These are some of the characters we have met, living in the minds of theorists and scribbled on dusty blackboards. Most of these characters will be forgotten, as data from experiments will awaken us from one dream and show the way to others. As a reader and citizen, and a current, would-be, or armchair scientist, you get to decide how much you care about what really happened at the Big Bang and what you want to do about it. The power to remember our past lies in your future.

ACKNOWLEDGMENTS

First and foremost, I'd like to thank my co-author, Niayesh Afshordi, for all the long hours and struggles of putting the book together. Thanks also to Joseph Calamia, Matt Lang, Jessica Wilson, and everyone else at the University of Chicago Press for making this manuscript so much better than its original version. I'd also like to thank friends and family who read through early versions of the manuscript either in part or in whole and offered invaluable feedback, especially my wife, Monica Halper—to whom I am grateful as well for tireless support, encouragement, and, most of all, love. I'm likewise thankful for the support and love of my brother Barry Halper, Joshua Salafsky, and Max Gold. David Yates and Christian Unterdechler should also be thanked here for both their encouragement and their invaluable and generous work creating the CGI images you see in this book. Thanks to Nick Franco for teaching me how to film an interview and for some of the images that appear here. And of course many thanks to all the scientists who agreed to be interviewed or who answered questions via email, including Stephen Hawking, Sir Roger Penrose, Alan Guth, Paul Steinhardt, Andy Albrecht, Anna Ijjas, Avi Loeb, Andrei Linde, Alex Vilenkin, David Spergel, David Kasier, Yasunori Nomura, Hans Halvorson, Thomas Hertog, Jonathan Halliwell, Slava Mukhanov, Jean-Luc Lehners, Will Kinney, Robert Brandenberger, Ali Nayeri, Stephen Alexander, Brian Keating, Gabriele Veneziano, Maurizio Gasperini, Anthony Aguirre, Abhay Ashtekar, Lee Smolin, Carlo Rovelli, Param Singh, Ivan Agullo, Francesca Vidotto, Martin Bojowald,

George Efstathiou, Paul Frampton, Paweł Nurowski, Krzysztof Meissner, Douglas Scott, Kevin Ludwick, Sean Carroll, Rafael Sorkin, Stav Zalel, Nikodem Poplawski, Ghazal Geshnizjani, John Moffat, João Magueijo, Richard Gott, Beth Gould, Rai Weiss, Neil Cornish, Bangalore Sathyaprakash, Bernard F. Shutz, Eiichiro Komatsu, Masashi Hazumi, and Don Page. Thanks to my astronomical mentors at University College London, Francisco Diego, Adam Burnley, Matt Page, Stephen Feeney, Roger Wesson, and Joe Zuntz. And to anyone I may have forgotten, I am sorry.

I'd also like to thank our two anonymous referees. For help in getting the ball rolling, I thank Patrick Walsh and Jim Al-Khalili; for advice on publishing, I thank Phillip Ball, Sharon Miller Gold, and Sabine Hossenfelder. For permissions, thanks to Tom Wilks and Lisa Pallatroni at *Scientific American*, the M. C. Escher company, and Allison Rein and Max Howell at the Niels Bohr Library. For advice on titles and much stimulating discussion, I thank the Oxford Socratic Society, especially David Redman, John Nelson, and Alex O'Connor. Lastly, I'd like to thank my parents, Mike and Stephanie for all their love and support and especially for my father's curious mind, which sparked my own intellectual journey.

Phil Halper
April 24, 2024

Almost exactly twenty years ago today, I was rushing to wrap up my PhD thesis, "The Other 99 Percent," to be printed and bound in six copies, on July 14, 2004. First, I needed to finish the last chapter (which was yet to be peer reviewed); then I needed to write an introduction; and finally I needed to add the acknowledgments. My classmate Hiranya Peris, who had just defended her own thesis, suggested that the last part was best done over a glass of wine. Given that I didn't drink, all I had was coffee to keep me up through the long night in the basement of the Princeton University Observatory, until I could thank everyone thoroughly and properly, in writing, before rushing to the printers (and getting some pancakes).

In many ways, these acknowledgments pick up where the last ones left off. The struggle to unravel the mysteries of the other 99 percent that makes up our universe, and the quest to understand what really happened at the Big Bang, are deep and vexing. Yet they are, after all, human struggles.

For the struggle to create this book, I should thank Eric Henney, who first triggered the idea for me and who later cheered the project on, as did my friend Sabine Hossenfelder. But the most unlikely hero of this story is my co-author, Phil, the former banker and current aurora chaser/YouTuber/amateur astronomer/animal pain researcher/cosmologist interviewer ... whose passion and curiosity are unparalleled. And, of course, it was Joe Calamia, our editor at the University of Chicago Press, who managed to masterfully land this plane. Thanks, Joe!

While it is reassuring that we finally get to tell the story of the battle over ideas about the Big Bang, the fact that there is a story to tell is thanks to those who made this journey possible. First and foremost among them is my PhD supervisor David Spergel, as well as many mentors over the years, most notably Robert Brandenberger, Bohdan Paczynski, Michael Strauss, Matias Zaldarriaga, Ramesh Narayan, Avi Loeb, Daniel Chung, Neil Turok, Brian McNamara, Lee Smolin, Justin Khoury, Robb Mann, and Rob Myers. And there were those who inspired me to think differently: Stephon Alexander, Ramit Dey, Ben Holder, Petr Hořava, Matt Johnson, Will Kinney, João Magueijo, Maria Mylova, Naritaka Oshita, Douglas Scott, Rafael Sorkin, and Dejan Stojkovic.

Perhaps no single event had a bigger impact on my academic journey and worldview than arriving at Perimeter Institute (PI). While it was Andrew Tolley, Claudia de Rham, Mark Wyman, Federico Piazza, Simone Speziale, and Justin Khoury who welcomed Ghazal and me, it was the entire PI community who kept this grand, bold experiment alive and thriving. And then it was my first and second home at the University of Waterloo, Physics and Astronomy, that brought me back to Earth and amongst some of the most talented students on the planet.

Amongst the many students who entertained my idiosyncratic ideas and taught me so much about the ways of the world are my former and current PhD students, many of whom are featured in this book: Chanda Prescod-Weinstein, Siavash Aslanbeigi, Farbod Kamiab, Mehdi Saravani, Jahed Abedi, Beth Gould, Yasaman Yazdi, Nosiphiwo Zwane, Natacha Altamirano, Mansour Karami, Qingwen Wang, Matthew Robbins, Krishan Saraswat, Conner Dailey, and Alice Chen, as well as Abhineet Agarwall and Razieh Pourhasan. We have had our challenges, but I am proud of each and every one of you.

And there was my student Chiamaka Okoli, who defended her PhD on "Dark Matter and Neutrinos in the Foggy Universe" but didn't live to attend her convocation. She inspired all who knew her by her strength and brilliance, but ultimately taught us in new ways how cosmology is such a human struggle.

I should, of course, thank my partner in crime, Ghazal, who stuck with me through this journey, both in life and in science, and my parents, Eftekhar and Hassan, as well as my sisters, Setayesh and Parastesh. And to my sons, Juyah and Auzaud, I hope you enjoy reading this sometime very, very soon.

Last but not least, this book would not have been possible without your support, the citizens across the globe whose tax money goes to support foundational research, and those who spend their day and night pondering deep questions about the true nature of the Big Bang. This work is dedicated to you and for you.

Now, time for some pancakes!

Niayesh Afshordi
April 24, 2024

APPENDIX

APPENDIX TABLE 1. Big Bang Models Cheat Sheet

Name	Include inflation?	Is the model cyclic?	Does it have an hourglass structure?	What's the source of the variations in the CMB temperature?	Is this a universe with a beginning?	Description	Chapter
Eternal inflation	Yes	No	In some versions	Quantum	Maybe	Our observable universe is one of infinitely many bubbles in a huge, rapidly expanding quantum fuzz	2
Hartle-Hawking no-boundary proposal	Yes	No	Maybe	Quantum	Maybe	Our universe is born from a 4-dimensional space without a time dimension	3
Tunneling from nothing	Yes	No	No	Quantum	Yes	Our universe is born from a quantum fluctuation in a space with no space	3
String gas cosmology	No	No	Maybe	Thermal	No	The Big Bang starts from the hot gas of tiny, vibrating strings	4

(Continued)

APPENDIX TABLE 1. Big Bang Models Cheat Sheet (*Continued*)

Name	Include inflation?	Is the model cyclic?	Does it have an hourglass structure?	What's the source of the variations in the CMB temperature?	Is this a universe with a beginning?	Description	Chapter
Emergent universe	Yes	No	No	Quantum	No	The universe is static, hot, and dense and then gradually starts to inflate	4
Pre–Big Bang	No	No	Yes	Quantum	No	A collapsing universe maps into an expanding universe, thanks to the mathematics of string theory	4
Ekpyrotic	No	Yes	No	Quantum	Maybe	The Big Bang is triggered by the collision of higher-dimensional membranes	4
Loop quantum cosmology	Yes	No	Yes	Quantum	No	Quanta of space, proposed by string theory's rival, bounce a collapsing universe into an expanding one	5
Conformal cyclic cosmology	No	Yes	No	Unclear	Maybe	The universe expands in cycles, but instead of collapsing, it forgets how big it was	6
Baum-Frampton	Yes	Yes	No	Quantum	No	Incipient phantom dark energy grows to create a new Big Bang	6
Higgs Bang	No	Yes	No	Unclear	No	An unstable Higgs field turns a collapsing universe into an expanding one	6

Name	Include inflation?	Is the model cyclic?	Does it have an hourglass structure?	What's the source of the variations in the CMB temperature?	Is this a universe with a beginning?	Description	Chapter
Mirror universe	No	No	Yes	Quantum	No	The future and past of the Big Bang singularity are mirror images of each other	6
Janus universe	Yes	No	Yes	Unclear	No	The direction of time reverses at the Big Bang	6
Slow contraction	No	Yes	No	Classical	Maybe	The universe contracts as in an Ekpyrotic model, but without the colliding membranes	6
De Sitter equilibrium	Yes	Yes	No	Quantum	No	Most of the universe is an empty desert, with universes like ours bubbling out as rare quantum fluctuations	6
Carroll-Chen	Yes	No	Yes	Quantum	No	The universe is in a state of eternal inflation, and the Big Bang is sparked by random downward fluctuations in entropy	6
Cosmological natural selection	Yes	No	No	Quantum	No	A universe is born inside every black hole, each with slightly different laws of physics	7
Torsion bounce	Yes	No	No	Quantum	No	A new twistiness in space-time leads to a Big Bounce inside a black hole	7

(Continued)

APPENDIX TABLE 1. Big Bang Models Cheat Sheet (*Continued*)

Name	Include inflation?	Is the model cyclic?	Does it have an hourglass structure?	What's the source of the variations in the CMB temperature?	Is this a universe with a beginning?	Description	Chapter
4D black hole	No	No	No	Thermal	No	Our universe is a membrane, expanding out of the horizon of a higher-dimensional black hole	7
Varying speed of light	No	Possibly	No	Unclear	Maybe	Light moves much faster as we reach the Big Bang	8
Bi-thermal Big Bang	No	No	No	Thermal	Maybe	Sound waves move much faster as we reach the Big Bang, but the universe remains hot	8
Holographic cosmology	Yes	No	No	Quantum	Maybe	The universe is born out of a timeless quantum theory	9
Space from Hilbert space	Yes	No	No	Quantum	No	Space is an illusion born out of correlations within a cosmic quantum computer	9
Gott and Li	Yes	No	No	Quantum	No	Time loops abound in the early universe but then evaporate	10
Periodic time cosmology	No	Yes	No	Conformal rescaling	No	The end is the beginning; the beginning is the end	10

NOTES

PREFACE

1 Sabine Hossenfelder, "Did the Big Bang Happen?" August 28, 2022, YouTube video, 16:58, https://www.youtube.com/watch?v=CAVUvq6BE1E.

CHAPTER 1

1 J. L. E. Dreyer, *A History of Astronomy from Thales to Kepler* (New York: Dover, 2003).
2 Dirk L. Couprie, "Anaximander (c. 610–546 B.C.E.)," *Internet Encyclopedia of Philosophy*, accessed July 12, 2024, https://iep.utm.edu/anaximander/.
3 Carlo Rovelli, *Anaximander and the Birth of Science* (New York: Penguin, 2023).
4 Alan Chodos, "June, ca. 240 B.C. Eratosthenes Measures the Earth," American Physical Society, June 1, 2006, https://www.aps.org/publications/apsnews/200606/history.cfm.
5 David Furley, *The Greek Cosmologists, Volume 1: The Formation of the Atomic Theory and Its Earliest Critics* (Cambridge: Cambridge University Press, 1987).
6 Lucretius, *Lucretius on Love and Sex: A Commentary on "De rerum natura" IV, 1030–1287*, trans. Robert Duncan Brown, Columbia Studies in the Classical Tradition vol. 15 (Leiden: E. J. Brill, 1987).
7 Thomas McEvilley, *The Shape of Ancient Thought: Comparative Studies in Greek and Indian Philosophies* (New York: Constable and Robinson, 2012).
8 Morris Edward Opler, *Myths and Tales of the Jicarilla Apache Indians: Memoirs of the American Folklore Society* (New York: American Folk-Lore Society, G. E. Stechert and Co., 1938): 406.
9 A. L. Basham, *The Wonder That Was India: A Survey of the Culture of the Indian Sub-Continent before the Coming of the Muslims* (London: Picador, 1954).

10 Although he did speculate that God could "vary the Laws of Nature and make Worlds of several sorts in several Parts of the universe," that speculation was mostly ignored.

11 For a lengthier discussion, see Simon Singh, *Big Bang: The Most Important Scientific Discovery of All Time and Why You Need to Know about It* (London: Fourth Estate, 2004).

12 Edwin Powell Hubble, *The Realm of the Nebulae* (New Haven, CT: Yale University Press, 1982).

13 Abbé G. Lemaître, "Contributions to a British Association Discussion on the Evolution of the Universe," *Nature* 128 (1931): 704–6.

14 Helge Kragh and Robert W. Smith, "Who Discovered the Expanding Universe?" *History of Science* 41, no. 2 (June 2003): 141–62, https://doi.org/10.1177/007327530304100202.

15 "Space Opened Up by Einstein," *Los Angeles Times*, February 5, 1931: A1–A2.

16 Clark Kimberling, "Emmy Noether and Her Influence," in *Emmy Noether: A Tribute to Her Life and Work*, eds. James W. Brewer and Martha K. Smith (New York: Marcel Dekker, 1981): 14.

17 Cormac O'Raifeartaigh, Brendan McCann, Werner Nahm, and Simon Mitton, "Einstein's Exploration of a Steady-State Model of the Universe," *European Physical Journal H* 39, no. 3 (2014): 353–67.

18 O'Raifeartaigh et al., "Einstein's Exploration."

19 Fred Hoyle, quoted in Malcolm S. Longair, *The Cosmic Century: A History of Astrophysics and Cosmology* (Cambridge: Cambridge University Press, 2006): 324.

20 Fred Hoyle, *Home Is Where the Wind Blows: Chapters from a Cosmologist's Life* (Mill Valley, CA: University Science Books, 1994).

21 The original script can be seen at "Hoyle on the Radio: Creating the 'Big Bang,'" St. John's College University of Cambridge, accessed April 25, 2023, https://www.joh.cam.ac.uk/library/special_collections/hoyle/exhibition/radio.

22 Helge Kragh, "Big Bang: The Etymology of a Name," *Astronomy and Geophysics* 54, no. 2 (April 2013): 2.28–2.30.

23 John Horgan, "The Return of the Maverick," *Scientific American* 272, no. 3 (1995): 46–47.

24 However, notable anomalies between predictions and observations of some elements persist to this day.

25 Hoyle, *Home Is Where the Wind Blows*.

26 Ernest Rutherford, quoted in Maurice Goldsmith, *Sage: A Life of J. D. Bernal* (London: Hutchinson, 1980).

27 Herbert Dingle, "Modern Aristotelianism," *Nature* 139 (1937): 784–86.

28 J. V. Narlikar, "The Evolution of Modern Cosmology as Seen through a Personal Walk across Six Decades." *European Physical Journal H* 43 (2018): 43–72.

29 A. McKellar, "Evidence for the Molecular Origin of Some Hitherto Unidentified Interstellar Lines," *Publications of the Astronomical Society of the Pacific* 52, no. 307 (1940): 187.
30 Fred Hoyle, "The Big Bang in Astronomy," *New Scientist*, November 19, 1981.
31 Ken Croswell, *The Universe at Midnight: Observations Illuminating the Cosmos* (New York: Simon and Schuster, 2002).
32 Pedro G. Ferreira, *The Perfect Theory* (New York: Houghton Mifflin Harcourt, 2014).
33 Woodruff T. Sullivan III, *Cosmic Noise: A History of Early Radio Astronomy* (Cambridge: Cambridge University Press, 2009).
34 Hoyle, *Home Is Where the Wind Blows*.
35 Barbara Perkins Gamow, as quoted by George Gamow in an oral-history interview with Charles Weiner in Boulder, CO, April 25, 1968, for the American Institute of Physics, https://www.aip.org/history-programs/niels-bohr-library/oral-histories/4325.
36 Fred Hoyle, in a BBC interview quoted in Adam Curtis, "A Mile or Two off Yarmouth," *Medium and the Message* (BBC blog), February 24, 2012, https://www.bbc.co.uk/blogs/adamcurtis/entries/512cde83-3afb-3048-9ece-dba774b10f89.
37 Helge Kragh, *Cosmology and Controversy: The Historical Development of Two Theories of the Universe* (Princeton, NJ: Princeton University Press, 1996).
38 E. A. Ohm, "Receiving System," *Bell System Technical Journal* 40, no. 4 (1961): 1065–94.
39 Jim Peebles, in his lecture on receiving the Nobel prize in Physics, delivered December 8, 2019, at the Aula Magna, Stockholm University; available online at James Peebles, "James Peebles Nobel Lecture: How Physical Cosmology Grew," *NobelPrize.org*, May 19, 2024, https://www.nobelprize.org/prizes/physics/2019/peebles/lecture/.
40 Evgeny M. Lifshitz and Isaak M. Khalatnikov, "Investigations in Relativistic Cosmology," *Advances in Physics* 12, no. 46 (1963): 185–249.
41 For any expert readers, this dependence of the fate of the universe on the value of omega becomes more convoluted in the presence of dark energy.
42 Alan H. Guth, *The Inflationary Universe: The Quest for a New Theory of Cosmic Origins* (New York: Random House, 1998), 18.

CHAPTER 2

1 John P. Preskill, "Cosmological Production of Superheavy Magnetic Monopoles," *Physical Review Letters* 43, no. 19 (1979): 1365; Alan H. Guth and S.-H. H. Tye, "Phase Transitions and Magnetic Monopole Production in the Very Early Universe," *Physical Review Letters* 44, no. 10 (1980): 631.
2 Alan H. Guth, *The Inflationary Universe: The Quest for a New Theory of Cosmic Origins* (New York: Random House, 1998).

3. Andrei Linde, interviewed by Alan Lightman, October 22, 1987, Niels Bohr Library and Archives, American Institute of Physics, College Park, MD, www.aip.org/history-programs/niels-bohr-library/oral-histories/34321.
4. Katsuhiko Sato, "First-Order Phase Transition of a Vacuum and the Expansion of the Universe," *Monthly Notices of the Royal Astronomical Society* 195, no. 3 (1981): 467–79.
5. Basil Hall Chamberlain, *The Kojiki: Japanese Records of Ancient Matters* (London: Forgotten Books, 2008).
6. The only (fundamental) scalar field that has been experimentally seen in nature is the Higgs field, discovered by Europe's Large Hadron Collider in 2012.
7. J. Richard Gott, "Creation of Open Universes from de Sitter Space," *Nature* 295 (1982): 304–7.
8. Guth, *Inflationary Universe*.
9. As reported by Jeffery Kahn, "LBL Physicist Discovers Fossil Relics of the Early Universe," *LBL Research Review* (Summer 1992): https://www2.lbl.gov/Science-Articles/Archive/cobe-discovery.html.
10. George Smoot, quoted in Zoë Corbyn, "George Smoot: We Mapped the Embryonic Universe," *Guardian*, April 20, 2104, https://www.theguardian.com/science/2014/apr/20/george-smoot-we-mapped-embryonic-universe-nobel-winning-big-bang-cosmos.
11. Michael D. Lemonick, *Echo of the Big Bang* (Princeton, NJ: Princeton University Press, 2003).
12. Lemonick, *Echo of the Big Bang*.
13. Guth, *Inflationary universe*.
14. Lemonick, *Echo of the Big Bang*.
15. George Smoot, "George Smoot: Birth of the Universe," interview by Monte Davis, March 1994, Smoot Group, https://aether.lbl.gov/www/personnel/OMNIinterviewSmMarch93.html.
16. Guth, *Inflationary Universe*.
17. L. M. Krauss and M. S. Turner, "The Cosmological Constant Is Back," *General Relativity and Gravitation* 27 (1995): 1137–44.
18. "WMAP 7-Year Results Released," National Aeronautics and Space Administration, last updated December 21, 2012, https://wmap.gsfc.nasa.gov/news/7yr_release.html.
19. Ryan Scranton, Andrew J. Connolly, Robert C. Nichol, Albert Stebbins, Istvan Szapudi, Daniel J. Eisenstein, Niayesh Afshordi, et al., "Physical Evidence for Dark Energy," *arXiv* preprint (2003): 0307335; Niayesh Afshordi, *The Other 99 Percent* (Princeton, NJ: Princeton University, 2004).
20. Dennis Overbye, "Astronomers Report Evidence of 'Dark Energy' Splitting the Universe," *New York Times*, July 22, 2003, https://www.nytimes.com/2003/07/22/us/astronomers-report-evidence-of-dark-energy-splitting-the-universe.html.

21 Steven Weinberg, "The Cosmological Constant Problem," *Reviews of Modern Physics* 61, no. 1 (1989).
22 R. H. Dicke, "Principle of Equivalence and Weak Interactions," *Reviews of Modern Physics* 29, no. 355 (1957).
23 Paul J. Steinhardt and Neil Turok, *Endless Universe: Beyond the Big Bang* (New York: Broadway, 2007).
24 Paul J. Steinhardt, "Cosmological Perturbations: Myths and Facts," *Modern Physics Letters A* 19, no. 13 (2004): 967–82.
25 Sabine Hossenfelder, "The Multiverse: Science, Religion, or Pseudoscience?" YouTube, September 11, 2022, MP4 video, 17:00, https://www.youtube.com/watch?v=QHa1vbwVaNU.
26 G. Ellis and J. Silk, "Scientific Method: Defend the Integrity of Physics," *Nature* 516 (2014): 321–23.
27 Phil Halper, "Before the Big Bang 4: Eternal Inflation and the Multiverse," YouTube, June 28, 2016, MP4 video, 59:53, https://www.youtube.com/watch?v=QqjsZEZMR7I&t=2529s.
28 Anna Ijjas, Paul Steinhardt, and Abraham Loeb, "Inflationary Paradigm in Trouble after Planck," *Physics Letters B* 723, nos. 4–5 (2013): 261–66. https://doi.org/10.1016/j.physletb.2013.05.023.
29 Stanford, "Stanford Professor Andrei Linde Celebrates Physics Breakthrough," YouTube, March 18, 2014, MP4 video, 2:38, https://www.youtube.com/watch?v=ZlfIVEy_YOA.
30 See, for example, Luboš Motl, "BICEP2 vs Planck: Nothing Wrong with Screen Scraping," *Reference Frame* (blog), May 14, 2014, archived at https://web.archive.org/web/20140516184813/http://motls.blogspot.nl/; and Luboš Motl, "BICEP2 and PRL: Journalists Prove That They're Trash," *Reference Frame* (blog), June 23, 2014, archived at https://web.archive.org/web/20140630052159/http://motls.blogspot.com/2014/06/bicep2-and-prl-journalists-prove-that.html#more.
31 Anna Ijjas, Paul J. Steinhardt, and Abraham Loeb, "Cosmic Inflation Theory Faces Challenges," *Scientific American* 316, no. 2 (2017): 32–39.
32 Paul J. Steinhardt, "The Inflation Debate," *Scientific American* 304, no. 4 (2011): 36–45.
33 Alan H. Guth, David I. Kaiser, Andrei D. Linde, Yasunori Nomura, Charles L. Bennett, J. Richard Bond, François Bouchet, et al., "A Cosmic Controversy," *Scientific American* 317, no. 1 (2017): 5–7.
34 Avi Loeb, *Extraterrestrial: The First Sign of Intelligent Life beyond Earth* (New York: Houghton Mifflin, 2021).
35 Debika Chowdhury, Jérôme Martin, Christophe Ringeval, Vincent Vennin, "Assessing the Scientific Status of Inflation after Planck," *Physical Review D* 100, no. 8 (2019): 083537.
36 Alan Harvey Guth and Yasunori Nomura, "What Can the Observation of Nonzero Curvature Tell Us?" *Physical Review D* 86 (2012): 023534.

37 Jaume Garriga, Alexander Vilenkin, and Jun Zhang, "Black Holes and the Multiverse," *Journal of Cosmology and Astroparticle Physics* 2016, no. 2 (February 2016): 64.
38 Edward R. Harrison, "Fluctuations at the Threshold of Classical Cosmology," *Physical Review D* 1, no. 10 (1970): 2726; Ya. B. Zeldovich, "A Hypothesis, Unifying the Structure and the Entropy of the Universe," *Monthly Notices of the Royal Astronomical Society* 160, no. 1 (1972): 1P–3P.

CHAPTER 3

1 See, for example, Adam Hills (@adamhillscomedy), "In Sydney, Paralympians were treated as equals. In London, they were treated as heroes. In Rio, they were treated to a party. Obrigado Rio," Twitter, September 20, 2016, https://x.com/adamhillscomedy/status/777840818929471488.
2 Werner Heisenberg, *Physics and Philosophy: The Revolution in Modern Science* (London: Allen and Unwin, 1958).
3 Claus Jönsson, "Elektroneninterferenzen an mehreren künstlich hergestellten Feinspalten," *Zeitschrift für Physik* 161 (August 1961): 454–74.
4 Yasunori Nomura, "Physical Theories, Eternal Inflation, and the Quantum Universe," *Journal of High Energy Physics* 2011, no. 11 (2011): 1–68; Raphael Bousso and Leonard Susskind, "Multiverse Interpretation of Quantum Mechanics," *Physical Review D* 85, no. 4 (2012): 045007.
5 James M. Bardeen, Brandon Carter, and Stephen W. Hawking, "The Four Laws of Black Hole Mechanics," *Communications in Mathematical Physics* 31 (1973): 161–70.
6 Niayesh Afshordi, personal communication with sources who prefer to remain anonymous.
7 Kitty Ferguson, *Stephen Hawking: His Life and Work: The Story and Science of One of the Most Extraordinary, Celebrated, and Courageous Figures of Our Time* (New York: Random House, 2011).
8 See Phil Halper, "Before the Big Bang V: The No Boundary Proposal," YouTube, November 7, 2017, MP4 video, 50:48, https://www.youtube.com/watch?v=Ry_pILPr7B8&t=3s.
9 James B. Hartle and Stephen W. Hawking, "Wave Function of the Universe," *Physical Review D* 28, no. 12 (1983): 2960.
10 Alexander Vilenkin, "Creation of Universes from Nothing," *Physics Letters B* 117, nos. 1–2 (1982): 25–28.
11 Lawrence M. Krauss, *A Universe from Nothing: Why There Is Something Rather than Nothing* (New York: Simon and Schuster, 2012).
12 David Albert, "On the Origin of Everything," *New York Times*, March 23, 2012, https://www.nytimes.com/2012/03/25/books/review/a-universe-from-nothing-by-lawrence-m-krauss.html.

13 Aron Wall, "Did the Universe Begin? VIII: The No Boundary Proposal," *Undivided Looking* (blog), July 13, 2103, http://www.wall.org/~aron/blog/did-the-universe-begin-viii-the-no-boundary-proposal/.
14 To our knowledge, the first version of an inflating universe with a reversal of the arrow of time was the one published in Anthony Aguirre and Steven Gratton, "Steady-State Eternal Inflation," *Physical Review D* 65, no. 8 (2002): 083507.
15 Job Feldbrugge, Jean-Luc Lehners, and Neil Turok, "No Smooth Beginning for Spacetime," *Physical Review Letters* 119, no. 17 (2017): 171301.
16 Jonathan J. Halliwell, James B. Hartle, and Thomas Hertog, "What Is the No-Boundary Wave Function of the Universe?" *Physical Review D* 99, no. 4 (2019): 043526.
17 Alexander Vilenkin and Masaki Yamada, "Tunneling Wave Function of the Universe," *Physical Review D* 98, no. 6 (2018): 066003.
18 Alice Di Tucci and Jean-Luc Lehners, "No-Boundary Proposal as a Path Integral with Robin Boundary Conditions," *Physical Review Letters* 122, no. 20 (2019): 201302.

CHAPTER 4

1 Albert Einstein, quoted in Ilse Rosenthal-Schneider, *Reality and Scientific Truth: Discussions with Einstein, von Laue, and Planck* (Detroit, MI: Wayne State University Press, 1980): 74.
2 Albert Einstein, *Eine neue Formale Deutung der Maxwellschen Feldgleichungen der Elektrodynamik* (Berlin: Akademie der Wissenschaften, 1916): 696.
3 *NOVA*, season 30, episode 12, "The Elegant Universe: Einstein's Dream," written, produced, and directed by Joseph McMaster, aired October 29, 2003, on PBS; transcript available at: https://www.scribd.com/document/185276/NOVA-The-Elegant-Universe-Transcript.
4 M. J. Duff, "A Layman's Guide to M-Theory," *arXiv*, last revised July 2, 1998, https://arxiv.org/abs/hep-th/9805177.
5 Laura Mersini-Houghton, *Before the Big Bang: The Origin of the Universe and What Lies Beyond* (London: Bodley Head, 2022).
6 William H. Kinney, "Limits on Entanglement Effects in the String Landscape from Planck and BICEP/Keck Data," *Journal of Cosmology and Astroparticle Physics* 2016 (November 2016).
7 Sabine Hossenfelder, "New Study Finds No Sign of Entanglement with Other Universes," *BackRe(Action)* (blog), June 18, 2016, http://backreaction.blogspot.com/2016/06/new-study-finds-no-sign-of-entanglement.html?showComment=1466261463362#c6591590283066480714. See also Peter Woit, "Quick Items," *Not Even Wrong* (blog), June 15, 2016, https://www.math.columbia.edu/~woit/wordpress/?p=8587.

8. Thomas Henry Huxley, address to the Liverpool Meeting of the British Association, *Nature*, September 15, 1870: 400–402, at 402.
9. Steven Weinberg, "Anthropic Bound on the Cosmological Constant," *Physical Review Letters* 59, no. 22 (1987): 2607.
10. Ali Nayeri, Robert H. Brandenberger, and Cumrun Vafa, "Producing a Scale-Invariant Spectrum of Perturbations in a Hagedorn Phase of String Cosmology," *Physical Review Letters* 97, no. 2 (2006): 021302.
11. Andrei Linde, as reported by Robert Brandenburger in an interview with Phil; Brandenburger tells the same story in an interview with David Zierler at the American Institute of Physics, April 22, 2021, https://www.aip.org/history-programs/niels-bohr-library/oral-histories/47043.
12. Nemanja Kaloper, Lev Kofman, Andrei Linde, and Viatcheslav Mukhanov, "On the New String Theory Inspired Mechanism of Generation of Cosmological Perturbations," *Journal of Cosmology and Astroparticle Physics* 2006, no. 10 (2006): 6, https://doi.org/10.48550/arXiv.hep-th/0608200.
13. R. H. Brandenberger and C. Vafa, "Superstrings in the Early Universe," *Nuclear Physics B* 316, no. 2 (April 10, 1989): 391–410.
14. For the more technically minded, Planck and BICEP 2 data were in tension at 3 sigma; this tension is resolvable if we assume a blue tensor tilt. 3 sigma represents three standard deviations from the mean, enough of a disparity to take seriously but not one high enough as the 5 sigma needed to declare a discovery. See Kendrick M. Smith, Cora Dvorkin, Latham A. Boyle, Neil Turok, Mark Halpern, Gary F. Hinshaw, and Benjamin Gold, "Quantifying the BICEP2-Planck Tension over Gravitational Waves," *Physical Review Letters* 113, no. 3 (2014): 031301.
15. Brian Keating, *Losing the Nobel Prize: A Story of Cosmology, Ambition, and the Perils of Science's Highest Honor* (New York: W. W. Norton and Company, 2018).
16. Brian Keating, "String Gas Cosmology to the Rescue! (336)," from Keating's *Into the Impossible* podcast, uploaded to YouTube on August 7, 2023, MP4 video, 1:47:22, https://www.youtube.com/watch?v=G3xy-bEDJCY&t=3243s.
17. For a biography of Ben Turok, see Shamielah Booley, "Ben Turok Biography," *South African History Online*, accessed January 5, 2024, https://www.sahistory.org.za/sites/default/files/archive_files/Ben%20Turok%20biography.pdf.
18. Neil Turok, "String-Driven Inflation," *Physical Review Letters* 60, no. 7 (1988): 549.
19. Paul J. Steinhardt and Neil Turok, *Endless Universe: Beyond the Big Bang* (New York: Doubleday, 2007).
20. Renata Kallosh, Lev Kofman, and Andrei Linde, "Pyrotechnic Universe," *Physical Review D* 64, no. 12 (2001): 123523.
21. Sean M. Carroll and Jennifer Chen, "Spontaneous Inflation and the Origin of the Arrow of Time," *arXiv*, October 27, 2004, https://arxiv.org/abs/hep-th/0410270.

22 This line was notably advocated by string-theorist-turned-philosopher Richard Dawid.

CHAPTER 5

1. Jörg Resag, *Feynman and His Physics: The Life and Science of an Extraordinary Man* (Cham: Springer, 2018). See also Leonard Mlodinow, "Physics: Fundamental Feynman," *Nature* 471 (2011): 296–97; and Lawrence M. Krauss, *Quantum Man: Richard Feynman's Life in Science* (New York: W. W. Norton, 2011): 234.
2. Shannon Palus, "Famed Physicist Richard Feynman Was Known as an Odd Genius. Was He Also an Abuser?" *Slate*, January 14, 2019, https://slate.com/technology/2019/01/richard-feynman-physical-abuse-science-wife-fbi.html.
3. Tim Adams, "Carlo Rovelli: 'Science Is Where Revolutions Happen,'" *Guardian*, October 1, 2016, https://www.theguardian.com/books/2016/oct/16/carlo-rovelli-interview-quantum-gravity-physics-science-is-where-revolutions-happen.
4. Matthew Reisz, "From Hippy Activist to Science Guru," *Times Higher Education*, September 29, 2022, https://www.timeshighereducation.com/depth/hippy-activist-science-guru-carlo-rovelli-why-wasting-time-key-scientific-discovery?c.
5. Phil Halper, "String Theory or Loop Quantum Gravity? David Gross vs Carlo Rovelli," YouTube, December 20, 2021, MP4 video, 1:43:01, https://www.youtube.com/watch?v=AUyylR5RPZw&t=191s.
6. Lee Smolin, *The Trouble with Physics: The Rise of String Theory, the Fall of a Science, and What Comes Next* (Boston, MA: Houghton Mifflin, 2007).
7. John Baez, "This Week's Finds in Mathematical Physics (Week 280)," *This Week's Finds* (blog), September 7, 2009, https://math.ucr.edu/home/baez/week280.html.
8. Charles J. Brainerd and Valerie F. Reyna, *The Science of False Memory*, Oxford Psychology Series (New York: Oxford University Press, 2005).
9. Martin Bojowald, "Absence of a Singularity in Loop Quantum Cosmology," *Physical Review Letters* 86, no. 23 (2001): 5227.
10. Abhay Ashtekar, Martin Bojowald, and Jerzy Lewandowski, "Mathematical Structure of Loop Quantum Cosmology," *arXiv* (April 2003): 233–68.
11. Alan Guth and Marc Sher, "The Impossibility of a Bouncing Universe," *Nature* 302 (April 1983): 505–6.
12. Abhay Ashtekar, Tomasz Pawlowski, and Parampreet Singh, "Quantum Nature of the Big Bang: Improved Dynamics," *Physical Review D* 74, no. 8 (2006): 084003.
13. Abhay Ashtekar, Alejandro Corichi, and Parampreet Singh, "Robustness of Key Features of Loop Quantum Cosmology," *Physical Review D* 77, no. 2 (January 2008): 024046.

14 Martin Bojowald, "Big Bang or Big Bounce?: New Theory on the Universe's Birth," *Scientific American*, October 1, 2008, https://www.scientificamerican.com/article/big-bang-or-big-bounce/.
15 Abhay Ashtekar and David Sloan, "Loop Quantum Cosmology and Slow Roll Inflation," *Physics Letters B* 694, no. 2 (2010): 108–12.
16 Linda Linsefors and Aurelien Barrau, "Duration of Inflation and Conditions at the Bounce as a Prediction of Effective Isotropic Loop Quantum Cosmology," *Physical Review D* 87, no. 12 (June 2013): 123509.
17 Martin Bojowald, "What Happened before the Big Bang?" *Nature Physics* 3 (2007): 523–25, https://doi.org/10.1038/nphys654.
18 Alejandro Corichi and Parampreet Singh, "Corichi and Singh Reply," *Physical Review Letters* 101, no. 20 (2008): 209002.
19 See, for example, Martin Bojowald, "Absence of a Singularity in Loop Quantum Cosmology," *Physical Review Letters* 86, no. 23 (2001): 5227; Martin Bojowald, "Critical Evaluation of Common Claims in Loop Quantum Cosmology," *Universe* 6 (2020): 36, https://doi.org/10.48550/arXiv.2002.05703.
20 Martin Bojowald, *Once before Time: A Whole Story of the Universe* (New York: Knopf, 2010): 91. On the next page, Bojowald asserts that an analyst of Ashtekar's ilk "at his humble best advances and crystalizes the laws underlying a physical theory, but at his greedy worst may take known results and refurbish them as his own" (92).
21 Martin Bojowald and Suddhasattwa Brahma, "Loops Rescue the No-Boundary Proposal," *Physical Review Letters* 121, no. 20 (2018): 201301.
22 Ivan Agullo, Dimitrios Kranas, and V. Sreenath, "Anomalies in the CMB from a Cosmic Bounce," *General Relativity and Gravitation* 53, no. 2 (2021): 17.
23 "Planck Reveals an Almost Perfect Universe," European Space Agency, March 21, 2013, https://www.esa.int/Science_Exploration/Space_Science/Planck/Planck_reveals_an_almost_perfect_Universe.
24 "Replay: Planck's Cosmic Microwave Background Map Media Briefing," European Space Agency, March 21, 2013, https://www.esa.int/ESA_Multimedia/Videos/2013/03/Replay_Planck_s_Cosmic_Microwave_Background_map_Media_Briefing.
25 Agullo, Kranas, and Sreenath. "Anomalies in the CMB from a Cosmic Bounce."
26 Francesca Vidotto, "Spinfoam Cosmology," *Journal of Physics: Conference Series* 314, no. 1 (2011): 012049.
27 James Riordon, "The Universe Began with a Bang, Not a Bounce, New Studies Find," *Scientific American*, May 24, 2023, https://www.scientificamerican.com/article/the-universe-began-with-a-bang-not-a-bounce-new-studies-find/.
28 See Rothna K. and Sreenath, V. "Non-Gaussianity in the Cosmic Microwave Background from Loop Quantum Cosmology," *arXiv* preprint

(January 2023), https://www.researchgate.net/publication/367165883_Non-Gaussianity_in_the_cosmic_microwave_background_from_loop_quantum_cosmology.

For Agullo's full account of this incident, see his interview on Phil's YouTube channel at Phil Halper, "Has the Big Bounce Been Ruled Out?" YouTube, December 16, 2023, MP4 video, 29:53, https://www.youtube.com/watch?v=a8A8TMMNFns&t=107s.

29 Rodolfo Gambini and Jorge Pullin, "Emergence of Stringlike Physics from Lorentz Invariance in Loop Quantum Gravity," *International Journal of Modern Physics D* 23, no. 12 (2014): 1442023.

30 ViXra Admin, "Super Yang-Mills vs Loop Quantum Gravity: The Same Bloody Thing," YouTube, July 9, 2013, MP4 video, 5:33, https://www.youtube.com/watch?v=W_TO2WESSA4&t=1s.

CHAPTER 6

1 A. Friedmann, "Über die Krümmung des Raumes," *Zeitschrift für Physik* 10 (1922): 377–86; Aleksandr Friedman, *Mir Kak Prostranstvo i Vremya* [The World as Space and Time] (Leningrad: Academia, 1923).

2 The astute reader will have noticed that what we called original order was of course arbitrary. We will come back to this point later, in chapter 11, where we discuss the low-entropy paradox.

3 Helge Kragh, "Cyclic Models of the Relativistic Universe: The Early History," *arXiv* preprint (2103): 1308.0932.

4 Edward Frenkel, "The Holy Grail of Quantum Physics on Your Kitchen Table [Excerpt]," *Scientific American*, September 27, 2013, https://www.scientificamerican.com/article/the-holy-grail-of-quantum-physics-on-your-kitchen-table-excerpt/.

5 Jason Palmer, "Cosmos May Show Echoes of Events before Big Bang," *BBC News*, November 27, 2010, https://www.bbc.co.uk/news/science-environment-11837869.

6 Lisa Grossman, "Theory of Recycled Universe Called into Question," *Wired*, May 13, 2011, https://www.wired.com/2011/05/no-cmb-circles/.

7 J. Zuntz, J. P. Zibin, C. Zunckel, and J. Zwart, "Non-Standard Morphological Relic Patterns in the Cosmic Microwave Background," *arXiv* preprint (2011): 1103.6262.

8 Phil Halper, "Before the Big Bang 2—Conformal Cyclic Cosmology," YouTube, January 24, 2013, MP4 video, 39:04, https://www.youtube.com/watch?v=sM47acQ7pEQ.

9 Daniel An, Krzysztof A. Meissner, and Paweł Nurowski, "Ring-Type Structures in the Planck Map of the CMB," *arXiv*, last revised August 24, 2016, https://arxiv.org/abs/1510.06537.

10 Daniel An, Krzysztof A. Meissner, and Paweł Nurowski, "Ring-Type Structures in the Planck Map of the CMB," *Monthly Notices of the Royal Astronomical Society* 473, no. 3 (2018): 3251–55.

11 R. Fernández-Cobos, A. Marcos-Caballero, and Enrique Martínez-González, "Radial Derivatives as a Test of Pre–Big Bang Events on the Planck Data," *Monthly Notices of the Royal Astronomical Society* 499, no. 1 (2020): 1300–11.

12 Phil Halper, "Multiverse or Cyclic Universe? Alan Guth vs Roger Penrose," YouTube, August 16, 2021, MP4 video, 1:38:28, https://www.youtube.com/watch?v=YhbULagUKhA&t=499s.

13 Robert R. Caldwell, Marc Kamionkowski, and Nevin N. Weinberg, "Phantom Energy: Dark Energy with $w < -1$ Causes a Cosmic Doomsday," *Physical Review Letters* 91, no. 7 (2003): 071301.

14 Lauris Baum and Paul H. Frampton, "Turnaround in Cyclic Cosmology," *Physical Review Letters* 98, no. 7 (2007): 071301.

15 Xin Zhang, "Comment on 'Turnaround in Cyclic Cosmology,'" *arXiv* preprint (2007): 0711.0667.

16 Paul Frampton, *Tricked! The Story of an Internet Scam* (self-published, 2014). For a more neutral account of Frampton's odyssey, see Maxine Swann, "The Professor, the Bikini Model and the Suitcase Full of Trouble," *New York Times*, March 10, 2013, https://www.nytimes.com/2013/03/10/magazine/the-professor-the-bikini-model-and-the-suitcase-full-of-trouble.html.

17 See "Former UNC Professor to Receive Back-Pay for Time Spent in Prison," *Chapelboro.com*, June 6, 2015, https://chapelboro.com/news/unc/former-unc-professor-to-receive-back-pay-for-time-spent-in-prison.

18 Matthew G. Brown, Katherine Freese, and William H. Kinney, "The Phantom Bounce: A New Oscillating Cosmology," *Journal of Cosmology and Astroparticle Physics* 2008, no. 3 (2008): 2.

19 See also Paul Steinhardt's interview by David Zierler, June 4, June 18, June 30, and July 8, 2020, Niels Bohr Library and Archives, American Institute of Physics, College Park, MD, www.aip.org/history-programs/niels-bohr-library/oral-histories/46757.

CHAPTER 7

1 William Cole, quoted in Marcia Bartusiak, *Black Hole: How an Idea Abandoned by Newtonians, Hated by Einstein, and Gambled on by Hawking Became Loved* (New Haven, CT: Yale University Press, 2015): 9.

2 Alasdair Wilkins, "The Forgotten Genius Who Discovered Black Holes Over 200 Years Ago," *Gizmodo*, December 23, 2010, https://web.archive.org/web/20210316003559/https://io9.gizmodo.com/the-forgotten-genius-who-discovered-black-holes-over-20-5717082.

3. Karl Schwarzschild, "On the Gravitational Field of a Sphere of Incompressible Fluid According to Einstein's Theory," *arXiv* preprint (1999): 9912033.
4. For a fuller account of Chandra's struggles with Eddington, see A. I. Miller, *Empire of the Stars: Obsession, Friendship, and Betrayal in the Quest for Black Holes* (Boston, MA: Houghton Mifflin, 2005).
5. "Meeting of the Royal Astronomical Society, Friday, 1935 January 11," *Observatory* 58 (1935): 33–41.
6. Spencer Weart, interview of Subrahmanyan Chandrasekhar, May 17, 1977, Niels Bohr Library, American Institute of Physics, College Park, MD, https://www.aip.org/history-programs/niels-bohr-library/oral-histories/4551-1.
7. Kip S. Thorne, John Archibald Wheeler, and Charles W. Misner, *Gravitation* (San Francisco, CA: Freeman, 2000).
8. For a fuller account of the history of the term, see Carlos A. R. Herdeiro and José P. S. Lemos, "The Black Hole Fifty Years After: Genesis of the Name," *arXiv* preprint (2018): 1811.06587.
9. Roger W. Romani, D. Kandel, Alexei V. Filippenko, Thomas G. Brink, and WeiKang Zheng, "PSR J0952−0607: The Fastest and Heaviest Known Galactic Neutron Star," *Astrophysical Journal Letters* 934, no. 2 (2022): L18.
10. Rafael D. Sorkin, "A Modified Sum-over-Histories for Gravity," presented at *Highlights in Gravitation and Cosmology: Proceedings of the International Conference on Gravitation and Cosmology*, Goa, India, December 14–19, 1987.
11. Richard Dawkins, "Why Darwin Matters," *Guardian*, February 8, 2008, https://www.theguardian.com/science/2008/feb/09/darwin.dawkins1.
12. Mitchell Langbert, Anthony J. Quain, and Daniel B. Klein, "Faculty Voter Registration in Economics, History, Journalism, Law, and Psychology," *Econ Journal Watch* 13, no. 3 (2016): 422–51.
13. Andrzej Trautman, "Spin and Torsion May Avert Gravitational Singularities," *Nature Physical Science* 242, no. 114 (1973): 7–8.
14. Nikodem Popławski, "Universe in a Black Hole in Einstein–Cartan Gravity," *Astrophysical Journal* 832, no. 2 (2016): 96.
15. Shantanu Desai and Nikodem J. Popławski, "Non-Parametric Reconstruction of an Inflaton Potential from Einstein–Cartan–Sciama–Kibble Gravity with Particle Production," *Physics Letters B* 755 (2016): 183–89.
16. Gia Dvali, Gregory Gabadadze, and Massimo Porrati, "Metastable Gravitons and Infinite Volume Extra Dimensions," *Physics Letters B* 484, nos. 1–2 (2000): 112–18.
17. Razieh Pourhasan, Niayesh Afshordi, and Robert B. Mann, "Out of the White Hole: A Holographic Origin for the Big Bang," *Journal of Cosmology and Astroparticle Physics* 2014, no. 4 (2014): 5.

CHAPTER 8

1. Andreas Albrecht and João Magueijo, "Time Varying Speed of Light as a Solution to Cosmological Puzzles," *Physical Review D* 59, no. 4 (1999): 043516.
2. João Magueijo, *Faster Than the Speed of Light: The Story of a Scientific Speculation* (New York: Random House, 2011).
3. G. Ellis, "Einstein Not Yet Displaced," *Nature* 422 (2003): 563–64.
4. See Phil Halper, "Before the Big Bang 8: Varying Speed of Light Cosmology (VSL)," YouTube, October 17, 2018, MP4 video, 59:35, https://www.youtube.com/watch?v=kbHBBtsrU1g.
5. Niayesh Afshordi, Daniel J. H. Chung, and Ghazal Geshnizjani, "Causal Field Theory with an Infinite Speed of Sound," *Physical Review D—Particles, Fields, Gravitation, and Cosmology* 75, no. 8 (April 2007): 083513.
6. Magueijo, *Faster Than the Speed of Light*.
7. John W. Moffat, "Superluminary Universe: A Possible Solution to the Initial Value Problem in Cosmology," *International Journal of Modern Physics D* 2, no. 3 (1993): 351–65.
8. Halper, "Before the Big Bang 8."
9. Howard Burton, First Principles: The Crazy Business of Doing Serious Science (Toronto: Key Porter Books, 2009).
10. Paul Wells, "Outside the Box," *Maclean's*, April 16, 2009, https://macleans.ca/news/canada/outside-the-box/.
11. Magueijo, *Faster Than the Speed of Light*.
12. João Magueijo "Speedy Sound and Cosmic Structure," *Physical Review Letters* 100, no. 23 (2008): 231302.
13. Mike Hale, "A Mind Geared to Physics, a Profile Set for Prime Time," *New York Times*, May 27, 2008, https://www.nytimes.com/2008/05/27/arts/television/27bang.html.
14. Petr Hořava, "Quantum Gravity at a Lifshitz Point," *Physical Review D* 79, no. 8 (2009): 084008.
15. Zeeya Merali, "Splitting Time from Space—The Evidence," *Scientific American*, November 24, 2009, https://www.scientificamerican.com/article/splitting-time-from-space-evidence/.
16. George Gamow, *The Creation of the Universe* (New York: Viking Press, 1952).
17. Itzhak Bars, John Terning, and Farzad Nekoogar, *Extra Dimensions in Space and Time* (New York: Springer, 2010).
18. Robbert Dijkgraaf, Ben Heidenreich, Patrick Jefferson, and Cumrun Vafa, "Negative Branes, Supergroups and the Signature of Spacetime," *Journal of High Energy Physics* 2018, no. 2 (2018): 1–63.
19. To be precise, our model says the fluctuations should be exactly −3.52 percent (negative values mean slightly red noise, with more power at lower frequencies). The latest and final Planck measurement for this deviation

(from 2018) was between −3.0 percent and −3.7 percent. We expect future experiments, such as the fourth-stage investigations of ground-based CMB observatories, to improve these constraints by a factor of three. So, depending on the actual performance of these experiments and their final measurements, our −3.52 percent prediction might be ruled out (or confirmed) within the next decade.

20 Maria Mylova, Marianthi Moschou, Niayesh Afshordi, and João Magueijo, "Non-Gaussian Signatures of a Thermal Big Bang," *Journal of Cosmology and Astroparticle Physics* 2022, no. 7 (2022): 5. In this paper, we showed the value of local, primordial non-Gaussianity to be exactly −1.5. Roughly speaking, this finding means that if we live in a region of the universe that is 1 percent denser than average, then the energy in primordial sound waves is 1.5 percent less than that of an average part of the universe. (Profuse apologies to the experts at this point, for oversimplification.)

21 "Unifying Tests of General Relativity," schedule for Burke Institute Workshop, California Institute of Technology, Pasadena, CA, July 19–21, 2016, http://www.tapir.caltech.edu/~unifying-gr-tests/Schedule/.

CHAPTER 9

1 "Astronomy Chairs," Harvard University Department of Astronomy, accessed February 25, 2024, https://astronomy.fas.harvard.edu/astronomy-alumni.

2 Otto Struve, as quoted in "History," Harvard University Department of Astronomy, accessed April 18, 2023, https://astronomy.fas.harvard.edu/history.

3 Alan Finder, Patrick D. Healy, and Kate Zernike, "President of Harvard Resigns, Ending Stormy 5-Year Tenure," *New York Times*, February 22, 2006, https://www.nytimes.com/2006/02/22/education/22harvard.html/.

4 The title of a talk which can be seen at Nomen Nominandum, "Don't Modify Gravity—Understand It! - Nima Arkani-Hamed (2013)," YouTube, January 8, 2016, MP4 video, 1:24:35, https://www.youtube.com/watch?v=_k_V8TNWTHg.

5 Richard Phillips Feynman and Ralph Leighton, *"What Do You Care What Other People Think?": Further Adventures of a Curious Character* (New York: W. W. Norton and Company, 2001).

6 For a more thorough account see Leonard Susskind, *The Black Hole War: My Battle with Stephen Hawking to Make the World Safe for Quantum Mechanics* (London: Hachette UK, 2008).

7 For more on Arkani-Hamed's story, see Natalie Walchover, "Visions of Future Physics," *Quanta*, September 22, 2015, https://www.quantamagazine.org/nima-arkani-hamed-and-the-future-of-physics-20150922.

8 Juan Maldacena, "The Large-N Limit of Superconformal Field Theories and Supergravity," *International Journal of Theoretical Physics* 38, no. 4 (1999): 1113–33.

9 SVAstronomyLectures, "The Black Hole Wars: My Battle with Stephen Hawking," YouTube, October 1, 2008, MP4 video, 1:34:51, https://youtu.be/KR3Msi1YeXQ.

10 Carlo Rovelli, "A Critical Look at Strings," *Foundations of Physics* 43 (2013): 8–20. Even more to Galileo's style, Rovelli wrote "A Dialog on Quantum Gravity" in 2003, modeling his critiques of AdS/CFT and string theory after Galileo's 1632 "Dialogue Concerning the Two Chief World Systems." See Carlo Rovelli, "A Dialog on Quantum Gravity," *International Journal of Modern Physics D* 12, no. 9 (2003): 1509–28.

11 Leonard Susskind, *The Cosmic Landscape: String Theory and the Illusion of Intelligent Design* (London: Hachette UK, 2008).

12 Shamit Kachru, Renata Kallosh, Andrei Linde, Juan Maldacena, Liam McAllister, and Sandip P. Trivedi, "Towards Inflation in String Theory," *Journal of Cosmology and Astroparticle Physics* 2003, no. 10 (2003): 13.

13 Daniel Jafferis, Alexander Zlokapa, Joseph D. Lykken, David K. Kolchmeyer, Samantha I. Davis, Nikolai Lauk, Hartmut Neven, and Maria Spiropulu, "Traversable Wormhole Dynamics on a Quantum Processor," *Nature* 612 (2022): 51–55.

14 Juan Maldacena, Alexey Milekhin, and Fedor Popov, "Traversable Wormholes in Four Dimensions," *arXiv* preprint (2018): 1807.04726.

CHAPTER 10

1 Martin Davis, "Gödel's Universe," *Nature* 435, no. 7038 (May 4, 2005): 19–20, https://doi.org/10.1038/435019a; "Kurt Gödel: Life, Work, and Legacy," Institute for Advanced Study, 2024, https://www.ias.edu/kurt-g%C3%B6del-life-work-and-legacy; Mark Balaguer, "Kurt Gödel," *Encyclopedia Britannica*, updated August 23, 2024, https://www.britannica.com/biography/Kurt-Godel; Sunny Labh, "The Mathematical Genius Who Starved Himself to Death," *Medium*, September 27, 2022, https://piggsboson.medium.com/the-mathematical-genius-who-starved-himself-to-death-68fd4bbee269.

2 Andrei Lossev and Igor D. Novikov, "The Jinn of the Time Machine: Nontrivial Self-Consistent Solutions," *Classical and Quantum Gravity* 9, no. 10 (1992): 2309.

3 Eric Poisson and Werner Israel, "Inner-Horizon Instability and Mass Inflation in Black Holes," *Physical Review Letters* 63, no. 16 (1989): 1663.

4 Stephen W. Hawking, "Chronology Protection Conjecture," *Physical Review D* 46, no. 2 (1992): 603.

5 Li-Xin Li, "Must Time Machines Be Unstable against Vacuum Fluctuations?" *Classical and Quantum Gravity* 13, no. 9 (1996): 2563.
6 Li-Xin Li and J. Richard Gott III, "Self-Consistent Vacuum for Misner Space and the Chronology Protection Conjecture," *Physical Review Letters* 80, no. 14 (1998): 2980.
7 Six years is the time it takes the spaceship to reach 99.99930 percent of the speed of light, starting from a resting position.
8 J. Richard Gott III and Li-Xin Li, "Can the Universe Create Itself?" *Physical Review D* 58, no. 2 (1998): 023501.
9 Roger Penrose, *Fashion, Faith, and Fantasy in the New Physics of the Universe* (Princeton, NJ: Princeton University Press, 2016).
10 William A. Hiscock, "Quantized Fields and Chronology Protection," *arXiv* preprint (2000): gr-qc/0009061.
11 Roberto Emparan and Marija Tomašević, "Holography of Time Machines," *Journal of High Energy Physics* 2022, no. 3 (2022): 1–33.
12 J. Richard Gott, interviewed in Jill Neimark, "J. Richard Gott on Life, the Universe, and Everything," *Science and Spirit*, 2007, archived at https://web.archive.org/web/20070928020457/http://www.science-spirit.org/article_detail.php?article_id=270.
13 However, considering virtual quantum particle/antiparticle pairs as time loops is subject to interpretation.
14 Something similar happened to the fluctuations during the bi-thermal Big Bang, as we described in chapter 8, but also happens in all other cosmological models, such as the early-time inflation of chapter 2, when structures are stretched to scales bigger than the cosmological horizon.

CHAPTER 11

1 See Ewen MacAskill, "George Bush: 'God Told Me to End the Tyranny in Iraq,'" *Guardian*, October 7, 2005, https://www.theguardian.com/world/2005/oct/07/iraq.usa; see also the text of a speech entitled "American Plots against Iran," delivered by Ayatollah Khomeini to the Iranian Central Insurance office staff at Qum on November 5, 1979, available online at Emam.com, http://emam.com/posts/view/15718/Speech.
2 For an overview of Nash's life, see Sylvia Nassar, "John Nash, His Life," in *The Abel Prize 2013–2017*, eds. Helge Holden and Ragni Piene (Cham: Springer, 2019): 35777.
3 William Lane Craig, "The Ultimate Question of Origins: God and the Beginning of the Universe," *Astrophysics and Space Science* 269 (1999): 721–38.
4 Sam Harris, during a University of Notre Dame debate accessible at University of Notre Dame, "The God Debate II: Harris vs. Craig," YouTube,

April 12, 2011, MP4 video, 2:06:54, https://www.youtube.com/watch?v=yqaHXKLRKzg.

5. David Bourget and David J. Chalmers, "Philosophers on Philosophy: The 2020 Philpapers Survey," *Philosopher's Imprint* 23, no. 11 (2023), https://philpapers. org/rec/BOUPOP-3.

6. Phillip Halper. "The Kalam Cosmological Argument: Critiquing a Recent Defence," *Think* 20, no. 57 (2021): 153–65.

7. See William Lane Craig, "Cosmos and Creator," *Origins and Design* 17, no. 2 (November 14, 1996), http://www.arn.org/docs/odesign/od172/cosmos172.htm.

8. William Lane Craig, *Kalam Cosmological Argument* (London: Palgrave Macmillan, 1979).

9. G. Geshnizjani, E. Ling, and J. Quintin, "On the Initial Singularity and Extendibility of Flat Quasi-de Sitter Spacetimes," *arXiv* preprint (2023): 2305.01676.

10. Alex Vilenkin, *Many Worlds in One: The Search for Other Universes* (New York: Hill and Wang, 2007).

11. Robert B. Stewart, ed., *God and Cosmology: William Lane Craig and Sean Carroll in Dialogue* (Minneapolis, MN: Fortress Press, 2016).

12. Stephon Alexander, Sam Cormack, and Marcelo Gleiser, "A Cyclic Universe Approach to Fine Tuning," *Physics Letters B* 757 (2016): 247–50.

13. See, for example, Phil Halper, "Physicists & Philosophers Debunk the Fine Tuning Argument," YouTube, November 17, 2022, MP4 video, 59:43, https://www.youtube.com/watch?v=jJ-fj3lqJ6M&t=1407s; and Memoirs of Professor Richard Dawkins, "Richard Dawkins - Steven Weinberg Discuss Science and Religion," YouTube, August 25, 2017, MP4 video, 1:14:17, https://www.youtube.com/watch?v=lv-CTPIfAas&t=1980s.

14. For a fuller discussion of the argument, see Hans Halvorson, "A Theological Critique of the Fine-Tuning Argument," in *Knowledge, Belief, and God: New Insights in Religious Epistemology*, eds. Matthew A. Benton, John Hawthorne, and Dani Rabinowitz (Oxford: Oxford University Press, 2018): 122–35.

15. T. McGrew, L. McGrew, and E. Vestrup, 2001, "Probabilities and the Fine-Tuning Argument: A Sceptical View," *Mind* 110, no. 440 (2001): 1027–38.

16. Richard Dawkins Foundation for Reason & Science, "'A Universe from Nothing' by Lawrence Krauss, AAI 2009," YouTube, October 21, 2009, MP4 video, 1:04:51, https://www.youtube.com/watch?v=7ImvlS8PLIo&t=2008s.

17. Lawrence M. Krauss, *A Universe from Nothing: Why There Is Something Rather Than Nothing* (New York: Simon and Schuster, 2012).

18. Stephen Hawking and Leonard Mlodinow, *The Grand Design* (London: Bantam, 2010).

19. Sean M. Carroll, "What If Time Really Exists?" *arXiv* preprint (2008): 0811.3772.

20. B. J. Alters, "Whose Nature of Science?" *Journal of Research in Science Teaching* 34 (1997): 39–55.
21. For a more through discussion, see Martin Curd and Jan A. Cover, eds., *Philosophy of Science: The Central Issues* (New York: Norton, 1998).
22. G. Irzik and R. Nola, "A Family Resemblance Approach to the Nature of Science for Science Education," *Science and Education* 20, nos. 7–8 (2011): 591–607.
23. João Magueijo, *Faster Than the Speed of Light: The Story of a Scientific Speculation* (New York: Random House, 2011).
24. FeynmanChaser, "Feynman Chaser - Imagination in a Straitjacket," YouTube, July 14, 2008, MP4 video, 2:49, https://www.youtube.com/watch?v=IFBtlZfwEwM.
25. Karl R. Popper, *The Open Society and Its Enemies* (Princeton, NJ: Princeton University Press, 2020).

CHAPTER 12

1. For a fuller account, see Janna Levin, *Black Hole Blues and Other Songs from Outer Space* (New York: Anchor, 2016).
2. Neil J. Cornish, David N. Spergel, Glenn D. Starkman, and Eiichiro Komatsu, "Constraining the Topology of the Universe," *Physical Review Letters* 92, no. 20 (2004): 201302.
3. Carlo Ungarelli and Alberto Vecchio, "Are Pre-Big-Bang Models Falsifiable by Gravitational Wave Experiments?" in *American Institute of Physics Conference Proceedings* 523, no. 1 (2000): 90–93.
4. Matteo Rini, "Researchers Capture Gravitational-Wave Background with Pulsar 'Antennae,'" *Physics* 16, no. 118 (June 29, 2023), https://physics.aps.org/articles/v16/118#.
5. Jim Simons, interview by Brady Haran (Numberphile2), YouTube, May 13, 2015, MP4 video, 1:00:42, https://www.youtube.com/watch?v=QNznD9hMEh0&t=2709s.
6. Dr. Brian Keating, "The Eternal Sky Episode 3," YouTube, May 9, 2018, MP4 video, 9:29, https://www.youtube.com/watch?v=Ft8SLtfIPiA.
7. "WMAP Produces New Results," National Aeronautics and Space Administration, last updated December 21, 2012, https://wmap.gsfc.nasa.gov/news/7yr_release.html.
8. Recall that a 5-sigma measurement indicates a result's statistical significance and is commonly taken as threshold for detection. It means that the chance of getting the same result from random noise is the same as throwing coins and getting twenty-two heads in a row.
9. Kitty Ferguson, *Stephen Hawking: A Life Well Lived* (New York: Random House, 2011).

10 Peter A. R. Ade, Nabila Aghanim, Charmaine Armitage-Caplan, Monique Arnaud, M. Ashdown, F. Atrio-Barandela, J. Aumont, et al., "Planck 2013 Results. XXII. Constraints on Inflation," *Astronomy and Astrophysics* 571 (2014): A22.

11 Jean-Luc Lehners and Paul J. Steinhardt, "Planck 2013 Results Support the Cyclic Universe," *Physical Review D* 87, no. 12 (2013): 123533.

12 Peter A. R. Ade, Nabila Aghanim, Charmaine Armitage-Caplan, Monique Arnaud, M. Ashdown, F. Atrio-Barandela, J. Aumont, et al., "Planck 2013 Results. XXIV. Constraints on Primordial Non-Gaussianity," *Astronomy and Astrophysics* 571 (2014): A24.

13 Yi Wang and Wei Xue, "Inflation and Alternatives with Blue Tensor Spectra," *Journal of Cosmology and Astroparticle Physics* 2014, no. 10 (2014): 75.

14 Bin Chen, Yi Wang, Wei Xue, and Robert Brandenberger, "String Gas Cosmology and Non-Gaussianities," *arXiv* preprint (2007): 0712.2477.

15 Xingang Chen, Abraham Loeb, and Zhong-Zhi Xianyu, "Unique Fingerprints of Alternatives to Inflation in the Primordial Power Spectrum," *Physical Review Letters* 122, no. 12 (2019): 121301.

16 Sunny Vagnozzi and Abraham Loeb, "The Challenge of Ruling Out Inflation via the Primordial Graviton Background," *Astrophysical Journal Letters* 939, no. 2 (2022): L22.

INDEX

Page numbers in italics refer to figures and tables.

Abedi, Jahed, 231
accelerating universe, 112, 157, 323
Adler Planetarium, 35
Agarwal, Abhineet, 230
Aguirre, Anthony, 59, 132
Agullo, Ivan, 148
Ahmadinejad, Mahmoud, 203
Albert, David, 88
Albrecht, Andreas: background of, 37; cyclic cosmology and, 179–81; de Sitter equilibrium and, 180–82, 284, 294, 299, *335*; multiverse and, 37–38, 41–42; Steinhardt and, 37; VSL and, 179–80, 212, 215
Alexander, Stephon, 296
Alexander the Great, 2, 6, 54
Allah, 3
"All You Zombies" (Heinlein), 261, 271
"Almost Perfect Universe, An" (Planck press release), 143
Alpher, Ralph, 17
Altamirano, Natacha, 206
Alters, Brian J., 302
American Astronomical Society, 48
Anaximander, 4, 89
Anaximenes, 89

Andromeda galaxy, 11, 21, 149
Anscombe, Elizabeth, 258
anti–de Sitter space (AdS): CFT and, 243–44, 248, 250, 258; holograms and, 243–46, 248, 250, 258
antiparticles, 33, 75, 78–80, 265, 274
Aperion, 89
arche, 3
Aristarchus, 305
Aristotle, 3, 6, 12
Arkani-Hamed, Nima, 122, 151, 238, 241; "Cosmological Collider Physics," 254
arXiv, 53, 219, 228
Ashtekar, Abhay: background of, 128; black holes and, 188; Bojowald and, 135–42; cutoff and, 137; Einstein and, 128–29, 151; Feynman and, 128; general relativity and, 129; Hartle and, 150; Lewandowski on, 128; loop quantum gravity and, 128–32, 135–42, 145–48, 151; mathematical rigor of, 141; Penrose and, 129; Singh and, 137–42, 146–47; singularities and, 135–40; Smolin and, 129–30, 132, 141, 188

Ashtekar variables, 129
Asimov, Isaac, 3
Aspect, Alain, 224
astrostatistics, 184
atheism, 21–22, 258, 273, 291, 299–301
atomic bomb, 24, 94, 171
atomic clocks, 284
atomists: competing worldviews and, 7; Democritus, 5; Greek, 6–7, 53–54, 74, 82; Heron of Alexandria, 82; Lucretius, 5; multiverse and, 53–54; Plato and, 6; quantum theory and, 74
atoms: black holes and, 185, 188, 190, 193–95; CMB and, 308, 311–12; cyclic cosmology and, 167, 171, 180; Hawking and, 72–73; holograms and, 254; multiverse and, 7, 32, 36, 45, 53–54, 63; Newton and, 7–8; plasma and, 17 (*see also* plasma); primeval, vii–viii, 8–13, 15–16, 24, 87; quantum theory and, 5, 28, 72–74, 82, 87, 94, 97–100, 193–95, 211, 254, 312; Rukeyser on, xii; speed of light and, 211; string theories and, 99–102; time and, 278; Vilenkin and, 87
Atum, 261

Babylonians, 3, 235
Background Imaging of Cosmic Extragalactic Polarization (BICEP): B-modes and, 63, *65*, 112, 117, 218, 313–20; CIB error of, 64–66; CMB and, 313–14, 316, 323; cyclic cosmology and, 62, 66, 112, 230, 303; Guth and, 62, 64–65; Kuo and, 63–64; Linde and, 63–64; multiverse and, 62–66; religion and, 289, 303; speed of light and, 218, 230; Steinhardt and, 62, 65–66, 303; string theories and, 111–12, 117, 124–25; 3 sigma and, 344n14; Turok and, 62, 303
Back to the Future (film), 265, 272
Bahcall, John N., 263
Balloon Observations of Millimetric Extragalactic Radiation and Geophysics (BOOMERanG), 51
Bardeen, James, 80
Barish, Barry, 309
Bars, Itzhak, 171, 228
Baum, Lauris, 166
Baum-Frampton model, *334*; comeback-empty principle and, 167–70, 181; cyclic cosmology and, 154, 166–67, 170, 181, 275, 284; Ludwick and, 168–69; religion and, 284, 294, 299; time and, 275; w parameter and, 166–67; Zhang and, 168
BBC, 16, 162
Beautiful Mind, A (film), 290
Before the Big Bang (Mersini-Houghton), 104
Bekenstein, Jacob: Big Bounce and, 131; black holes and, 79–81, 107, 131, 198, 204, 238–42, 255; CMB and, 324; entropy and, 79–81, 107, 131, 198, 204, 237–42; holograms and, 235–42, 255; religion and, 287; singularities and, 79–81; string theories and, 107
Belinski, Vladimir, 24
Bernard, Jean-Philippe, 63
Bessel, Friedrich, xiii
Bethe, Hans, 190
Bible, 3, 6, 21–22, 187, 207, 235
Big Bang: Big Bounce and, 127–52; bi-thermal, 224–30, 232, 289, 304, 318, *336*; black holes and, 183–208; CMB and, 307–28; concept of, vii–xiii; cyclic cosmology and, 153–82; final problem and, 325–28; geometric ripples from, 308–10;

Hawking and, 71–73 (*see also* Hawking, Stephen); historical perspective on, 1–30; holograms and, 235; Hoyle's invention of term, 16; as mankind's earliest memory, vii–viii; as mirage, x; models of, *333–36*; multiverse and, 31–70; pre–Big Bang and, 112–17 (*see also* pre–Big Bang model); religion and, 22, 281–86; singularities of, 28 (*see also* singularities); speed of light and, 209–33; as "standard model," 48; steady-state model and, 13–23; string theories and, 99–126; time and, 261–80

Big Bang Observer (BBO), 310–12, 319, 327

"Big Bang or Big Bounce?" (Bojowald), 140

Big Bang Theory, The (TV show), 30, 131, 256

Big Bounce: Bekenstein and, 131; black holes and, 128, 131, 137, 140, 150, *335*; CMB and, 143–50; COBE and, 143; cosmic forgetfulness and, 140–41; cosmological constant and, 133, 136–38; curvature and, 133, 138, 148; density and, 137, 139, 143; Dicke and, 151; Ekpyrosis and, 134; electrons and, 131, 141; expansion and, 134–38, 141, 149–51; galaxies and, 149; Gamow and, 151; Gaussian distribution and, 147–49; general relativity and, 127–33, 138; geometry and, 131, 133, 136; gravity and, 127–36, 140–42, 146, 149–51; Guth and, 139, 143; Hartle and, 15–20, 142–50; inflation and, 134–36, 140, 143, 146–47; internal strife over, 140–42; loop quantum gravity and, 127–52; NASA and, 143; numerics and, 139–40; Penrose and, 129; photons and, 143; Planck and, 143–44, 148–49; plasma and, 143–44; pre–Big Bang model and, 140–41, 144; probability and, 140, 145; quantum gravity and, 127–36, 140–51; quantum theory and, 127–51; radiation and, 131; Singh and, 137–42, 146–47; singularities and, 128, 132, 135–40, 142, 146; space-time and, *93*, 127, 133, 138, 143, 148; Spergel and, 143; Steinhardt and, 143; string gas cosmology and, 107–8; temperature and, 143–44, 147, 149; Turok and, 142; Veneziano and, 151; wave functions and, 141; WMAP and, 143

Big Crunch, 166, 200

Big Rip, 167, 328

Big Squeeze, 22

black-body spectrum, 45–46, 225–26

Black Eyed Peas, 155

"Black Hole at the Beginning of Time, The" (Afshordi, Mann, and Pourhasan), 206

black holes: accretion discs and, 48; Ashtekar and, 188; atoms and, 185, 188, 190, 193–95; Bekenstein and, 79–81, 107, 131, 198, 204, 238–42, 255; Big Bang models and, *335–36*; Big Bounce and, 128, 131, 137, 140, 150; Brandenberger and, 206, 211, 217; causal set theory and, 193–95; Chandrasekhar and, 184–85, 189; CMB and, 199, 202, 205, 207, 308–9, 320, 324; collisions of, 117, 176, 231, 308–9; cosmological constant and, 157, 165–66, 170, 175, 182; cosmological natural selection and, 187–96, 296, 322; creation and, 187, 201; curvature and, 185; cyclic cosmology and, 157, 160–61, 167, 171, 174, 176, 182; dark energy and,

black holes (*continued*)
193–96, 200–202; dark matter and, 193, 201; Darwinism and, 188–91, 196; density and, 185, 195–200; Eddington and, 185; entropy and, 198–201, 204–5, 207, 242; event horizon of, 79–81, 161, 198, 200, 204, 228, 242, 248–49, 267–68, 320; expansion and, 196–200, 205; flatness problem and, 199, 201; Friedmann and, ix; galaxies and, 186, 205; general relativity and, 186; geometry and, 183, 194, 204; gravity and, 24, 75, 79, 81, 100, 140, 150, 176, 184–202, 205, 214, 224, 228, 231, 240, 242, 248, 252; Hawking and, 80–83, 92, 107, 131, 161, 176, 186, 195, 197–98, 204, 207, 238–44, 248, 251, 268, 320, 324; Heisenberg and, 195; holograms and, 238–44, 248–52, 255; horizon problem and, 201; Hubble and, 196; inflation and, 187, 190–91, 198–99, 205; information paradox and, 230–33; LIGO and, 117, 176, 231–32; loop quantum gravity and, 128, 131, 137, 140, 150, 187, 192–94, 198; Michell and, 183–86, 208; multiverse and, 48, 68; Penrose and, 24, 79, 157, 160–61, 176, 185–86, 197, 200; Planck and, 197–99; quantum theory and, 75, 187, 191–98, 205; radiation and, 81, 205; religion and, 283; Rovelli and, 188, 192; Schwarzschild radius and, 184; Sciama and, 197; singularities and, 24–25, 79, 83, 92, 128, 176, 185–87, 192, 199–201, 204–5, 214, 228, 248–49, 268; Smolin and, 187–96, 202; space-time and, 187, 192–97, 200; speed of light and, 214, 219, 224, 228–32; Spergel and, 162, 202; standard model and, 188–89; string theories and, 100, 107, 117; Susskind and, 191; symmetry and, 185, 198, 204; temperature and, 185, 199, 202, 204–5; thermodynamics and, 200–201; time and, 267–68, 275, 277; torsion bounce and, 196–99, 322, *335*; vacuum and, 191, 198, 200, 205; Vilenkin and, 190–91, 196; Weinberg and, 195–96; white holes, 199–208; WMAP and, 202

Black Hole War, The (Susskind), 240
blogs, 64, 92
blueshift, 12
blue spectrum, 12, 111, 320–23
B-modes: BICEP and, 63, 65, 112, 117, 218, 313–20; CMB and, 313–20; multiverse and, 63, 65; speed of light and, 218; string theories and, 112, 117
Bohr, Niels, 76, 78, 87, 159, 249
Bojowald, Martin: Ashtekar and, 135–42; "Big Bang or Big Bounce?," 140; cosmic forgetfulness and, 140–41; loop quantum gravity and, 135–42, 146–47; *Once before Time*, 141; singularities and, 135–40
Boltzmann, Ludwig, 180
Bombelli, Rafael, 82
Bondi, Hermann, 15–16, 19–20, 44
Borde, Arvind, 58, 123, 292–93
Borde-Guth-Vilenkin (BGV) theorem, 58, 123–24, 292–93
bosons, 100, 123, 145, 170
Bousso, Raphael, 132
Boyle, Latham, 174
Brahma, 153
Brandenberger, Robert: background of, 106–7; black holes and, 206, 211, 217; CMB and, 320; holograms and, 247; Linde and, 109; loop quantum gravity and, 151;

string gas cosmology and, 106–13, 120, 211, 320; Vafa and, 106–12

branes: colliding, 119–24, 134, 154, 172–73, *334–35*; DGP gravity and, 201–2, 204; Ekpyrosis and, 118–23; higher-dimensional, 103, 118–19; Hořava-Witten theory and, 118–19; inflation and, 110; membranes and, 103, 225, 229, *334–36*; M-theory and, 103, 117–18, 154; Stoics and, 122; string theories and, 103, 110, 118–25, 134, 173, 201, 205, 219, 225, 228–29, 242, 246

Brief History of Time, A (Hawking), 273

Brown, Gerald, 190

bubbles: Big Bang models and, *333*; collisions and, 59–60; Hawking and, 71; multiverse and, 37, 42–44, 53, 59–60, 68; string theories and, 123, 144; vacuums and, 37, 42–43, 53, 59

Buddhism, 6–7, 153

Burke, Bernard, 23

Burton, Howard, 216–17

Bush, George W., 286

Caldwell, Robert, 166

Cannon, Annie Jump, 236

"Can the Universe Create Itself?" (Gott and Li), 270–71

Carrey, Jim, 120

Carroll, Sean: Big Bang model of, *335*; Chen and, 123, 179–80, 255, 284, 294, 299, *335*; cyclic cosmology and, 178–80; holograms and, 255–59; Penrose and, 178–80, 291, 294; religion and, 284, 291–95, 299, 301; time and, 272

Cartan, Élie, 128, 197

Carter, Brandon, 25–26, 80

Catholicism, 13, 21–22, 183, 290

Cauchy horizon, 267–72

causal patches, 167–68

causal set theory, 193–95

Cepheid variables, 11, 13, 21, 50

Chandrasekhar, Subrahmanyan, 50, 140, 162, 184–85, 189

chaos, 3, 6, 37, 41

ChatGPT, 187

Chen, Jennifer: Big Bang model of, *335*; Carroll and, 123, 179–80, 255, 284, 294, 299, *335*; cyclic cosmology and, 179–80; holograms and, 255; string theories and, 123

Chen, Xingang, 323–24

Chern-Simons quantum field theories, 315

Chibisov, Gennady, 40

Chowdhury, Debika, 68

Christianity, 6, 21, 287–90, 293, 297–98

Chung, Daniel, 213

Circle Limit (Escher), 159–60

Classical and Quantum Gravity (journal), 268

Clauser, John, 224

closed time-like curve (CTC), 262–63, 267–73

Cole, William, 183

come-back-empty principle, 167–70, 181

competition: Big Bang theories and, x–xi; black holes and, 187, 189, 191; CMB and, 284, 288, 296, 301, 303, 306, 314, 326; creation models and, 7; cyclic cosmology and, 164; general relativity, 223; gravitational waves, 230; Great Debate, 11; Hawking and, 85; Hoyle vs. Ryle, 18–21; multiverse and, 41, 43, 47, 63, 65; religion and, 281, 284, 288, 296, 301, 303, 306; string theories and, 108, 110, 121, 187; VSL and, 225

complex plane, 82–83

Comte, Auguste, xiii
Conformal Field Theory (CFT), 243–44, 248, 250, 258
Connecticut Yankee in King Arthur's Court, A (Twain), 261
conservation, 14, 86, 161
Copenhagen interpretation, 78
Coriano, Claudio, 253
Corichi, Alejandro, 139
Coriolis force, 109
Cornish, Neil, 310
cosmic infrared background (CIB), 64
Cosmic Landscape, The (Susskind), 245
cosmic microwave background (CMB): atoms and, 308, 311–12; BBO and, 310–12, 319, 327; Bekenstein and, 324; BICEP and, 313–14, 316, 323; Big Bounce and, 143–50; black holes and, 199, 202, 205, 207, 308–9, 320, 324; B-modes and, 313–20; Brandenberger and, 320; COBE, 86 (*see also* Cosmic Microwave Background Explorer (COBE)); creation and, 23; curvature and, 324–25; cyclic cosmology and, 156, 160–65, 174, 176, 181–82; dark energy and, 310, 324; DECIGO, 311–12, 319, 327; density and, 324; Dicke and, 22–23, 53; discovery of, 17–18; Ekpyrosis and, 320, 322–23; electromagnetism and, 312; expansion and, 307, 317, 321–27; final problem and, 325–28; flatness problem and, 36–37, 53, 199; galaxies and, 18, 29, 53, 149, 205, 227, 250, 263, 284, 307, 317, 321–25; Gamow and, 308; Gaussian distribution and, 147, 315, 321–25; geometry and, 308–10; Gott and, 263, 325; gravitational waves and, 279, 308–13, 316–27; ground observations of, 313–17; Guth and, 143, 321; Hawking and, 47, 319–20; holograms and, 250–54; horizon problem and, 29, 36, 225–26; Hoyle and, 17–18; inflation and, 314–28; Kelvin temperature of, 23, 29, 36; LIGO and, 309–13, 325; Linde and, 321, 324; LISA and, 309–12, 327; Loeb and, 323–24; loop quantum gravity and, 143; Magueijo and, 289, 318; McKellar and, 18, 23; Mersin-Houghton on, 104–5; multiverse and, 36, 39, 45–47, 51–53, 55, 59–63, 66, 69; NASA and, 45, 47, 52, 142, 212, 310, 317, 319, 322; non-Gaussianity and, 148–49, 229, 254, 279, 289, 315, 321–25; as oldest light in cosmos, 29, 307–8; omega and, 29; Penrose and, 60, 235–36, 320; Penzias and, 23, 60, 263, 312; photons and, 36, 51, 53, 55, 307; Planck and, 308–18, 322–23, 327; plasma and, 307–10, 316; pre-Big Bang model and, 310, 320; quantum theory and, 310, 312, 315, 321–22, 326; radiation of, 23, 307–32; religion and, 47, 284, 289; Sciama and, 160; like seeing God, 47; singularities and, 326–28; Skenderis and, 323; Smolin and, 322; Smoot and, 47; space observations of, 317–21; speed of light and, 212, 218, 224–29; Spergel and, 310, 314–17, 321; Steinhardt and, 315, 320, 323–24; string theories and, 104, 108–9, 117, 125; temperature and, 18, 23, 29, 36, 46–47, 51, 55, 143–44; 147, 149, 162, 182, 199, 202, 205, 224–26, 229, 252, *289*, 308, 318, 320, *333–36*; time and, 263, 275–79,

307–8; uniformity of, 29, 36, 39, 47; Veneziano and, 320, 326; Wilson and, 23, 60, 263, 312; WMAP and, *163*, 308–18, 322

Cosmic Microwave Background Explorer (COBE): Big Bounce and, 143; cyclic cosmology and, 156; DMR and, 46–47; FIRAS and, 46; first results of, 46; God and, 45–54; holograms and, 246; multiverse and, 45–54, 66; NASA and, 45–46, 52, 143; string theories and, 321; Vilenkin and, 86; WMAP and, *46*, 52, 143, 312, 314

"Cosmological Collider Physics" (Maldacena and Arkani-Hamed), 254

cosmological constant: Big Bounce and, 133, 136–38; as "biggest blunder," viii, 10, 27, 34, 49–50; black holes and, 157, 165–66, 170, 175, 182; cyclic cosmology and, 157, 165–66, 170, 175, 182; Einstein and, viii, 10, 15–17, 27, 49, 112, 133, 136, 170, 175, 214; expansion and, 10, 15–16, 49, 136–37, 165–66, 170, 175, 196, 200, 214, 296; holograms and, 246; multiverse and, 49–51, 54; religion and, 296–97, 305; space-time and, 10, 133, 192, 296; speed of light and, 213–14, 230; string theories and, 105, 112; symmetry and, 50

cosmological natural selection, *335*; black holes and, 187–96; religion and, 296; Smolin and, 187–96, 296, 322

"Cosmos May Show Echoes of Events before Big Bang" (Palmer), 162

COVID-19 pandemic, xi, 31, 114, 280

Cox, Brian, 60

Craig, William Lane, 273, 290–96

creation: Aperion and, 4; arche and, 3; atomists and, 5–7, 53–54, 74, 82; Babylonian, 3; bi-thermal, 230; black holes and, 187, 201; chaos and, 3, 6, 37, 41; Christian, 6; CMB and, 23; conservation law and, 86; continual, 187; cosmological theories and, 299–302; Egyptian, 3; Einstein field and, 16, 44; Empedocles and, 4–5; galaxies and, 2, 15–16, 39, 44, 115, 153, 167, 322; Genesis and, 6, 21, 187, 235; God and, 3, 6–8, 22; Greek, 6; Hawking and, 72, 300–301; Hindu, 7; holograms and, 251; Islamic, 7, 22; Japanese, 41; kosmoi, 4–5, 7, 30, 54; Mulla Sadra and, 22; no-boundary proposal and, 91–94; from nothing, 6–7, 73, 88, 94, 291, 300; perpetual, 22; quantum theory and, 88–91, 94, 230, 305; religion and, 291, 300–301, 305; spontaneous, 86–87, 91, 300–301

Curtis, Herbert, 11

curvature: Big Bounce and, 133, 138, 148; black holes and, 185; CMB and, 324–25; cyclic cosmology and, 172–73; Dicke and, 29, 34; expansion and, 10, 14, 138, 173–74, 324–25; Friedmann and, 24; in higher dimensions, 10; multiverse and, 34–36, 68; omega and, 29–30, 34–36, 48–51; religion and, 284, 289–91, 299; speed of light and, 228; string theories and, 115

Cuscuton: holograms and, 251; Hořava gravity and, 223–24, 283, 305; speed of light and, 213–19, 223–24, 230, 251, 283, 305

cutoffs, 137–38, 281–82

cyanogen, 18

cyclic cosmology: Albrecht and, 179–81; atoms and, 167, 171, 180;

cyclic cosmology (*continued*)
Baum-Frampton model and, 154, 166–67, 170, 181, 275, 284; BICEP and, 62, 66, 112, 230, 303; Big Crunch and, 166, 200; Big Rip and, 167; black holes and, 157, 160–61, 167, 171, 174, 176, 182; Carroll and, 178–80; causal patches and, 167–68; CMB and, 156, 160–65, 174, 176, 181–82; COBE and, 156; come-back-empty principle and, 167–70, 181; conformal (CCC), 157–67, 170, 181–82, 227, 243, 275–78, 282–83, 294, 320, *334*; cosmological constant and, 157, 165–66, 170, 175, 182; curvature and, 172–73; dark energy and, 156, 166–70, 175–76, 181–82; dark matter and, 160; density and, 166, 168, 174–77, 181; dualities and, 174; Ekpyrosis and, 6, 153, 171–74; electrons and, 156; Empedocles and, 4–5; entropy and, 154–56, 160–61, 166–67, 177–82; equilibrium and, 180–82; expansion and, 154–60, 165–67, 170–81; galaxies and, 153, 167, 180; geometry and, 157, 160, 163–64, 176; gravity and, 156, 162, 166, 175–76, 181; Guth and, 165, 174, 178; Hawking and, 154, 156, 161, 163; Hubble and, 310–11; Ijjas and, 172–77, 180–81; inflation and, 154, 160, 165–69, 172–81; LIGO and, 176; Little Rip and, 168; loop quantum gravity and, 167, 321–22, 326; monopoles and, 174; NASA and, 163; Penrose and, 154–67, 171–81; phantom menace and, 166–67; photons and, 36; Planck and, 175; pre–Big Bang model and, 160; probability and, 154; quantum theory and, 159–61, 167, 175–79; radiation and, 157, 161, 167; Sciama and, 157, 160; singularities and, 154–57, 171, 173, 176, 182; spectrums and, 156, 174, 176; standard model and, 170; steady-state model and, 160; Steinhardt and, 157, 171–77, 181; symmetry and, 173; temperature and, 162, 182; thermodynamics and, 155, 179–80; Turok and, 170–72, 174, 303; vacuum and, 170–71, 177, 180; WMAP and, 163

Danu, 3
Darius the Great, 5
Dark (TV show), 269
dark energy: Big Bang models and, *334*; black holes and, 193–96, 200–202; CMB and, 310, 324; come-back-empty principle and, 167–70; cyclic cosmology and, 156, 166–70, 175–76, 181–82; gravity and, 27, 52, 57, 166, 176, 181, 193, 200–202, 222, 245, 279, 287, 310; holograms and, 245–46; multiverse and, 51–53, 57; phantom, 166–70, 181–82, 324, *334*; religion and, 282, 297; speed of light and, 214–15, 219; standard model and, *101*; string theories and, 112, 121; symmetry and, 53; time and, 275, 278–79
dark matter: Bekenstein and, 238, 287; black holes and, 193, 201; cyclic cosmology and, 160; erebons and, 160; holograms and, 237–38, 256; Moffat and, 238, 287; speed of light and, 214–15; standard model and, *101*; w parameter and, 166–67
Darwin, Charles: black holes and, 188–91, 196; grave of, 81; Hawking and, 81; multiverse and, 57; religion and, 296, 300

Dawkins, Richard, 188, 300
Deci-Hertz Interferometer Gravitational Wave Observatory (DECIGO), 311–12, 319, 327
deGrasse Tyson, Neil, 264
Delle Rose, Luigi, 253
Democritus, 5–6
density: Big Bounce and, 137, 139, 143; black holes and, 185, 195–200; CMB and, 324; constant, 15–16, 40, 44, 49, 137, 166, 200; cyclic cosmology and, 166, 168, 174–77, 181; galaxies and, 2, 15–16, 19, 29, 40, 44; Hawking and, 72, 83, 89; holograms and, 245–46; Hoyle and, 16; infinite, 2, 72, 83, 185, 200, 285, 291; multiverse and, 32, 34, 40, 44, 49, 52–53; Planck and, 175, 197–98; religion and, 285, 291; singularities and, 2, 24, 72, 83, 114, 176, 185, 200, 291; speed of light and, 225, 230; steady-state model, 19, 44; string theories and, 114, 120, 123; temperature and, 2, 24, 72, 120, 143, 185, 225, 291
Desai, Shantanu, 199
Descartes, René, 83
de Sitter, Willem, 180
de Sitter equilibrium, 180–82, 284, 294, 299, *335*
deuterium, 17, 26
Dewitt, Bryce, 187
DGP gravity, 201–2, 204
Dicke, Bob: background of, 22; Big Bounce and, 151; CMB and, 22–23, 53; curvature and, 29, 34; flatness problem and, 29–30, 34–35, 53; Gamow and, 22–23; Guth and, 30, 34; loop quantum gravity and, 151; multiverse and, 34–35, 53; omega and, 29–30, 34
Differential Microwave Radiometers (DMR), 46–47

Dijkgraaf, Robbert, 228–29
Dingle, Herbert, 17
Dirac medal, 62
Doctor Who (TV show), 263
"Does History Repeat Itself?" (Afshordi and Gould), 277–78
Doppler effect, 12
Doroshkevich, Andrei, 23
double-slit experiment, 76–78
Dowker, Fay, 195
dualities: cyclic cosmology and, 174; holograms and, 243, 250, 252; particle-wave, 73; religion and, 289; string theories and, 113–16; T, 113–14, 116, 250, 275; time and, 275; Veneziano and, 113–14
Duhem-Quine thesis, 303, 315
Dvali, Giorgi (Gia), 60, 110, 201–2, 204, 220
Dykaar, Hannah, 231

Eddington, Arthur: black holes and, 185; eclipse journey of, 8, 99, 284; general relativity and, 284; primeval atom and, 8; string theories and, 99
Edge of All We Know, The (film), 88
Efstathiou, George, 61
Egyptians, 3, 261, 286
Einstein, Albert: Ashtekar and, 128–29, 151; Big Bang models and, *333*; biggest blunder of, viii, 10, 27, 34, 49–50; cosmological constant and, viii, 10, 15–17, 27, 49, 112, 133, 136, 170, 175, 214; creation field and, 16, 44; expansion and, 10, 13–16, 49–50, 116, 136, 138, 151, 170, 175, 197, 209, 211, 214, 224; final problem and, 325; Gamow and, 86, 151; general relativity and, viii, 9–10, 27, 34, 79, 86, 99, 127–33, 138, 210, 220, 229, 237, 261; God and,

Einstein, Albert (*continued*)
87; gravity and, 9, 24, 26, 32, 34, 50, 79, 86, 97, 99, 127, 129, 133, 151, 193, 196–97, 209, 214, 220–24, 232, 238, 240, 256, 269, 291, 326; Hubble and, 13, 16, 50; Newton and, 10, 14, 133, 238; quantum theory and, 9, 34, 74, 79, 86–87, 97–100, 104, 127–30, 133, 136, 138, 151, 161, 167, 193, 196, 209, 211, 220–24, 232, 235, 240, 255–56, 312, 326; Smolin and, 129–30, 133, 220; space-time and, viii, 9–10, 32, 99, 127, 133, 138, 193, 197, 262; special relativity and, 9, 223, 265; time and, 261–69, 273; unstable universe and, 10

Einstein-Cartan gravity, 197

Ekpyrosis: BGV theorem and, 124; Big Bang models and, *334–35*; Big Bounce and, 134; branes and, 118–23; Carrey and, 120; Carroll-Chen model and, 123; CMB and, 320, 322–23; cyclic cosmology and, 153, 171–74; Guth and, 122–23; Hawking and, 120; Hořava-Witten theory and, 118–19; Linde and, 120–21; loop quantum gravity and, 134; pre-Big Bang model and, 120, 123, 125; press attention of, 120; speed of light and, 217, 219, 226, 230; string theories and, 6, 101, 117–25, 134, 153, 171–74, 217, 219, 226, 230, 320, 322–23, *334–35*; time and, 123; Turok and, 117–22; Vilenkin and, 123

electromagnetism: blue wavelengths, 12; CMB and, 312; light wave behavior and, 8, 32; multiverse and, 62; red wavelengths and, 12; spectrum of, 12 (*see also* spectrum); string theories and, 100, 117

electron mass, 131, 156

electrons: Big Bounce and, 131, 141; cosmic rays and, 63; cyclic cosmology and, 156; double-slit experiment and, 76–78; Feynman and, 76, 82; holograms and, 250; multiverse and, 45, 63; plasma and, 17; quantum theory and, 74–79; scattering, 45; string theories and, 102; time and, 265; Vilenkin and, 96

Ellis, George: Hawking and, 58, 108; multiverse and, 58–59; Sciama and, 25, 212; speed of light and, 212–13, 226, 233; string theories and, 108

emergent universe, 108, *334*

Emparan, Roberto, 273

Empedocles, 4–5, 7, 10

entropy: Bekenstein and, 79–81, 107, 131, 198, 204, 237–42; Big Bang models and, *335*; Big Bounce and, 131, 137, 140, 149–50; black holes and, 198–201, 204–5, 207, 242; Carroll-Chen model and, 123, *335*; CMB and, 324; cyclic cosmology and, 154–56, 160–61, 166–67, 177–82; event horizon and, 79–80, 161, 177, 198, 242, 248; Hawking and, 80–81, 93, 107, 131, 154, 161, 198, 204, 207, 237–42, 248, 324; high, 154–56, 166, 239, 294; holograms and, 237–42, 248, 255; low, 123, 150, 154, 156, 160, 166–67, 177–82, 294, 297; as measure of disorder, 79–80; no-boundary condition and, 93, 95; paradox of, 161, 177–82, 239–40, 294; quantum theory and, 81–82, 95, 137, 140, 161, 179, 239–40, 242, 248, 255, 282, 294; religion and, 282, 294, 297; singularities and, 79–82, 154, 156, 182, 205, 248, 282, 294; string gas cosmology and, 123, 282; string theories and, 107, 123; temperature and, 80–81, 182, 204–5; thermodynamics and, 80–81, 123, 155, 180, 200, 238–40, 294; time and, 93, 107, 123, 150, 155–56, 160, 166, 168, 178, 180, 198, 200–201, 242, 277, 282, 294

Enūma Eliš, 3

EP = EPR, 255–56, 259

equilibrium: cyclic cosmology and, 180–82; de Sitter, 180–82, 284, 294, 299, *335*; horizon problem and, 36, 211, 226; multiverse and, 36, 45; religion and, 284, 294, 299; speed of light and, 211, 225–27; string theories and, 107–8, 123; thermal, 45, 107–8, 211, 225–27

erebons, 160

Erhard, Werner, 239

Eriksen, Hans, 162

Escher, M. C., 159–60

ether: Michelson-Morley experiment and, 8–9, 309; speed of light and, 8–9, 214, 224, 232–33

Euclid satellite, 279

European Association for Nuclear Research (CERN), 113, 170–71, 307, 327

European Space Agency (ESA), 309, 318–19, 327

event horizon: black holes and, 79–81, 161, 198, 200, 204, 228, 242, 248–49, 267–68, 320; entropy and, 79–80, 161, 177, 198, 242, 248

Everett, Hugh, 78, 256

expansion: accelerating universe and, 112, 157, 323; Big Bang models and, *333–36*; Big Bounce and, 134–38, 141, 149–51; black holes and, 196–200, 205; Bombelli and, 82; Bondi and, 16; bubble collisions and, 59–60; CMB and, 307, 317, 321–27; cosmological constant and, 10, 15–16, 49, 136–37, 165–66, 170, 175, 196, 200, 214, 296; curvature and, 10, 14, 138, 173–74, 324–25; cyclic cosmology and, 154–60, 165–67, 170–81; Dicke and, 22, 29–30; Einstein and, 10, 13–16, 49–50, 116, 136, 138, 151, 170, 175, 197, 209, 211, 214, 224; exponential, 37–39, 43, 48–49, 54, 57, 59, 69, 116, 122, 165, 175, 211, 225, 283, 325; Gamow and, 16; Gold and, 16; Hawking and, 25, 71, 92, 96–97; historical perspective and, 243–46; holograms and, 243–46, 257; Hoyle and, 15–16; Hubble and, 12–13, 16, 50, 196, 232; Janus universe and, 93; Lehners and, 96–97; Lemaître and, 11–13, 15, 25, 293; magnetic monopoles and, 28; multiverse and, 32–34, 37–39, 43–44, 48–51, 54, 57, 59, 62, 69; no-boundary condition and, 71–73, 83–85, 89–97, 150, 299–301, 322, *333*; Planck and, 30, 60–62; protons and, 17; quantum theory and, 30, 32, 39, 57, 94, 96, 115, 120, 134–38, 141, 149, 151, 175, 179, 196, 209, 211, 217, 224, 226, 243, 257, 283, 302, *333*; redshift and, 12, 14, 52, 225, 279; religion and, 283, 293, 296, 304; singularities and, 2, 25–28, 37, 57, 69, 113–14, 138, 154, 156, 200, 205, 209, 257, 293; Slipher data and, 11–12; speed of light and, 209, 211, 214, 217, 224–27, 232; string theories and, 107–16, 119–23; time and, 275–76, 279

explanatory power, 282–83, 304

Faraday, Michael, 8

Far Infrared Absolute Spectrophotometer (FIRAS), 46

Feeney, Stephen, 59
Feldbrugge, Job, 94
Fermi gamma-ray observatory, 233
Ferreira, Pedro, 19
Feynman, Richard: Ashtekar and, 128; definition of science by, 305–6; double-slit experiment and, 76; electrons and, 76, 82; Hawking and, 82, 240; *Lectures on Physics*, 128; no-boundary condition and, 83; path integral of, 82–83; quantum theory and, 76, 240, 266; textbook error of, 128; thought experiment of, 76; wave functions and, 76, 82–83; Wheeler and, 82
Fiat lux, 22
firewall paradox, 161, 230, 240, 249, 283
First Principles (Burton), 216–17
5 sigma, 63, 145, 303, 318, 355n8
flatness problem: black holes and, 199, 201; CMB and, 36–37, 53, 199; Dicke and, 29–30, 34–35, 53; Guth and, 36–37; multiverse and, 34–37, 53, 56; no-boundary condition and, 299; omega and, 34–36; religion and, 297, 299; speed of light and, 227; string theories and, 110, 121–22; time and, 278
Flauger, Raphael, 64
Foundational Questions Institute Conference, 131–32
4D black hole, *336*
Fowler, Alfred, 185
Frampton, Paul, 166–69. *See also* Baum-Frampton model
Friedmann, Alexander, viii–iv, 12, 23–24, 52, 154
From Lenin to Gorbachev (Crowley and Vaillancourt), 273
Fuller, Buckminster, 129
Furley, David, 4

Gabadadze, Gregory, 201–2, 204
galaxies: Andromeda, 11, 21, 149; Big Bounce and, 149; black holes and, 186, 205; CMB and, 18, 29, 53, 149, 205, 227, 250, 263, 284, 321–22, 325; creation and, 2, 15–16, 39, 44, 115, 153, 167, 322; cyclic cosmology and, 153, 167, 180; density and, 2, 15–16, 19, 29, 40, 44; distribution of, 18–19, 26, 39, 250, 284, 321–22; formation of, 2; holograms and, 238, 250; Mersin-Houghton on, 104–5; Milky Way, 4, 11, 44, 149, 276, 303; multiverse and, 39–40, 44, 48, 50, 53, 63, 65; radio waves and, 18–19, 63; redshift and, viii, 12, 14, 52, 225, 279; religion and, 282, 284, 286; Slipher data and, 11–12; speed of light and, 211–12, 219, 227; steady-state model and, 13, 16, 18–21, 44; string theories and, 99, 105, 113, 115; time and, 263, 267, 275, 277–78
Galileo, 64
Gambini, Rudolfo, 151
Gamow, Barbara, 20–21
Gamow, George: Alpher and, 17; Big Bounce and, 151; Big Squeeze and, 22; CMB and, 308, 328; Dicke and, 22–23; Einstein and, 86, 151; Genesis and, 21; Hilbert and, 221; Hoyle and, 16–18, 20; loop quantum gravity and, 151; speed of light and, 221
Garwin, Dick, 309
Gasperini, Maurizio, 115, 320
Gaussian distribution: Big Bounce and, 147–49; CMB and, 315, 321–25; holograms and, 254; multiverse and, 55–56, 61; non-Gaussianity and, 148–49, 229, 254, 279, 289, 315, 321–25, 351n20;

religion and, 289; speed of light and, 229; time and, 279
Gell-Mann, Murray, 84
general relativity: Ashtekar and, 129; Big Bounce and, 127–33, 138; black holes and, 186; Eddington eclipse journey and, 8, 99, 284; Einstein and, viii, 9–10, 27, 34, 79, 86, 99, 127–33, 138, 210, 220, 229, 237, 261; gravity and, 9–10, 28–29, 34, 79, 86, 99, 127–33, 220, 223, 230, 242, 252; Hawking and, 79, 86; holograms and, 237, 240, 242, 252; Lemaître and, 12; multiverse and, 34; omega and, 29; probability and, 28; quantum mechanics contradiction and, 99–100; religion and, 282–83, 292; scale and, 27–28; speed of light and, 210, 220, 223, 229–30; string theories and, 99; time and, 261, 264, 267–71; unification of, 9–10; universal stability and, 10; Wheeler and, 79, 82, 186, 188, 195
Genesis, Bible book of, 6, 21, 187, 207, 235
Genghis Khan, 2
geocentrists, 64
geometry: Big Bounce and, 131, 133, 136; black holes and, 183, 194, 204; CMB and, 308–10; cyclic cosmology and, 157, 160, 163–64, 176; gravity and, 10, 34, 55, 68, 131, 133, 136, 176, 183, 194, 220, 255–56, 308, 310; holograms and, 235, 255–57, 259; as illusion, 259; multiverse and, 34–35, 48, 51, 55, 68; no-boundary proposal and, 84; Pythagoras and, 4; quantum theory and, 84, 88, 90, 92, 131, 133, 136, 176, 194, 220, 235, 255, 257, 259, 310; ripples and, 308–10; space-time and, 10, 55, 84, 92, 133, 220, 259, 308–10; speed of light and, 220, 229

Gibbons, Gary, 39–40
Glashow, Sheldon Lee, 32, 104
God, 128, 279; Aristotle on, 6; atheist scientists and, 273; belief in one, 211; Bible and, 3, 6, 21–22, 187, 207, 235; Carroll and, 293; CMB and, 47; COBE and, 45–54; Craig and, 273, 291, 293, 295; creation and, 3, 6–8, 22; disproving creator and, 299–302; Egyptians and, 261; Einstein and, 87; fine-tuning and, 295–99; Gott and, 273; Halvorson and, 297–99; Harris and, 291; Hawking and, 72, 84, 273, 300–301; Hinduism and, 153; multiverse and, 47, 58; Plato on, 6; religion and, 287, 291; tower of Babel and, 235; Vilenkin and, 89
Gödel, Kurt, 262–63
Gold, Thomas, 15–20, 44
Good, I. J., 186–87
Gott, Rich: Big Bang model of, 336; "Can the Universe Create Itself?," 270–71; CMB and, 263, 325; deGrasse Tyson and, 264; final problem and, 325; Hartle-Hawking model and, 269–70; Hawking and, 28; Li and, 268–73, 325, 336; multiverse and, 42; religion and, 287; Strauss and, 264; string theories and, 268; time and, 263–64, 268–76; *Welcome to the Universe*, 264
Gould, Beth: black holes and, 207; "Does History Repeat Itself?," 277; holograms and, 251, 253, 258–59; "New Views on the Cosmological Big Bang," 280; time and, 274–75, 277, 279–80
Gould, Stephen Jay, 188

Grand Design, The (Hawking and Mlodinow), 91–92, 300
Grand Unified Theories, 32, 120
Gravitation (Wheeler, Misner, and Thorne), 79
gravitational constant, 183
gravitational waves: black holes and, 201; CMB and, 308–13, 316–27; cyclic cosmology and, 162, 176, 181; DECIGO and, 311–12, 319, 327; LIGO and, 117, 176, 231–33, 279, 309–13, 325; multiverse and, 55, 62–63, 65, 68; Planck and, 95; religion and, 284, 289, 302; speed of light and, 213, 227, 230–33; string theories and, 111–12, 117, 124–25; time and, 279
gravitons, 100, 102, 324
gravity: as attractive force, viii; Big Bounce and, 127–36, 140–42, 146, 149–51; black holes and, 24, 75, 79, 81, 100, 140, 150, 176, 184–202, 205, 214, 224, 228, 231, 240, 242, 248, 252; CMB and, 308–13, 316–27; cyclic cosmology and, 156, 162, 166, 175–76, 181; dark energy and, 27, 52, 57, 166, 176, 181, 193, 200–202, 222, 245, 279, 287, 310; Einstein and, 9, 24, 26, 32, 34, 50, 79, 86, 97, 99, 127, 129, 133, 151, 193, 196–97, 209, 214, 220–24, 232, 238, 240, 256, 269, 291, 326; event horizon and, 79–81, 161, 198, 200, 204, 228, 242, 248–49, 267–68, 320; general relativity and, 9–10, 28–29, 34, 79, 86, 99, 127–33, 220, 223, 230, 242, 252; geometry and, 10, 34, 55, 68, 131, 133, 136, 176, 183, 194, 220, 255–56, 308, 310; Hawking and, 25–27, 57, 71, 75, 79, 81, 86–87, 91, 95, 97, 120, 176, 197, 240, 242–43, 248, 268–69, 292, 326; holograms and, 238–52, 255–58; loop quantum, 135 (*see also* loop quantum gravity); multiverse and, 32–36, 50, 52, 55, 57, 62–63, 65, 68; Newton and, 7–8, 71, 81, 128, 133, 183, 238; Penrose and, 25–26, 57, 79, 156, 162, 176, 181, 197, 227, 243, 279, 282, 292, 320, 325–26; Planck time and, 30; quantizing, 127; quantum, 28, 310 (*see also* quantum gravity); religion and, 282–84, 287–92, 302, 305–6; Schwarzschild radius and, 184; singularities and, 24, 28, 57, 75, 79, 114, 128, 132, 136, 146, 176, 185–86, 192, 200, 209, 214, 227–28, 247–48, 291, 326; space-time and, 9–10, 28, 32, 50, 52, 55, 62, 87, 99, 127, 130, 133, 166, 192–93, 197, 200, 219–21, 227, 231, 242, 248–50, 258, 269, 284, 306–12, 325; speed of light and, 210, 213–33; string theories and, 99–102, 105, 111–17, 120, 124–25; time and, 269, 275, 279

Great Debate, 11
Green's function, 95–96
Greenspan, Alan, 132
Gross, David, 132
Groundhog Day (film), 267, 269
Gurzadyan, Vahe, 162, 164
Guth, Alan, vii; BICEP and, 62, 64–65; BGV theorem and, 58, 123–24, 292–93; Big Bounce and, 139, 143; CMB and, 143, 321; cyclic cosmology and, 165, 174, 178; Dicke and, 30, 34; Ekpyrosis and, 122–23; flatness problem and, 36–37; holograms and, 244; horizon problem and, 36, 211; Linde and, 37–38; loop quantum gravity and, 139, 143; multiverse and, 32–49, 53–54, 57–58, 61–69; Nayeri and, 108–9; omega and, 30, 34, 49;

religion and, 292–94, 297–99; Sher and, 139; speed of light and, 211; Steinhardt and, 36–37; string theories and, 108, 122–23; time and, 271, 279; Tye and, 32–33; Vilenkin and, 42, 44, 49, 58, 68, 123, 292–93, 299

Guven, Jemal, 43, 271

Hajian, Amir, 162
Hale telescope, 18–19
Halliwell, Jonathan, 90–91, 95–96
Halvorson, Hans, 297–98
Harris, Sam, 291
Harrison, Edward, 69
Hartle, James: Big Bang model of, *333*; Big Bounce and, 15–20, 142–50; Hawking and, 82–85, 88, 90–93, 96, 150, 269–70, 299, *333*; loop quantum gravity and, 150; no-boundary condition and, 82–85, 88–96, 142, 150, 253, 269–70, 299, *333*; time and, 253; Wick rotation and, 83
Harvard Computers, 11, 236–37
Harvard-Smithsonian Center for Astrophysics, 236
Hawking, Jane, 72
Hawking, Stephen: atoms and, 72–73; Big Bang model of, *333*; black holes and, 80–83, 92, 107, 131, 161, 176, 186, 195–98, 204, 207, 238–44, 248, 251, 268, 320, 324; *A Brief History of Time*, 273; bubbles and, 71; CMB and, 47, 319–20; creation and, 72, 300–301; cyclic cosmology and, 154, 156, 161, 163, 171, 176; Darwinism and, 81; density and, 72, 83, 89; Ekpyrosis and, 120; Ellis and, 58, 108; entropy and, 80–81, 93, 107, 131, 154, 161, 198, 204, 207, 237–42, 248, 324; expansion and, 25, 71, 92, 96–97; Feynman and, 82, 240; final problem and, 326; general relativity and, 79, 86; God and, 72, 84, 273, 300–301; Gott and, 28; *The Grand Design*, 91–92; gravity and, 25–27, 57, 71, 75, 79, 81, 86–87, 91, 95, 97, 120, 176, 197, 240–43, 248, 268–69, 292, 326; Hartle and, 82–85, 88, 90–93, 96, 150, 269–70, 299, *333*; holograms and, 237–44, 248, 251, 254; inflation and, 71, 95; Linde and, 38–39, 58, 120; loop quantum gravity and, 130–31, 143, 150; Newton and, 81; no-boundary condition and, 71–73, 83–85, 89–97, 150, 299–301, 322, *333*; at Paralympics, 71–72; Penrose and, 25–27, 30, 38, 57, 68, 86, 107, 154, 156, 161, 171, 176, 186, 197, 243, 254, 292, 326; Perry and, 87–88; probability and, 74–78, 82, 91; protons and, 92; quantum theory and, 72–97; radiation and, 81; religion and, 292, 299–301; Sciama and, 26, 72; singularities and, 25–26, 30, 57, 69, 72–73, 83, 92, 107, 154, 156, 171, 176, 186, 248, 254, 269–70, 326; space-time and, 84, 92–93; temperature and, 80–81, 320; *The Theory of Everything*, 25, 72; thermodynamics and, 80–81, 238–40; time and, 268–70, 273; U-turn of, 72–73, 154; vacuum and, 75, 79–80, 86–88, 96; at Vatican conference, 84; Vilenkin and, 73, 86–91, 95–96, 292, 299; Wick rotation and, 83
Heinlein, Robert, 261, 271
Heisenberg, Werner: black holes and, 195; Bohr and, 76, 78; quantum theory and, 74–76, 78; string theories and, 99; time and, 275; uncertainty principle of, 74–75, 86, 99, 195, 275

helium, 2, 17, 26, 236
Herman, Robert, 17, 22, 45
Herodotus, 3
Heron of Alexandria, 82
Hertog, Thomas, 93
Hidden Figures (Shetterly), 237
Higgins, William, xiii
Higgs Bang, 171, 181, *334*
Higgs boson, 123, 145, 170
Higgs Centre for Theoretical Physics, 172
Higgs field, 170–71, 175, 283, *334*
Higgs mechanism, 197
Hilbert, David, 14, 128, 221, 257, *336*
Hill, Colin, 64
Hinduism, 3, 6–7, 41, 153, 304
Hiscock, William, 271–72
holograms: AdS/CFT, 243–44, 248, 250, 258; atoms and, 254; Bekenstein and, 235–42, 255; Big Bang models and, *336*; birth of holography, 238–41; black holes and, 238–44, 248–52, 255; Brandenberger and, 247; Carroll and, 255–59; CMB and, 250–54; COBE and, 246; cosmological constant and, 246; creation and, 251; Cuscuton and, 251; dark energy and, 245–46; dark matter and, 237–38, 256; density and, 245–46; dualities and, 243, 250, 252; electrons and, 250; entropy and, 237–42, 248, 255; EP = EPR and, 255–56, 259; expansion and, 243–46, 257; future of, 254–59; galaxies and, 238, 250; Gaussian distribution and, 254; general relativity and, 237, 240, 242, 252; geometry and, 235, 255–57, 259; gravity and, 238–52, 255–58; Guth and, 244; Hawking and, 237–44, 248, 251, 254; holographic cosmology, 247–54; Ijjas and, 254; inflation and, 246–47, 250–54, 257; KKLMMT construction, 246; Leavitt and, 237; Linde and, 246; Loeb and, 254; loop quantum gravity and, 253; Maldacena and, 241–46, 252–55, 259; monopoles and, 244; NASA and, 237; Penrose and, 243, 254; Planck and, 245, 252–53; quantum gravity and, 242, 247–48, 251–52, 255–58; quantum mechanics and, 235, 239–42, 248–49, 255–58; radiation and, 240, 248; Rovelli and, 245; singularities and, 247–49, 254, 257; Skenderis and, 247–53, 258; space-time and, 254–59; Spergel and, 246; standard model and, 252–53; Steinhardt and, 254; Susskind and, 240, 243–49, 255, 259; symmetry and, 257–58; temperature and, 252, 254; thermodynamics and, 238–40; Turok and, 247, 250; vacuum and, 243, 250, 258; wave functions and, 256–58; Weinberg and, 244
"Holographic Cosmology" (workshop), 247
Hořava gravity, 326; Cuscuton and, 223–24, 283, 305; speed of light and, 219–24, 227, 230–33, 282
Hořava-Witten theory, 118–19, 219
horizon problem: black holes and, 201; CMB and, 29, 36, 225–26; equilibrium and, 36, 211, 226; Guth and, 36, 211; speed of light and, 211, 213, 224–26; string theories and, 108, 120; time and, 278
Hossenfelder, Sabine, xii–xiii, 58
Hoyle, Fred: "Big Bang" coined by, 16; Bondi and, 15–16, 19–20, 44; CMB and, 17–18; density and, 16; expansion and, 15–16; Gamow and, 16–18, 20; Gold and, 15–20, 44;

multiverse and, 44; Penrose and, 24; radio broadcasts of, 24; religion and, 21–22; Ryle and, 18–21; speed of light and, 216; steady-state model and, 44, 216, 319

Hubble, Edwin: black holes and, 196; CMB and, 310–11; distance measurement and, 11–12, 50; Einstein and, 13, 16, 50; expansion and, 12–13, 16, 50, 196, 232; multiverse and, 50; redshift and, viii, 12; speed of light and, 232

Hubble constant, 14

Hubble Telescope, 311

Humason, Milton, 12, 50

Huxley, Thomas Henry, 105

hydrogen, 2, 16–17, 26, 236, 254

Ijjas, Anna: cyclic cosmology and, 172–77, 180–81; holograms and, 254; multiverse and, 61, 65–69; religion and, 287

imaginary numbers, 82–83, 92

imaginary time, 92, 270

Inception (film), 160

inflation: accelerating universe and, 112, 157, 323; beginning theories of, 31–42; BICEP error and, 64–66; Big Bang models and, *333–36*; Big Bounce and, 134–36, 140, 143, 146–47; black holes and, 187, 190–91, 198–99, 205; branes and, 110; bubble universes and, 123; CMB and, 314–28; cyclic cosmology and, 154, 160, 165–69, 172–81; entropy and, 131, 137, 140, 149–50; eternal, 42–45, 54, 59, 61, 64, 71, 78, 90, 106, 122–25, 178, 181, 190–91, 218, 251, 295–96, 324, *333*, *335*; fundamental forces and, 32; Hawking and, 71, 95; holograms and, 246–47, 250–54, 257; as mainstream cosmology, 31; multiverse and, 31–69; no-boundary condition and, 71–73, 83–85, 89–97, 150, 299–301, 322, *333*; Planck's proof of, 60–62; quantum theory and, 39–40, 47, 57, 75, 78, 87, 90–91, 95, 100, 104, 115, 120, 124, 134, 136, 140, 146, 176, 179, 198, 205, 209–11, 217–20, 247, 250–52, 257, 268, 282–83, 288, 292, 299, 304–5, 321–22, 326; religion and, 282–92, 295–99, 302–6; space-time and, 41, 219, 284, 292, 296; speed of light and, 209–13, 217–20, 223–27, 230; string gas cosmology and, 101, 108–12, 115–16, 120, 123, 125, 134, 199, 205, 211, 282, 320, 323; string theories and, 100–101, 104–12, 115–25; superinflation and, 136, 140; time and, 263, 268, 270–71, 275–79; Vilenkin and, 42–44, 49, 58, 68, 87, 90–91, 94, 123, 190, 218, 292, 299

information paradox, 161, 230–33, 240, 249, 283

interferometers, 320; DECIGO, 311–12; LIGO, 117, 176, 231–32, 279, 309–13, 325; Michelson-Morley experiment and, 8–9

Interstellar (film), 248–49

Ionian League, 5

Iran hostage crisis, 202, 286

Iranian Islamic revolution, 216, 241, 286

Iran-Iraq war, 1, 203

Iraq, 1, 203, 286–87, 290

Islam, 1; creation and, 3; European massacres and, 118; Golden Age of, 7; Helli, 221; Iranian Islamic revolution, 216, 241, 286; jinn, 266; Mulla Sadra, 22; Quran, 207, 266, 281; Shia, 287–88; Sunni, 288; terrorism and, 286; theocracy of, 286

island universes, 11

Jackson, Mary, 237
Jacobson, Ted, 220
Jainism, 153
James Webb Space Telescope (JWST), 307
Janus universe, 93, 171, 178–79, *335*
Jet Propulsion Laboratory, 322
Jews, 6, 79, 85, 184, 288
jinn particles, 266, 274–75
Jodrell Bank telescope, 21
Johnson, Katherine, 237
Jönsson, Claus, 76

Kachru, Shamit, 246
Kaiser, David, 61, 66
Kalam cosmological argument, 291–92
Kallosh, Renata, 38, 63–64, 120, 246
Kant, Immanuel, 11
Keating, Brian, 111–12
Khoury, Justin, 119
Kibble, Tom, 197
kinematics, 133
Kinney, Will, 104
Kiowa Apache, 6–7
Kirchhoff, Gustav Robert, 45
Kite Runner, The (Hosseini), 274
KKLMMT construction, 246
Klein, Felix, 14
Kofman, Lev, 120
Komatsu, Eiichiro, 317–18
kosmoi, 4–5, 7, 30, 54
Kovac, John, 62, 64
Kragh, Helge, 16
Krauss, Lawrence, 49, 87–88, 299–301
Kuo, Chao-Lin, 63–64

Labyrinth (film), 160
Landau, Lev, 17, 102
Lange, Andrew, 51–52
Laniakea, 278
Large Hadron Collider, 71, 82, 113, 164, 173, 307, 327

large language models, 187
"Large-N Limit of Superconformal Field Theories and Supergravity, The" (Maldacena), 241–42
Laser Interferometer Gravitational-Wave Observatory (LIGO), *313*; black holes and, 117, 176, 231–32; CMB and, 309–13, 325; cyclic cosmology and, 176; gravitational waves and, 117, 176, 231–33, 279, 309–13, 325; speed of light and, 231–33; time and, 279
Laser Interferometer Space Antenna (LISA), 309–12, 327
Late Show with David Letterman (TV show), 53
Lazaridis, Mike, 217
Leavitt, Henrietta Swan: Cepheid variables and, 11–13, 21, 50; as Harvard Computer, 11, 237; multiverse and, 50; redshift and, viii
Lebanon, 286
Lectures on Physics (Feynman), 128
Lehners, Jean-Luc, 94–97, 171
Lemaître, Georges: as Catholic priest, 13, 22, 290; expansion and, 11–13, 15, 25, 293; general relativity and, 12; historical perspective and, 12–13, 15–16, 20, 22, 24–25, 328; Penrose and, 25; Pius XII and, 22; primeval atom and, vii–viii, 12–13, 15–16, 24, 87; religion and, 13, 22, 290, 293; Vilenkin and, 87
Lemonick, Michael, 47
Lewandowski, Jerzy, 128, 137
Li, Li-Xin: Big Bang model of, *336*; "Can the Universe Create Itself?," 270–71; final problem and, 325; Gott and, 268–73, 325, *336*; Hartle-Hawking model and, 269–70; "Must Time Machines Be Unstable against Vacuum Fluctuations?," 268

Lifshitz, Evgeny, 24, 221, 223, 282–83, 326
light-years, 11, 277–78, 319, 321
Linde, Andrei: BICEP and, 63–64; Brandenberger and, 109; CMB and, 321, 324; Ekpyrosis and, 120–21; Guth and, 37–38; Hawking and, 38–39, 58, 120; KKLMMT construction and, 246; Kuo and, 63–64; multiverse and, 37–39, 42, 58, 62–66; speed of light and, 213; string gas cosmology and, 109, 115, 120; string theories and, 106, 109–10, 115, 120–21
Lisbon earthquake, 183
LiteBIRD, 317–19
lithium, 17
Little Rip, 168
Loeb, Avi: CMB and, 323–24; holograms and, 254; multiverse and, 61, 65–69
loop quantum gravity: Ashtekar and, 128–32, 135–42, 145–48, 151; Big Bang model of, *334*; Big Bounce and, 127–52; black holes and, 128, 131, 137, 140, 150, 187, 192–94, 198; Bojowald and, 135–42, 146–47; Brandenberger and, 151; CMB and, 143–50, 321–22, 326; cosmic forgetfulness and, 140–41; cutoff and, 137; Dicke and, 151; Ekpyrosis and, 134; final problem and, 326; Foundational Questions Institute Conference and, 131–32; Gamow and, 151; Guth and, 139, 143; Hawking and, 130–31, 143, 150; holograms and, 253; kinematics of, 133; need for, 97; numerics and, 139–40; Planck satellite and, 148–49; religion and, 282, 284, 288, 294, 299, 305; Rovelli and, 130, 132–33, 150–51; Singh and, 137–42, 146–47; singularities and, 128, 132, 135–40, 142, 146; Smolin and, 129–33, 141, 146; speed of light and, 220, 223, 229; vs. string theory, 132; superinflation and, 136, 140; symmetry and, 134, 147; Veneziano and, 151
Lorentz invariance, 219
Losing the Nobel Prize (Keating), 112
Lossev, Andrei, 266, 274
Lovell, Bernard, 21
Lucretius, 5, 53
Ludwick, Kevin, 168–69
Lydia, 2

Magueijo, João: bi-thermal Big Bang and, 224–30; CMB and, 289, 318; Imperial College and, 217; Moffat and, 215; Mukhanov and, 217–19; speed of light and, 211–20, 224–25, 287; "Speedy Sound and Cosmic Structure," 218; VSL and, 211–20, 224–25
Maldacena, Juan: "Cosmological Collider Physics," 254; EP = EPR and, 255–56, 259; holograms and, 241–46, 252, 254–55, 259; KKLMMT construction and, 246
Mandé people, 43
Mann, Robb, 203, 206
Mansouri, Reza, 210
many-worlds interpretation, 78, 256, 272
Margulis, Lynn, 188
Marx, Karl, 290
Mathematica, 138
Mather, John, 46
Max Planck Institute, 310
Maxwell, James Clerk, 8
McAlister, Liam, 246
McFadden, Paul, 250–52, 258–59
McGrew, Lydia, 298
McGrew, Tim, 298
McKellar, Andrew, 18, 23, 52–53

Medians, 2
Meissner, Krzysztof, 164
membranes, 103, 225, 229, *334–36*.
  See also branes
Merali, Zeeya, 206, 220
Mersini-Houghton, Laura, 104–5
methylidyne, 18
Michell, John, 183–86, 208, 232
Michelson, Albert A., 8–9, 309
Microwave Anisotropy Probe
  (MAP), 52
Midnight's Children (Rushdie), 128
Milani, Denise, 168–69
Miletus, 4–5
Milky Way, 4, 11, 44, 149, 276, 303
mirror universe, 171, *335*
Misner, Charles, 79
Mlodinow, Leonard, 91–92
Mochoscou, Mary, 229
Moffat, John: background of, 216; dark matter and, 238; Magueijo and, 215; "Superluminary Universe," 215; VSL and, 215–20, 238, 287
monopoles: cyclic cosmology and, 174; holograms and, 244; magnetic, 28; multiverse and, 32–33, 36–37; string theories and, 108, 120; time and, 278–79
Morley, Edward W., 8–9, 309
Motl, Luboš, 64
M-theory, 103, 117–18, 154
Mukhanov, Slava: black holes and, 187; CMB and, 318; cyclic cosmology and, 172; holograms and, 252; multiverse and, 40; speed of light and, 217–18, 225
Mulla Sadra, 22
multiverse: Albrecht and, 37–38, 41–42; atoms and, 7, 32, 36, 45, 53–54, 63; BICEP and, 62–66; black holes and, 48, 68; B-modes and, 63, 65; bubbles and, 37, 42–44, 53, 59–60, 68; CMB and, 36, 39, 45–47, 51–53, 55, 59–63, 66, 69; COBE and, 45–54, 66; cosmic inflation and, 31–42; cosmological constant and, 49–51, 54; curvature and, 34–36, 68; dark energy and, 51–53, 57; Darwinism and, 57; density and, 32, 34, 40, 44, 49, 52–53; Dicke and, 34–35, 53; electromagnetism and, 62; electrons and, 45, 63; Ellis and, 58–59; equilibrium and, 36, 45; eternal inflation and, 42–45; expansion and, 32–34, 37–39, 43–44, 48–51, 54, 57, 59, 62, 69; flatness problem and, 34–37, 53, 56; galaxies and, 39–40, 44, 48, 50, 53, 63, 65; Gaussian distribution and, 55–56, 61; general relativity and, 34; geometry and, 34–35, 48, 51, 55, 68; God and, 47, 58; Gott and, 42; gravitational waves and, 55, 62–63, 65, 68; gravity and, 32–36, 50, 52, 55, 57, 62–63, 65, 68; Guth and, 32–49, 53–54, 57–58, 61–69; Hoyle and, 44; Hubble and, 50; Ijjas and, 61, 65–69; inflation and, 31–69; Leavitt and, 50; Linde and, 37–39, 42, 58, 62–66; Loeb and, 61, 65–69; monopoles and, 32–33, 36–37; NASA and, 45–47, 52, 67; omega and, 34–36, 48–51; Penrose and, 38, 57, 60, 68–69; photons and, 36, 45, 51, 53, 55; Planck and, 45–46, 60–66; plasma and, 40, 62; pre-Big Bang model and, 59; probability and, 56–57, 61–63; protons and, 41; quantum theory and, 32–34, 39–40, 43–49, 57, 62; radiation and, 39–42, 45, 51, 69; rebellion and, 54–59, 66–69; singularities and, 37, 57, 69; Smoot and, 46–48; Soviet Union and, 37–40; space-time

and, 32, 41, 47, 62; spectrum and, 45–46, 52, 55, 69; Spergel and, 47–48, 52, 63–64; standard model and, 48, 61; steady-state model and, 44; Steinhardt and, 36–38, 42, 54–55, 61–62, 65–69; temperature and, 32–33, 36, 43–47, 51, 55; thermodynamics and, 45; Turok and, 48, 54, 62; vacuum and, 33–43, 49–54, 59; Vilenkin and, 42–44, 49, 58, 68, 73; Weinberg and, 32, 38, 53, 58, 67; WMAP and, 35, 46, 52–53, 55, 59–60, 63, 67

"Must Time Machines Be Unstable against Vacuum Fluctuations?" (Li), 268

Mylova, Maria, 229

Narlikar, Jayant, 319
NASA: Big Bounce and, 143; CMB and, 45, 47, 52, 142, 212, 310, 317, 319, 322; COBE and, 45–46, 52, 143; cyclic cosmology and, 163; holograms and, 237; MAP and, 52; multiverse and, 45–47, 52, 67; speed of light and, 212
Nash, John, 290
National Academy of Sciences, 11
NATO, 118
Nature (journal), 206, 212
Nayeri, Ali, 108–10
neutron stars, 117, 188–92, 197, 233, 313
New Atheists, 299
Newman, Ezra "Ted," 163–64
New Testament, 288
Newton, Isaac: Ashtekar and, 128; atoms and, 7–8; clockwork universe and, 74; conservation laws and, 14; Einstein and, 10, 14, 133, 238; gravity and, 7–8, 71, 81, 128, 133, 183, 238; Hawking and, 81; *Principia*, 7; quantum theory and, 74, 76; unstable universe and, 10
Newtonian mechanics, 14
"New Views on the Cosmological Big Bang" (Gould), 280
New York Times, 8, 53, 88, 219
Nietzsche, Friedrich, 280
Nobel Prize: Aspect and, 224; Barish and, 309; Clauser and, 224; Feynman and, 128; Gell-Man and, 84; Glashow and, 32, 104; Gross and, 132; Guth and, 62, 64, 321–22; Keating on, 112; Kibble and, 197; Landau and, 17; LIGO team and, 117, 232; Linde and, 62, 321–22; Misner and, 79; Nash and, 290; Penrose and, 156; Penzias and, 23; Planck and, 64, 66–67; Raman and, 185; Salam and, 32, 216; Smoot and, 46; Sorkin and, 195; Steinhardt and, 62; Strickland and, 232, 237; 't Hooft and, 240; Thorne and, 79, 244, 309; Weinberg and, 32, 53; Weiss and, 309; Wilson, 23; Zeilinger and, 224
no-boundary condition: entropy and, 93, 95; expansion and, 71–73, 83–85, 89–97, 150, 299–301, 322, *333*; Feynman and, 83; flatness and, 299; Green's function and, 95–96; Hartle and, 82–85, 88–96, 142, 150, 253, 269–70, 299, *333*; Hawking and, 71–73, 83–85, 89–97, 150, 299–301, 322, *333*; inflation and, 71–73, 83–85, 89–97, 150, 299–301, 322, *333*; Picard-Lefschetz technique and, 94; rethinking, 91–94; time and, 253, 260; Vatican conference and, 84; wave functions and, 83–84, 89, 91, 95, 97; Wick rotation and, 83
Noether, Emmy, 14–15, 86
Nomura, Yasunori, 61, 66–68

North Korea, 286
NOVA (TV show), 102
Novikov, Igor, 23, 187, 266, 274
Nu, 3
nucleus, 74, 94
Nuffield Workshop, 40–41, 45
null energy condition, 175, 287
Nurowski, Pawel, 163–64

Occam's razor, 207, 304
Ohm, E. A., 23
Olbers, Heinrich Wilhelm, 8
Om, 41
omega: CMB and, 29; curvature and, 29, 30, 34–36, 48–51; Dicke and, 29–30, 34; flatness problem and, 34–36; Guth and, 30, 34, 49; multiverse and, 34–36, 48–51; Planck time and, 30
Omni magazine, 128
Once before Time (Bojowald), 141
Open Society and Its Enemies, The (Popper), 306
Oppenheimer (film), 171
Oppenheimer, Robert, 24–25, 185
Oracle of Delphi, 89
"Out of the White Hole" (Pourhasan, Afshordi, and Mann), 205–6
Ovrut, Burt, 118–19
oxygen, 254

Page, Don, 244
Parker, Leonard, 39
Parmenides, 4
Pathria, Raj, 186–87
Pauli, Wolfgang, xiii, 1–2
Pawlowski, Tomasz, 137, 139
Payne-Gaposchkin, Cecilia, 236, 256
Peebles, Jim, 23
peer review, 53, 58, 66, 164
Penrose, Oliver, 24
Penrose, Roger: Ashtekar and, 129; Big Bounce and, 129; black holes and, 24, 79, 157, 160–61, 176, 185–86, 197, 200; Carroll and, 178–80, 291, 294; CCC and, 157–67, 170, 181–82, 227, 243, 275–76, 282, 294, 320; CMB and, 60, 235–36, 320; cosmic censorship and, 200; cyclic cosmology and, 154–67, 171–81; final problem and, 325–26; gravity and, 25–26, 57, 79, 156, 162, 176, 181, 197, 227, 243, 279, 282, 292, 320, 325–26; Hawking and, 25–27, 30, 38, 57, 68, 86, 107, 154, 156, 161, 171, 176, 186, 197, 243, 254, 292, 326; holograms and, 243, 254; Hoyle and, 24; multiverse and, 38, 57, 60, 68–69; religion and, 282–83, 291–94, 297; Sciama and, 24–26, 157, 160, 197; singularities and, 24–26, 30, 57, 69, 79, 107, 114, 154, 156–57, 171, 176, 185–86, 200, 227, 254, 291, 294, 326; speed of light and, 227–28; string theories and, 107, 114; time and, 271, 275–79; U-turn of, 154; Wheeler and, 79
Penrose Staircase, *159*, 160
Penzias, Arno, 23, 60, 263, 312
Perimeter Institute: Abedi and, 231; Arkani-Hamed and, 241; Bars and, 228; Black Hole Bistro, 171, 193, 203; Boyle and, 174; Burton and, 216–17; Dykaar and, 231; "Holographic Cosmology" workshop at, 247; Khoury and, 119; Lehners and, 94; McFadden and, 251; Moffat and, 215–20, 238, 287; Poplawski and, 199; Smolin and, 188, 193, 223; Turok and, 94, 119, 171–72, 174, 217, 228, 247
periodic time cosmology, 276, 279, 325, *336*
Perry, Malcolm, 87–88
Persian Gulf, 210

Persians, 1–2, 5, 172, 274
phantom menace, 166–67
Philo, 6
Philosophiæ Naturalis Principia Mathematica (*Principia*) (Newton), 7
photons: Big Bounce and, 143; CMB and, 36, 51, 53, 55, 307; cyclic cosmology and, 36; multiverse and, 36, 45, 51, 53, 55; Penrose on, 24; string theories and, 100, 102
Physical Review Letters (journal), 269
Picard-Lefschetz technique, 94
Pius XII, 22
Planck, Max: black-body radiation and, 45; quantum theory and, 45, 74–75, 79, 175
Planck density, 175, 197–98, 199
Planck length, 75, 79, 198, 245
Planck satellite: CMB and, 46, 60–66, 143–44, 225, 252–53, *289*, 308, 312–18, 350n19; conformal rescaling and, 278; gravitational waves and, 95; loop quantum cosmology and, 148–49; proof of inflation and, 60–61, 104, 322–23; religion and, 289; 3 sigma and, 344n14
Planck time, 30, 75, 79
plasma: Big Bounce and, 143–44; CMB and, 307–10, 316; electrons and, 17; multiverse and, 40, 62; primeval, 17, 26; speed of light and, 225–26, 230; string theories and, 108, 111
Plato, 6, 54, 89
Podolsky, Boris, 255
Poisson, Eric, 268
Poliakoff, Serge, 216
Poplawski, Nikodem, 139, 197–99, 202, 322
Popper, Karl, 302–3, 306
Porrati, Massimo, 201–2, 204
positrons, 78–79, 250, 265

Pourhasan, Razieh, 203
pre–Big Bang model, *334*; Big Bounce and, 140–41, 144; CMB and, 310, 320; cyclic cosmology and, 160; multiverse and, 59; string theories and, 112–17, 120, 123, 125
Predestination (film), 261
Preskill, John, 32, 244, 279
primeval atom: Big Bang models and, *333–36*; historical perspective on, 8–16, 24; Lemaître and, vii–viii, 12–16, 24, 87
probability: Big Bounce and, 140, 145; cyclic cosmology and, 154; Hawking and, 74–78, 82, 91; multiverse and, 56–57, 61–63; quantum theory and, 28, 74–78, 82, 140, 267; religion and, 296–98; Smoot and, 47; string theories and, 122; time and, 267
protons: expansion and, 17; Hawking and, 92; imaginary time and, 92; multiverse and, 41; quantum theory and, 76; standard model and, *101*
proto-science, 305–6
pseudoscience, 302–5
Pullin, Jorge, 151
Pythagoras, 4

quantum gravity: Big Bounce and, 127–36, 140–51; black holes and, 187, 192–95; CMB and, 310, 321; cyclic cosmology and, 175–77; Hawking and, 75, 79, 97; holograms and, 242, 247–48, 251–52, 255–58; multiverse and, 57; religion and, 283, 288; speed of light and, 209–10, 217–28, 232; string theories and, 100, 105, 115, 120; time and, 268
"Quantum Gravity at a Lifshitz Point" (Hořava), 221

quantum mechanics: Big Bounce and, 129–30, 141, 151; black holes and, 193, 196, 198; CMB and, 326; contradiction of, 99–100; cyclic cosmology and, 161, 176, 178; Hawking and, 72–78, 81, 83, 91–97; holograms and, 235, 239–42, 248–49, 255–58; multiverse and, 45, 49; Pauli effect and, 1–2; Planck time and, 30; religion and, 283, 304; scale and, 28; speed of light and, 220, 224; string theories and, 99, 124; time and, 265–66, 269, 272, 274

quantum theory: antiparticles and, 33, 75, 78–80, 265, 274; atoms, 5, 28, 72–74, 82, 87, 94, 97–100, 193–95, 211, 254, 312; Big Bounce, 127–51; black holes, 75, 187, 191–98, 205; Chern-Simons, 315; CMB and, 310, 312, 315, 321–22, 326; Copenhagen interpretation, 78; creation and, 88–91, 94, 230, 305; cyclic cosmology and, 159–61, 167, 175–79; double-slit experiment, 76–78; Einstein and, 9, 34, 74, 79, 86–87, 97–100, 104, 127–30, 133, 136, 138, 151, 161, 167, 193, 196, 209, 211, 220–24, 232, 235, 240, 255–56, 312, 326; electrons and, 74–79; energy packets, 74; entropy and, 79–82, 95, 137, 140, 161, 179, 239–40, 242, 248, 255, 282, 294; expansion and, 30, 32, 39, 57, 94, 96, 115, 120, 134–38, 141, 149, 151, 175, 179, 196, 209, 211, 217, 224, 226, 243, 257, 283, 302, *333*; Feynman and, 76, 240, 266; geometry and, 84, 88, 90, 92, 131, 133, 136, 176, 194, 220, 235, 255, 257, 259, 310; Hawking and, 72–97; Heisenberg and, 74–76, 78; holograms and, 235, 239–44, 247–59; inflation and, 39–40, 47, 57, 75, 78, 87, 90–91, 95, 100, 104, 115, 120, 124, 134, 136, 140, 146, 176, 179, 198, 205, 209–11, 217–20, 247, 250–52, 257, 268, 282–83, 288, 292, 299, 304–5, 321–22, 326; Lucretius and, 5; many-worlds interpretation and, 78; multiverse and, 32–34, 39–40, 43–49, 57, 62; Newton and, 74, 76; no-boundary proposal and, 91–94; Pauli effect and, 1; Planck and, 39, 45, 74–75, 79, 104, 148–49, 175, 253; probability and, 28, 74–78, 82, 140, 267; protons and, 76; real world and, 89–91; religion and, 281–84, 288, 292, 294, 297–301, 304–5; Schrödinger and, 73, 257; singularities and, 28, 57, 72–73, 75, 83, 132, 136, 138, 142, 146, 176, 192, 209, 227–30, 247–49, 254, 257, 282, 292–93, 326, *335*; speed of light and, 209–11, 217, 219–33; strange world of, 73–79; string theories and, 99–105, 115, 120, 124; supercooling and, 33–34, 43, 225, 312; Susskind and, 78; thermodynamics and, 45, 81, 240, 294; time and, 265–69, 272–74; Turok and, 94–95, 120, 142, 217, 247, 250; Vilenkin and, 73, 86–88, 91, 191, 196, 299; wave functions and, 76, 78, 82–84, 91, 95, 97, 141, 256–58
quantum tunneling, 75–76, 85–89, 92, 192, *333*
quarks, *101*, 190
Quine, Willard, 303
Quinn, Philip, 21–22
Quran, 207, 266, 281

radiation: Big Bounce and, 131; black holes and, 81, 205; CMB, 23, 307–28; cyclic cosmology and, 157, 161, 167; Hawking and, 81;

holograms and, 240, 248; multiverse and, 39–42, 45, 51, 69; string theories and, 119
radioactivity, 12, 32, 43, 87, 100
radio astronomy, 19, 23
Radio Television Series of Serbia, 118
radio waves, 18–19, 46, 313, 324
Raman, Chandrasekhara Venkata, 185
real time, 92–94, 150
Red Army, 274
redshift: galaxies and, 12, 14, 52, 225, 279; Hubble and, viii, 12; Leavitt and, viii
red spectrum, 12, 52, 111, 279, 320
Rees, Martin, 58, 67, 324
religion: atheism and, 21–22, 258, 273, 291, 299–301; Baum-Frampton model and, 284, 294; Bekenstein and, 287; BICEP and, 289, 303; *The Big Bang Theory* and, 131; black holes and, 283; Buddhism, 6–7, 153; Carroll and, 284, 291–95, 299, 301; Catholicism, 13, 21–22, 183, 290; Christianity, 6, 21, 287–90, 293, 297–98; CMB and, 284, 289; cosmological constant and, 296–97, 305; cosmological natural selection and, 296; cosmology within, 290–95; Craig and, 273, 290–96; creation and, 291, 299–302, 305; curvature and, 284, 289–91, 299; dark energy and, 282, 297; Darwinism and, 296, 300; density and, 285, 291; dogma and, 220, 257, 285, 287, 300–301, 304; dualities and, 289; entropy and, 282, 294, 297; equilibrium and, 284, 294, 299; expansion and, 283, 293, 296, 304; flatness problem and, 297, 299; galaxies and, 282, 284, 286; Gaussian distribution and, 289; general relativity and, 282–83, 292; God and, 287, 291; Gott and, 287; gravity and, 282–84, 287–92, 302, 305–6; Guth and, 292–94, 297–99; Hawking and, 292, 299–301; Hinduism, 3, 6–7, 41, 153, 304; Ijjas and, 287; inflation and, 282–92, 295–99, 302–6; Islam, 1, 3, 7, 22, 118, 207, 216, 221, 241, 266, 286–88; Jainism, 153; Jewish, 6, 79, 85, 184, 288; Lemaître and, 13, 22, 290, 293; loop quantum gravity and, 282, 284, 288, 294, 299, 305; Marx on, 290; New Testament, 288; Penrose and, 282–83, 291–94, 297; Planck and, 289; probability and, 296–98; pseudoscience and, 302–5; quantum theory and, 281–84, 288, 292, 294, 297–301, 304–5; Rovelli and, 291; science and, 285–306; singularities and, 282, 291–94, 300; Smolin and, 296; Smoot and, 47; space-time and, 284, 292, 296; Steinhardt and, 283, 287, 303; thermodynamics and, 294; Torah, 288; vacuum and, 296, 299; Vatican conference and, 84; Vilenkin and, 292–93, 299–300; Weinberg and, 297
Renaissance Capital, 315
Research in Motion (RIM), 217
Rosen, Nathan, 255
Rosenberg, Jack, 239, 244
Rovelli, Carlo: background of, 130; black holes and, 188, 192; holograms and, 245; loop quantum gravity and, 130, 132–33, 150–51; religion and, 291
Royal Astronomical Society, 17, 185
Rubakov, Valeria, 38
Rubin Observatory, 279
Rukeyser, Muriel, xi–xii

Rushdie, Salman, 128
Russell, Henry Norris, 185
Russo-Persian wars, 274
Rutherford, Ernest, 17
Ryle, Martin, 18–21
Ryu, Shinsei, 255

Sachs, Mendel, 86
Sachs, Rainer Kurt, 85–86
Sagan, Carl, 162, 189
Salam, Abdus, 32, 216
Sato, Katsuhiko, 38
Saudi Arabia, 286
Schrödinger equation, 73, 257
Schwarzschild radius, 184
Sciama, Dennis: black holes and, 197; Carter and, 26; CMB and, 160; cyclic cosmology and, 157, 160; Ellis and, 25, 212; Hawking and, 26, 72; Kibble and, 197; Penrose and, 24–26, 157, 160, 197; temperature and, 72; Vilenkin and, 86
Scientific American, 66, 140, 148, 206, 220
Scott, Douglas, 163–65
Sen, Amitabha, 129
September 11, 2001, attacks, 286
Serbia, 118
Shapley, Harlow, 11, 236
Sher, Marc, 139
Shetterly, Margot Lee, 237
Shutz, Bernard, 311
Silk, Joe, 58
Silverstein, Eva, 246–47
Simons, Jim, 315–16
Singh, Parampreet: Ashtekar and, 137–42, 146–47; loop quantum gravity and, 137–42, 146–47; Smolin and, 192; supercomputer simulations of, 146–47
singularities, x; Ashtekar and, 135–40; Bekenstein and, 79–81;

Big Bang models and, *335*; Big Bounce and, 128, 132, 135–40, 142, 146; black holes and, 24–25, 79, 83, 92, 128, 176, 185–87, 192, 199–201, 204–5, 214, 228, 248–49, 268; Bojowald and, 135–40; CMB and, 326–28; cyclic cosmology and, 154–57, 171, 173, 176, 182; density and, 2, 24, 72, 83, 114, 176, 185, 200, 291; entropy and, 79–82, 154, 156, 182, 205, 248, 282, 294; expansion and, 2, 25–28, 37, 57, 69, 113–14, 138, 154, 156, 200, 205, 209, 257, 293; farewell to, 135–40; Friedmann and, 23–24; gravity and, 24, 28, 57, 75, 79, 114, 128, 132, 136, 146, 176, 185–86, 192, 200, 209, 214, 227–28, 247–48, 291, 326; Hawking and, 25–26, 30, 57, 69, 72–73, 83, 92, 107, 154, 156, 171, 176, 186, 248, 254, 269–70, 326; holograms and, 247–49, 254, 257; loop quantum gravity and, 128, 132, 135–40, 142, 146; multiverse and, 37, 57, 69; no-boundary proposal and, 83; Penrose and, 24–26, 30, 57, 69, 79, 107, 114, 154–57, 171, 176, 185–86, 200, 227, 254, 291, 294, 326; quantum theory and, 28, 57, 72–73, 75, 83, 132, 136, 138, 142, 146, 176, 192, 209, 227–30, 247–49, 254, 257, 282, 292–93, 326, *335*; religion and, 282, 291–94, 300; speed of light and, 209, 214, 218, 227–30; string theories and, 107, 109, 113–16, 119, 125; temperature and, 2, 24, 72, 107, 125, 182, 185, 205, 254, 291; time and, 267–70
Skenderis, Kostas: CMB and, 323; holograms and, 247–53, 258; McFadden and, 250–52; time and, 274, 279

Slipher, Vesto, 11–12
slow contraction, 174–75, 182, 299, 335
Smolin, Lee: Ashtekar and, 129–30, 132, 141, 188; black holes and, 187–96, 202; CMB and, 322; cosmological natural selection and, 187–96, 296, 322; Einstein and, 129–30, 133, 220; Fuller and, 129; loop quantum gravity and, 129–33, 141, 146; Perimeter Institute and, 188, 193, 223; religion and, 296; speed of light and, 220, 223; *The Trouble with Physics*, 132, 187, 223
Smoot, George, 46–48
Snyder, Hartland, 24, 185
solar eclipses, 3, 8–9, 99, 284
Sorkin, Rafael, 105, 193
Souradeep, Tarun, 319
Soviet Union: Belinski and, 24; Doroshkevich and, 23; Gamow and, 16; Hosseini and, 274; Lifshitz and, 24; Linde and, 37; multiverse and, 37–40; Nash and, 290; Novikov and, 23; Starobinsky and, 37, 81; Vilenkin and, 73, 85; Zel'dovich and, 81
space-time: background independence and, 127–28; Big Bang models and, 335; Big Bounce and, 127, 133, 138, 143, 148; black holes and, 187, 192–97, 200; CMB and, 308; cosmological constant and, 10, 133, 192, 296; curvature and, 10, 24 (*see also* curvature); Einstein and, viii, 9–10, 32, 99, 127, 133, 138, 193, 197, 262; emergence of, 254–59; geometry and, 10, 55, 84, 92, 133, 220, 259, 308–10; Gott-Li model and, 270; gravity and, 9–10, 28, 32, 50, 52, 55, 62, 87, 99, 127, 130, 133, 166, 192–93, 197, 200, 219–21, 227, 231, 242, 248–50, 258, 269, 284, 306–12, 325; Hawking and, 84, 92–93; holograms and, 242, 254–59; inflation and, 41, 219, 284, 292, 296; multiverse and, 32, 41, 47, 62; religion and, 284, 292, 296; ripples in, 39, 55, 62, 77, 120, 143, 226, 231, 308–10, 313, 317; special relativity and, 9; speed of light and, 219, 229, 231, 265; string theories and, 115, 267; vacuum of, 33, 39, 43, 59, 75, 87–88, 170, 180, 195, 200, 219, 221, 243, 250, 270, 296, 299; Vilenkin and, 87, 292; warping of, 10, 197, 262–63; wormholes and, 89, 255, 259, 263, 274
special relativity, 9, 223, 265
spectroscope, 45
spectrum: black-body, 45–46, 225–26; blue, 12, 111, 320–23; CMB and, 279, 320–23; cyclic cosmology and, 156, 174, 176; Hawking and, 82; multiverse and, 45–46, 52, 55, 69; Planck and, 289; red, 12, 52, 111, 279, 320; speed of light and, 217–18, 226, 253; string theories and, 108–11, 116
speed of light: Albrecht and, 212, 215; atoms and, 211; BICEP and, 218, 230; bi-thermal Big Bang and, 224–30; black holes and, 214, 219, 224, 228–32; B-modes and, 218; CMB and, 212, 218, 224–29; cosmological constant and, 213–14, 230; curvature and, 228; Cuscuton and, 213–19, 223–24, 230, 251, 283, 305; dark energy and, 214–15, 219; dark matter and, 214–15; density and, 225, 230; Ekpyrosis and, 217, 219, 226, 230; Ellis and, 212–13, 226, 233; equilibrium and, 211, 225–27; ether and, 8–9, 214, 224, 232–33; expansion and, 209, 211,

speed of light (*continued*)
214, 217, 224–27, 232; flatness problem and, 227; galaxies and, 211–12, 219, 227; Gamow and, 221; Gaussian distribution and, 229; general relativity and, 210, 220, 223, 229–30; geometry and, 220, 229; gravity and, 210, 213–33; Guth and, 211; Hořava gravity and, 219–24, 227, 230–33; horizon problem and, 211, 213, 224–26; Hoyle and, 216; Hubble and, 232; inflation and, 209–13, 217–20, 223–27, 230; LIGO and, 231–33; Linde and, 213; loop quantum gravity and, 220, 223, 229; Magueijo and, 211–20, 224–25, 287; Michelson-Morley experiment and, 8–9; Mukhanov and, 217–19; NASA and, 212; Penrose and, 227–28; Planck and, 225, 350n19; plasma and, 225–26, 230; quantum gravity and, 209–10, 217–28, 232; quantum mechanics and, 220, 224; singularities and, 209, 214, 218, 227–30; Smolin and, 220, 223; space-time and, 219, 229, 231, 265; spectrum and, 217–18, 226, 253; Spergel and, 232; standard model and, 222; steady-state model and, 216; Steinhardt and, 228; symmetry and, 218–19; temperature and, 211, 224–26, 229; time and, 265; Turok and, 217, 228; vacuum and, 219–22; Vilenkin and, 218; VSL, 179, 209–27, 230–33, 265, 282–83, 299, 305, 323; WMAP and, 212, 218, 225

Spergel, David: Big Bounce and, 143; black holes and, 202; CMB and, 310, 314–17, 321; COBE and, 47–48, 52, 143, 246, 314; cyclic cosmology and, 162; holograms and, 246; Komatsu and, 317–18; multiverse and, 47–48, 52, 63–64; Schumer and, 48; speed of light and, 232; time and, 263

SphereX, 322–23
spin torsion, 292
"standard candles," 11, 13, 50
standard model: black holes and, 188–89; cyclic cosmology and, 170; holograms and, 252–53; multiverse and, 48, 61; speed of light and, 222; string theories and, *101*, 103

Starobinsky, Alexei, 37, 57, 81, 289, 318, 323
Star Wars (film), 212
steady-state model: Big Bang and, 13–23; Bondi and, 44; cyclic cosmology and, 160; density and, 44; galaxies and, 13, 16, 18–21, 44; Gold and, 44; Hoyle and, 44, 216, 319; multiverse and, 44; speed of light and, 216

Steinhardt, Paul: Albrecht and, 37; BICEP and, 62, 65–66, 303; Big Bounce and, 143; CMB and, 315, 320, 323–24; cyclic cosmology and, 157, 171–77, 181; holograms and, 254; multiverse and, 36–38, 42, 54–55, 61–62, 65–69; religion and, 283, 287, 303; speed of light and, 228; string theories and, 118–24; time and, 275

Stoics, 6–7, 54, 119, 122, 153
Stojkovic, Dejan, 194
Strauss, Michael, 263–64
Strickland, Donna, 232, 237
string gas cosmology, *333*; Big Bounce and, 107–8; Brandenberger and, 106–13, 120, 211, 320; Ekpyrosis and, 120, 123, 125; entropy and, 123, 282; inflation and, 101, 108–12, 115–16, 120,

123, 125, 134, 199, 205, 211, 282, 320, 323; Linde and, 109, 115, 120; Nayeri and, 108–9; Vafa and, 106–12

string theories: anthropic argument and, 105–6; atoms and, 99–102; background independence and, 127–28; Bekenstein and, 107; BICEP and, 111–12, 117, 124–25; black holes and, 100, 107, 117; B-modes and, 112, 117; branes and, 103, 110, 118–23, 125, 134, 173, 201, 205, 219, 225, 228–29, 242, 246; bubbles and, 123, 144; CMB and, 104, 108–9, 117, 125; COBE and, 321; cosmological constant and, 105, 112; curvature and, 115; dark energy and, 112, 121; debates over, 131–34; density and, 114, 120, 123; development of, 101–6; dualities and, 113–16; Eddington and, 99; Ekpyrosis, 6, 101, 117–25, 134, 153, 171–74, 217, 219, 226, 230, 320, 322–23, *334–35*; electromagnetism and, 100, 117; electrons and, 102; Ellis and, 108; entropy and, 107, 123; equilibrium and, 107–8, 123; expansion and, 107–16, 119–23; flatness problem and, 110, 121–22; Foundational Questions Institute Conference and, 131–32; galaxies and, 99, 105, 113, 115; general relativity and, 99; Gott and, 268; gravitational waves and, 111–12, 117, 124–25; gravity and, 99–102, 105, 111–17, 120, 124–25; Guth and, 108, 122–23; Heisenberg and, 99; horizon problem and, 108, 120; inflation and, 100–101, 104–12, 115–25; Linde and, 106, 109–10, 115, 120–21; loop quantum gravity, 127–52 (*see also* loop quantum gravity); monopoles and, 108, 120; M-theory and, 103, 117–18, 154; Penrose and, 107, 114; photons and, 100, 102; Planck and, 104; plasma and, 108, 111; pre–Big Bang model and, 112–17, 120, 123, 125; probability and, 122; quantum theory and, 99–105, 115, 120, 124; radiation and, 119; singularities and, 107, 109, 113–16, 119, 125; space-time and, 115, 267–68; spectrum and, 108–11, 116; standard model and, *101*, 103; Steinhardt and, 118–24; superpartners and, 103; Susskind and, 102; temperature and, 107, 120, 125; Theory of Everything, 102–3, 132; thermodynamics and, 123; Turok and, 117–22; vacuum and, 114; Veneziano and, 101–2, 112–17, 120; Vilenkin and, 123; Weinberg and, 105

String Wars, 131–34
Strominger, Andrew, 107
Struve, Otto, 236
Sullivan, Woodruff T., III, 19
Summers, Larry, 237
supercooling, 33–34, 43, 225, 312
superinflation, 136, 140
supernovae, 50, 310
superpartners, 103
supersymmetry, 103, 126, 134, 173, 258
Susskind, Leonard: black holes and, 191; *The Black Hole War*, 240; *The Cosmic Landscape*, 245; EP = EPR and, 255–56, 259; holograms and, 240, 243–49, 255, 259; quantum theory and, 78; string theories and, 102
Swampland conjecture, 112, 246–47, 295

symmetry: black holes and, 185, 198, 204; conservation laws and, 14; cosmological constant and, 50; cyclic cosmology and, 173; dark energy and, 53; holograms and, 257–58; loop quantum gravity and, 134, 147; perfect, 24–25, 185, 204; speed of light and, 218–19; super, 103, 126, 134, 173, 258

Takayanagi, Tadashi, 255
Tata Institute for Fundamental Research, 128
Taylor, Marika, 251
T-duality, 113–14, 116, 250, 275
Tegmark, Max, 132
temperature: Big Bang models and, *333–36*; Big Bounce and, 143–44, 147, 149; black holes and, 185, 199, 202, 204–5; CMB and, 18, 23, 29, 36, 46–47, 51, 55, 143–44, 147, 149, 162, 182, 199, 202, 205, 224–26, 229, 252, 289, 308, 318, 320, *333–36*; cyclic cosmology and, 162, 182; density and, 2, 24, 72, 120, 143, 185, 225, 291; entropy and, 80–81, 182, 204–5; Friedmann and, 24; Hawking and, 80–81, 320; holograms and, 252, 254; McKellar and, 18; multiverse and, 32–33, 36, 43–47, 51, 55; Sciama and, 72; laws of thermodynamics, 80; singularities and, 2, 24, 72, 107, 125, 182, 185, 205, 254, 291; speed of light and, 211, 224–26, 229; string theories and, 107, 120, 125
Tenchi-kaibyaku, 41
Thales, 3, 8
Theory of Everything, 102–3, 132
Theory of Everything, The (film), 25, 72
thermal equilibrium, 45, 107–8, 211, 225–27

thermodynamics: black holes and, 200–201; Boltzmann and, 180; cyclic cosmology and, 155, 179–80; entropy and, 80–81, 123, 155, 180, 200, 238–40, 294; first law of, 80; Hawking and, 80–81, 238–40; holograms and, 238–40; multiverse and, 45; quantum theory and, 45, 81, 240, 294; religion and, 294; second law of, 80, 123, 155, 179, 200–201, 277, 294; string theories and, 123; time and, 277
't Hooft, Gerard, 240
Thorne, Kip, 79, 244, 263, 309
time: antiparticles and, 33, 75, 78–80, 265, 274; arrow of, *93*, 155–56; atoms and, 278; Baum-Frampton model and, 275; black holes and, 267–68, 275, 277; Carroll and, 272; Cauchy horizon and, 267–72; CMB and, 263, 275–79, 307–8; CTC and, 262–63, 267–73; cyclic cosmology and, 153–82; dark energy and, 275, 278–79; dualities and, 275; Einstein and, 261–69, 273; Ekpyrosis and, 123; electrons and, 265; entropy and, 93, 107, 123, 150, 155–56, 160, 166, 168, 178, 180, 198, 200–201, 242, 277, 282, 294; eternal inflation, 42–45, 54, 59, 61, 64, 71, 78, 90, 106, 122–25, 178, 181, 190–91, 218, 251, 295–96, 324, *333*, *335*; expansion and, 275–76, 279; flatness problem and, 278; galaxies and, 263, 267, 275, 277–78; Gaussian distribution and, 279; general relativity and, 261, 264, 267–71; Gödel and, 262–63; Gott and, 263–64, 268–76; Gould and, 274–75, 277, 279–80; gravity and, 269, 275, 279; Guth and, 271, 279; Hartle and, 253, 260; Hawking and, 268–70, 273; Heisenberg and, 275; historical, 273–80; horizon

problem and, 278; imaginary, 92, 270; inflation and, 263, 268, 270–71, 275–79; jinn particles and, 266, 274–75; LIGO and, 279; Lossev and, 266, 274; monopoles and, 278–79; no-boundary condition and, 253, 260; Novikov and, 23, 187, 266, 274; omega and, 30; pendulums and, 155; Penrose and, 271, 275–79; Planck and, 30, 75, 278; probability and, 267; quantum theory and, 265–69, 272–74; real, 92–94, 150; second law of thermodynamics and, 155, 277; singularities and, 267–70; Skenderis and, 274, 279; space-time, 262 (*see also* space-time); speed of light and, 265; Spergel and, 263; Steinhardt and, 275; travel through, 27–28, 259–74; Turok and, 275; vacuum and, 265, 268–72; van Stockum and, 262; warping of, 10, 197, 262–63; wormholes and, 89, 255, 259, 263, 274
Time Machine, The (Wells), 261
Time magazine, 263
Times of London, 8
Tipler, Frank, 263
Todd, Paul, 157
Tomašević, Marija, 273
Torah, 288
torsion bounce, 196–99, 322, *335*
tower of Babel, 235
Tricked! (Frampton), 169
Trivedi, Sandip, 246
Trouble with Physics, The (Smolin), 132, 187, 223
Tryon, Edward, 86–87
Turing, Alan, 186–87
Turkey, 2–3
Turner, Michael S., 49, 51
Turok, Neil: background of, 117–18; BICEP and, 62, 303; Big Bounce and, 142; cyclic cosmology and, 170–72, 174, 303; Ekpyrosis and, 117–22; holograms and, 247, 250; Lehners and, 94, 96, 171; multiverse and, 48, 54, 62; Perimeter Institute and, 94, 119, 171–72, 174, 217, 228, 247; Picard-Lefschetz technique and, 94; quantum theory and, 94–95, 120, 142, 217, 247, 250; speed of light and, 217, 228; string theories and, 117–22; time and, 275
Twain, Mark, 261
Tye, Sze-Hoi Henry, 32–33, 110, 244

UAE, 286
Uddalaka Aruni, 6
Ungarelli, Carlo, 310
Universe from Nothing, A (Krauss), 88, 299
Unruh, Bill, 136

vacuums: antiparticles and, 33, 75, 78–80, 265, 274; black holes and, 191, 198, 200, 205; bubbles and, 37, 42–43, 53, 59; cyclic cosmology and, 170–71, 177, 180; energy in, 34, 37, 42–43, 49, 51, 53–54, 75, 86, 114, 170–71, 180, 195, 200, 219, 222, 243, 258; Hawking and, 75, 79–80, 86–88, 96; holograms and, 243, 250, 258; multiverse and, 33–43, 49–54, 59; religion and, 296, 299; space-time and, 33, 39, 43, 59, 75, 87–88, 170, 180, 195, 200, 219, 221, 243, 250, 270, 296, 299; speed of light and, 219–22; string theories and, 114; supercooling and, 33–34, 43, 225, 312; time and, 265, 268–72
Vafa, Cumrun: CMB and, 320; final problem and, 326; holograms and, 241–42, 246; speed of light

Vafa, Cumrun (*continued*) and, 228; string theories and, 106–12; Swampland conjecture, 112, 246–47, 295

Vagnozzi, Sunny, 324

Vandersloot, Kevin, 138

van Stockum, Willem Jacob, 262

varying speed of light (VSL), *336*; Albrecht and, 179–80, 212, 215; as alternative paradigm, 179, 209–20, 223–27, 230, 233, 265, 282–83, 299, 305, 323; bi-thermal Big Bang and, 224–30; black-hole information paradox and, 230–33; Cuscuton and, 213–19, 223–24, 230, 251, 283, 305; Magueijo and, 211–20, 224–25; Moffat and, 215–20, 238, 287; Mukhanov and, 217–19

Vatican conference, 84

Vaughan, Dorothy, 237

Vecchio, Alberto, 310

Veneziano, Gabriele: Big Bounce and, 151; CERN and, 113; CMB and, 320, 326; dualities and, 113–14; final problem and, 326; loop quantum gravity and, 151; string theories and, 101–2, 112–17, 120

Vidotto, Francesca, 146, 192

Vilenkin, Alex: atoms and, 87; background of, 85; BGV theorem and, 58, 123–24, 292–93; black holes and, 190–91, 196; COBE and, 86; Ekpyrosis and, 123; electrons and, 96; God and, 89; Guth and, 42, 44, 49, 58, 68, 123, 292–93, 299; Hawking and, 73, 86–91, 95–96, 292, 299; inflation and, 42–44, 49, 58, 68, 87, 90–91, 94, 123, 190, 218, 292, 299; multiverse and, 42–44, 49, 58, 68, 73; Perry and, 87–88; Planck and, 95; Plato and, 89; quantum theory and, 73, 86–88, 91, 191, 196, 299; religion and, 292–93, 299–300; Sachs and, 85–86; Sciama and, 86; Soviet Union and, 73, 85; spacetime and, 87, 292; speed of light and, 218; string theories and, 123; Tryon and, 86, 86–87

virtual particles, 39, 49, 80, 86, 221–22, 265

Wall, Aron, 92–93

Warner, Nick, 38

Watch Mr. Wizard (TV show), 30

"Wave-Function of the Universe" (Hartle and Hawking), 84

wave functions: Big Bounce and, 141; Feynman and, 76, 82–83; Gell-Man and, 84; Green's function and, 95–96; Hartle-Hawking model and, 84; holograms and, 256–58; interference and, 76–78; no-boundary condition and, 83–84, 89, 91, 95, 97; peak/trough cancellation and, 82; quantum theory and, 76, 78, 82–84, 91, 95, 97, 141, 256–58

W boson, 100

Weber, Joseph, 308–9

Wehus, Ingunn Kathrine, 162

Weinberg, Eric, 38

Weinberg, Steven: black holes and, 195–96; holograms and, 244; multiverse and, 32, 53, 58, 67; religion and, 297; string theories and, 105

Weiss, Rai, 309, 311

Welcome to the Universe (Gott, deGrasse Tyson, and Strauss), 264

Wells, H. G., 261

Wheeler, John, 79, 82, 186, 188, 195

white holes, 199–208

Wick rotation, 83

Wilkinson, David, 47, 52

Wilkinson Microwave Anisotropy Probe (WMAP): Big Bounce and, 143; black holes and, 202; CMB and, 308–18, 322; COBE and, *46*,

52, 143, 312, 314; cyclic cosmology and, 163; multiverse and, 35, 46, 52–53, 55, 59–60, 63, 67; speed of light and, 212, 218, 225
Wilks, Tom, 134–35
Wilson, Robert, 23, 60, 263, 312
Wired (magazine), 162
Witten, Ed, 103, 106, 117–18, 219, 241, 247
Wittgenstein, Ludwig, 258, 303–4
World War I, viii
World War II, 15, 38, 85, 262
wormholes, 89, 255, 259, 263, 274

Xianyu, Zhong-Zhi, 323–24

Yamada, Masaki, 96
YouTube, x–xii, 59, 112, 151, 155, 291

Zaldarriaga, Matias, 64
Zalel, Stav, 195
Z boson, 100
Zeilinger, Anton, 224
Zel'dovich, Yakub, 69, 81
Zhang, Xin, 167–68
Zwane, Nosiphiwo, 196

72ff Big Bang not an account of the explosion of the universe, but of our classical understanding...

73f no-boundary proposal (Heisenberg uncertainty)

74 Planck length

75 vanishingly small is not zero

88 Laws before things

96 Not our knowledge but the real world has uncertainty

(✽) 101ff String theory and the real world